Chrysophytes: aspects and problems

Chrysophytes: aspects and problems

Edited by
JØRGEN KRISTIANSEN
University of Copenhagen
and
ROBERT A. ANDERSEN
DePaul University
Chicago, Illinois

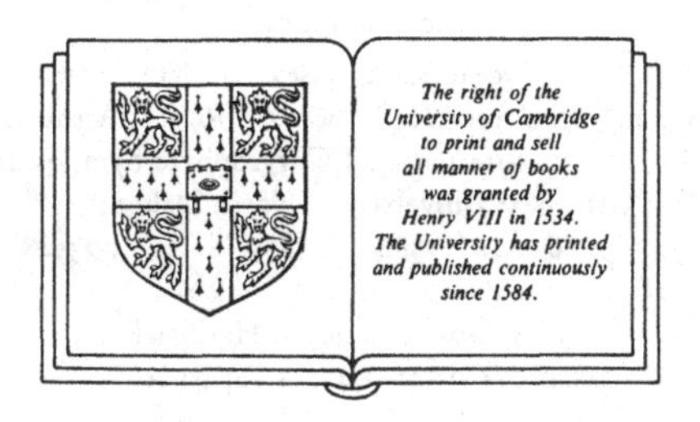
The right of the
University of Cambridge
to print and sell
all manner of books
was granted by
Henry VIII in 1534.
The University has printed
and published continuously
since 1584.

CAMBRIDGE UNIVERSITY PRESS

Cambridge
London New York New Rochelle
Melbourne Sydney

CAMBRIDGE UNIVERSITY PRESS
Cambridge, New York, Melbourne, Madrid, Cape Town, Singapore,
São Paulo, Delhi, Dubai, Tokyo, Mexico City

Cambridge University Press
The Edinburgh Building, Cambridge CB2 8RU, UK

Published in the United States of America by Cambridge University Press, New York

www.cambridge.org
Information on this title: www.cambridge.org/9780521180078

First published 1986
First paperback edition 2010

A catalogue record for this publication is available from the British Library

Library of Congress Cataloguing in Publication data
Main entry under title:
Chrysophytes: aspects and problems.
Papers from the first International Chrysophyte
Symposium, held at the University of North Dakota,
Aug. 11–16, 1983.
Includes index.
1. Chrysophyceae–Congresses. I. Kristiansen,
Jørgen. II. Andersen, Robert A. (Robert Arthur),
1948– . III. International Chrysophyte Symposium
(1st: 1983: University of North Dakota)
QK569.C62C48 1986 589.4′8 85-25548

ISBN 978-0-521-32090-0 Hardback
ISBN 978-0-521-18007-8 Paperback

Contents

Preface

The growing interest in the study of chrysophytes resulted in the First International Chrysophyte Symposium, which was held in Grand Forks (United States of America), at the University of North Dakota, August 11–16, 1983, in continuation of the Annual Meeting of the Phycological Society of America. At this symposium, 41 contributions on all aspects of the study of chrysophytes were presented. Of these, 21 papers have been collected in the present volume, and these contributions reflect important trends in today's investigations on chrysophytes.

One main complex of problems addressed in Part I is the content and delimitation of "chrysophytes." How should this group be structured? On closer study it appears more and more heterogeneous, and perhaps some of its members do not belong here at all. What are the relations to other algae, such as diatoms and brown algae, and to zooflagellates?

Electron microscopy has penetrated into almost every field of chrysophyte study, and taxonomy (Part II) is no exception. Ultrastructural evidence sometimes makes it necessary to split established families. For example, the once united silica-scaled species are now classified as two only distantly related families. Light microscopic characters are still used for arranging species in evolutionary lines, but in many cases electron microscopy is necessary in order to describe new species and to redefine established ones.

The chrysophytes have also been favorite organisms for cell biologists (Part III). Electron microscopy in combination with biochemical methods has made it possible to study the intricate problem of how silica scales are deposited in the scale case surrounding the *Synura* cell. As investigators probe at an ever-smaller scale, biochemical and molecular methods become increasingly more important, such as in the study of organelles and their function or the study of DNA in chloroplasts.

Equally important, but in a direction of increasingly larger scale, are studies to help explain how organisms interact with their environment (Parts IV and V). The occurrence of the chrysophytes in nature is regulated at least in part by changes in water chemistry, light, and temperature. Especially important are studies of life histories as correlated with environmental factors. However, occurrence is also governed by interspecific relations, sometimes even as symbiotic partnerships. Ecology and distribution of chrysophytes have been the object of many investigations, but severe species identification problems have hitherto hindered such studies. In the silica-scale-bearing groups these problems to a great extent are now being solved, and the methods of defining and recognizing chrysophyte assemblages have also improved.

The paleobiology (Part VI) of chrysophytes provides, just as it does for all groups of organisms, information that leads to a better understanding of today's living forms. An increasing number of Recent fossil scales and cysts can now be referred to present-day living species with known ecology, and these serve as environmental indicators for understanding aquatic ecosystems of the past. However, many problems remain, such as the identity of Cambrian scales – do they belong to ancient chrysophytes? The growing interest in fossil cysts makes it advantageous to introduce a standardized procedure for describing new forms.

Even if the contributions in this book reflect many viewpoints in the study of chrysophytes today, several additional themes might as well have been treated. In order to give an impression of the complete range of the Symposium, the Symposium papers not published here are listed by author and title.

The Chrysophyte Symposium was a success, and the editors hope that the publication of this treatise will be equally well received. We thank the Phycological Society of America which sponsored the Symposium and supported the publication of this volume. We also thank our editorial committee and the many anonymous reviewers who have striven for excellence in this publication. The First International Chrysophyte Symposium and this publication were supported in part by a United States National Science Foundation Grant BSR 82-16920 to R.A.A.

JK & RAA

Titles of other papers from the symposium

David H. Adam: Chrysophyte cysts: a geologist's approach.

R.A. Andersen: Ultrastructure of the flagellar apparatus of *Synura uvella* Ehr.

G. Cronberg: Scaled chrysophytes from the tropics.

D.A. Egloff, T.J. Cowles, and D. Stoecker: Effects of *Olisthodiscus luteus* Carter (Chrysophyceae) on feeding and swimming of the marine rotifer *Synchaeta cecilia* Rousselot.

L.R. Hoffman and M. Vesk: Ultrastructure of zoospores in *Hydrurus foetidus.*

B.A. Jacobsen: Scale-bearing Chrysophyceae from West Greenland.

H.J. Kling: Chrysophyceae in Central Arctic lakes.

Aa. Kristiansen: Locality types and occurrence in Denmark of marine benthic Chrysophyceae and Prymnesiophyceae.

K.H. Nicholls: *Mallomonas labrinthina* sp. nov. and *Paraphysomonas sediculosa* sp. nov., two new silica-scaled chrysophytes.

J.D. Pickett-Heaps: A comparison of scale formation in chrysophytes with valve formation in diatoms.

J.S. Prince: Comparative ultrastructure of coccolith, naked and scaly cells of *Coccolithus neohelis.*

C.D. Sandgren and H.J. Carney: A flora of fossil chrysophycean cysts from the recent sediments of Frains Lake, Michigan, U.S.A.

J.P. Smol: Chrysophycean microfossils in lacustrine deposits.

H.A. Thomsen: Fine structural studies on the flagellate genus *Bicosoeca* – a review.

S.D. van Valkenburg: The synchronization of silicoflagellate cultures.

M. Vesk and L.R. Hoffman: Mitosis and cell division in *Hydrurus foetidus.*

M. Vesk, S.W. Jeffrey, and J.L. Stauber: *Pelagococcus subviridis* from the East Australian current.

J.L. Wee and M. Gabel: The distributional relationship between pH, temperature, and assemblages of silica-scaled chrysophytes in the Tall-Grass Prairie of the USA.

D.E. Wujek: Ultrastructure of the flagellated chrysophyte *Chrysosphaerella.*

Contributors

Jane Aldrich, The Standard Oil Co., 3092 Broaday Avenue, Cleveland, Ohio 44115, U.S.A. (Chapter 11)

Carol Wagner Allison, University of Alaska Museum, Fairbanks, Alaska 99701, U.S.A. (Chapter 19)

Robert A. Andersen, Department of Biological Sciences, DePaul University, 1036 Belden Avenue, Chicago, Illinois 60614, U.S.A. (Chapter 8)

Chantal Billard, Laboratoire d'Algologie fondamentale et appliquée, Université de Caen, 39 rue Desmoueux, 14000-Caen, France (Chapter 3)

Steven Bressler, Department of Botany KB-15, University of Washington, Seattle, Washington 98195, U.S.A. (Chapter 11)

Rose Ann Cattolico, Department of Botany KB-15, University of Washington, Seattle, Washington 98195, U.S.A. (Chapter 11)

Jan S. Chock, East Woods School, 31 Yellow Cote Road, Oyster Bay, New York 11771, U.S.A. (Chapter 12)

Annette W. Coleman, Division of Biology and Medicine, Brown University, Providence, Rhode Island 02912, U.S.A. (Chapter 10)

Gertrud Cronberg, Institute of Limnology, University of Lund, Box 3060, S-220 03 Lund, Sweden (Chapters 20 and 21)

R.L. Cubel, Department of Biology and Living Resources, University of Miami, Miami, Florida 33149, U.S.A. (Chapter 14)

Monika Dürrschmidt, Institut für Pflanzenökologie, Justus-Liebig Universität, Heinrich-Buff-Ring 38, D-6300 Giessen, Federal Republic of Germany (Chapter 7)

Duncan Ersland-Talbot, Agrigenetics Research Park, 5649 East Buckeye Road, Madison, Wisconsin 53716, U.S.A. (Chapter 11)

Paulette Gayral, Laboratoire d'Algologie fondamentale et appliquée, Université de Caen, 39 rue Desmoueux, 14000-Caen, France (Chapter 3)

David J. Hibberd, Axle Tree Cottage, Starvecrow Lane, Peasmarsh, Rye, East Sussex TN31 6XL, England (Chapters 2 and 5)

Jerry W. Hilgert, Institute of Northern Forestry, U.S.D.A., U.S. Forest Service, 308 Tanana Drive, Fairbanks, Alaska 99701, U.S.A. (Chapter 19)
Jørgen Kristiansen, Institut for Sporeplanter, University of Copenhagen, Øster Farimagsgade 2D, 1353 Copenhagen K, Denmark (Chapter 16)
Akira Kurata, Lake Biwa Research Institute, 1-10, Uchide-hama, Otsu, Shiga Prefrecture, 520 Japan (Chapter 13)
Barry S.C. Leadbeater, Department of Plant Biology, University of Birmingham, Edgbaston, Birmingham, B15 2TT, England (Chapter 9)
Richard L., Meyer, Department of Botany and Microbiology, University of Arkansas, Fayetteville, Arkansas 72701, U.S.A. (Chapter 6)
Scott Newman, Department of Botany KB-15, University of Washington, Seattle, Washington 98195, U.S.A. (Chapter 11)
David J. Patterson, Department of Zoology, University of Bristol, Bristol, B58 1UG, England (Chapter 4)
Frances R. Pick, Department of Biology, York University, 4700 Keele Street, Toronto, Ontario, M3J 1P3, Canada (Chapter 14)
Hans R. Preisig, Institute of Systematic Botany, University of Zurich, Zollikerstr. 107, CH-8008, Zurich, Switzerland (Chapter 5)
Michael Reith, Department of Botany, University of Toronto, Toronto, Ontario M5S 1A1, Canada (Chapter 11)
R.M.M. Roijackers, Laboratory of Hydrobiology, Agricultural University, De Dreijen 12, 6703 BC Wageningen, The Netherlands (Chapter 17)
Frank E. Round, Department of Botany, University of Bristol, Bristol, B58 1UG, England (Chapter 1)
Craig D. Sandgren, Department of Biological Science, University of Wisconsin at Milwaukee, P.O. Box 413, Milwaukee, Wisconsin 53201, U.S.A. (Chapters 15 and 21)
Peter A. Siver, Department of Biology, Western Connecticut State University, Danbury, Connecticut 06810, U.S.A. (Chapter 12)
Absjørn Skogstad, Department of Biology, Division of Limnology, University of Oslo, P.O. Box 1027, Blindern, N-Oslo 3, Norway (Chapter 18)

PART I

Definition and relationships

1

The Chrysophyta – a reassessment

FRANK E. ROUND
Department of Botany
University of Bristol,
Bristol B58 1UG, England

Introduction

In a symposium of this nature, many problems of detail will be discussed. However, it is worth returning to some of the basic precepts of the group Chrysophyta, examining these, and raising problems that appear from modern studies of individual genera. Recently, interest is reviving in the overall classification and phylogeny of algal groups, and the "chrysophytes" play a distinctive role in these discussions. The group is traditionally one in which "protozoal" representatives have been closely allied with autotrophic forms, and the rationale of this merits consideration. In addition, the chrysophytes occupy a relatively isolated position among the algae, having apparently not developed along the usual algal lines to the same extent as other groups. This is especially true regarding the evolution of sexuality. Few other algal groups have been so fractured as have the chrysophytes and this process, involving a reassessment of the status of the widely recognized anomalous genera, will continue, although perhaps within the overall confines of a "heterokontophyte" series of organisms. Depending on whose classification one follows, the Chrysophyta can be a very restricted group with a single class Chrysophyceae or a group (Chrysophycophyta) encompassing six (Bold & Wynne 1978) or five (Ettl 1980) classes. The problems within the group are accentuated by the difficulty of obtaining reliable morphological data at the light microscopic level. Therefore, almost all decisions on critical genera rely on electron microscopy (EM). In addition, excellent EM fixation is required to determine, for example, details of the flagellar basal system. It is perhaps significant that data on the chrysophytes lags far behind the "chlorophytes" in this aspect. However, it is just these kinds of data that are leading to the exciting reevaluation of the latter series. The chrysophytes do, however, *appear* to be less complicated in terms of flagellar root structure, but this may be due only to lack of data.

History

Historically, the chrysophytes started merely as a loose cluster of mainly flagellate species, and it was not until 1914 that Pascher drew them together as the Chrysophyceae and placed them into a division Chrysophyta. Early workers relied on careful cytological techniques, and although these data were regarded with suspicion, they did reveal the "hairy" flagella, the heterokont condition, and many interesting structural details. The peak of this activity was the superb review of chrysophyte genera by Bourrelly (1957) with its detailed classification scheme and evolutionary discussion. The genera were also described, but with less discussion, by Bourrelly in his 1968 book. Bourrelly's detailed classification has been used as a basis by many, but, even in the 1960s, ultrastructural research was beginning to uncover additional features, resulting in the removal of the "haptophytes" (Parke 1961, Christensen 1962). The splitting of the "xanthophyte" series into two was made by Hibberd and Leedale (1972), and, in key papers, Hibberd (1976, 1980) discussed the basic structure of the chrysophyte and "prymnesiophyte" lines. Meanwhile, many details concerning the species bearing siliceous scales began to appear, starting with the classic paper by Manton (1955). The classification of these, based on scale structure, had been begun by light microscopists, but it was now put on a much firmer basis, and new species were recognized by detail that is almost only visible by electron microscopy. Throughout the history of chrysophytes, the bipartite cyst has figured prominently, but it is only in recent years that detailed studies have been made (Sheath, Hellebust, & Sawa 1975, Hibberd 1977, and most recently by Sandgren, 1980a,b). There was an exciting period when electron microscopy revealed scale types that had to be linked with whole cells and when coccoid filamentous forms were described and linked up to flagellate stages and into the classification. This process is nowhere near complete, and many groups are obviously little studied [cf. the recent description of a large number of new species of *Paraphysomonas* by Thomsen et al. (1981); Preisig & Hibberd (1982a,b, 1983)]. The outcome of these most recent studies is the realization that the type of flagellation, which was a prime feature of Bourrelly's classification, is perhaps not applicable when electron microscopic detail is included. Another, and perhaps more unfortunate, outcome of ultrastructural work has been the emphasis placed on a few genera with anomalous features. *Olisthodiscus* and *Sphaleromantis* ["a single peculiar species of uncertain status within a group to which in fact it may not belong" (Manton & Harris 1966)] immediately spring to mind. These genera and others (e.g., *Dictyocha, Pterosperma,* and *Pedinella*) have been retained in the set of species with what one might term the *Ochromonas* structure, and this has led to a heterogeneous cluster that is probably misleading. Evolution is more likely to have given rise to many small sets of organisms, in this case perhaps to be loosely conceived of as a heterokont group but not as a diversified chrysophyte group. Historically, there is a dearth of information at the biochemical/genetical level, and when this is forthcoming many of the existing problems may disappear.

Recent valuable reviews are those of Hibberd (1976, 1980) and Pienaar (1980), and it is not my intention to cover the same ground but rather to discuss some general features.

Concepts

In many accounts and especially in textbooks, the Chrysophyta include three classes, viz., Chrysophceae, Xanthophyceae, and Bacillariophyceae. The origin of this taxonomic treatment is Pascher (1914), and, considering Pascher's general approach, it is difficult to see the logic in the proposal since by 1931 he was proposing (Pascher 1931) that each class should have a series of life forms from flagellate through coccoid, simple filament, and branched filament to thalloid. Clearly, the diatoms (Bacillariophyceae) do not have such a range; in fact, they are purely coccoid. The Xanthophyceae do exhibit a wide range of form, although not to the same degree as some classes of algae. The Chrysophyceae are exceptional, with many flagellate and coccoid members, but relatively few with higher organization. So why were the three classes taken under the umbrella of the Chrysophyta? The usual explanation is that in some form or other there is often a bipartite wall and coupled with this a deposit of silica into the wall matrix or into discrete components (scales, valves, cysts, etc.). The flagellate stage is heterokont, but, when the amalgamation was made, the diatom motile gamete and many other flagellate forms were relatively unknown or incompletely known. In any case, the diatom motile sperm is not heterokont. (It may have been during early evolution.) Moestrup (1982) discussed the use of the term heterokont, which originally meant "unequal in length," and later described one mastigoneme-bearing flagellum and one smooth flagellum. By any criteria, the evidence for a combined series was very scant. Comparison with other well-established groups, e.g., Rhodophyta, and Phaeophyta (the latter admittedly heterokont in a loose sense), reveals a lack of consistency in the concept of the Chrysophyta; either the concept itself is incorrect or the consistency within other groups is fortuitious. Let us examine briefly each group. The "xanthophytes" in the old sense have already been fragmented by the removal of a number of genera to the Eustigmatophyceae (Hibberd & Leedale 1972; Hibberd 1981), and there have been several suggestions that the "siphonous" group might form a distinct subset – Vaucheriophyta of Maekawa (1960). Whether one leaves these as classes of the newly designated Tribophyceae [(Christensen 1979, Hibberd 1981) Xanthophyceae excl. Eustigmatophyceae] or as separate phyla is hardly relevant in the present discussion. Apart from heterokont flagellation, there are few features allying them to the Chrysophyceae or Bacillariophyceae: The rarely reported "cysts" require reexamination, and the bipartite nature of some walls is not an homologous feature. Bipartite walls occur also in the "chlorophyte" series. [To designate the groups as Bivalvophyta as Hirose, quoted in Takahashi (1964), did is unnatural and has never been taken up.] The Bacillariophyceae are unique among the heterokont series in their total consistency of form, which is undoubtedly bound to the rigidity of cell structure within the confines of the multipartite siliceous wall covering that allows little scope for variation (Round 1981). There are no anomalous sections that could be split off as in the other two groups. There have been rare suggestions that they might form a coccoid order of the chrysophyte series, but there are already simple coccoid chrysophytes (e.g., in the Chrysapiales, Chrysosphaerales, and Stichogloeales), and greater evidence of affinity would be needed to substantiate such a move. Their spore structures (auxospores) are not bipartite but multipartite. However, some of the simplest diatom resting spores do show features reminiscent of one subset within the Chrysophyceae. The Chrysophceae in the classic sense have, like the Xanthophyceae, suffered subdivision into Prymnesiophyceae and Chrysophyceae, and this appears eminently sensible with a cluster of features characterizing the former group, e.g., organic scales, equal flagella, haptonema, calcium carbonate deposits, and a predominately marine distribution [see Hibberd (1976) for details]. However, this does not leave a homogeneous group but rather a cluster of subsets, many with rather superficial claims for inclusion in the Chrysophyceae *sensu stricto*, i.e., based on Pascher's (1914) concept in which the endogenous cyst features as a key element. Unlike other algal phyla, there is a paucity of hard biochemical evidence to characterize the group as a whole or any of its subsets. The universal occurrence of "leucosin" is marginally permissible, but this (or very similar β-1,3-glycosides) occurs widely in heterokont algae. It is imperative that some of the biochemical pathways of chrysophytes be examined and compared with those of prymnesiophytes, diatoms, etc. We are left with one overwhelming feature that cannot be ignored and that is the cyst or statospore. By its mode of formation (internally in the cytoplasm) and its single, plugged aperture, the cyst is so distinctive that it must form a unifying attribute – it is possible of course, but unlikely, that this arose more than once in evolution. That the cyst unites forms with and without siliceous scales is significant and probably denotes a primitive character. Sandgren (1980b) refers to the cysts as "conserved through evolutionary time." If this is so, are the remaining non-cyst-forming genera derived by loss of such a structure or are they out of place? We ought to ask, what is the origin of the cyst? Is it a feature of the primitive, colorless eukaryote that on the serial endosymbiotic hypothesis must have predated the pigmented derivative? If this is its origin, then ought we not to find such cysts in modern colorless flagellates? If it is not a feature of the colorless ancestor, how did it arise in the life cycle after the incorporation of the plastids? It is not obviously sexually formed (see Sheath et al. 1975), but appears at intervals in the life cycle [see Sandgren (1980a) for details and possible sexual events after cyst formation]. That mitosis occurs prior to cyst formation in *Dinobryon* and *Uroglena* (Sheath et al. 1975; Sandgren 1980a,b) is slightly reminiscent of the necessity for nuclear division in diatoms before valve formation can occur. However, replication of the nucleus does not occur in the *"Chromulina"* type, and there is also a suggestion that in the Synuraceae other features are found (Sandgren 1980b), thus weakening this argument. Pascher (1924) referred to a filamentous chrysophyte that forms a cyst; however, the only illustration I have been able to locate is in a palmelloid piece of *Phaeothamnion,* and this is hardly filamentous (Pascher 1925). Later (Pascher 1932), he comments that cyst formation occurs only when the filament breaks down into a palmelloid mucilage mass. In fact, this seems to be a process in which swarmers are forming (the cytoplasm contains numerous contractile vacuoles), and the arrest of these leads to cyst formation. Pascher does continue with the comment that he has also seen cysts forming inside the cells (*innerhalb behäuteter Zellen*), but this is not illustrated.

Classification

The Chrysophyta were first treated systematically by Pascher (1914), and up to this time the diatoms (Bacillariophyceae) had always formed a well-defined taxon and indeed are one of the few groups with a completely internal consistency. The Heterokontae had been removed by Luther (1899) to become the Xanthophyceae (Allorge 1930) and most recently Tribophyceae (Hibberd 1981). Before 1914, the Chrysophyceae were simply a loose collection of phytoflagellates in the family Chrysomonadina (Klebs 1892). In 1960, Chadefaud argued in favor of clustering the chlorophyll *c*-containing algae into the Chromophycophytes [Chromophyta of Christensen (1962, 1980) and Bourrelly (1968) or Heterokontophyta of Leedale (1974)]. The latest proposal is that of Cavalier-Smith (1981), who proposes a nine-kingdom system, one of which is named Chromophyta: Phyta is not a suitable ending for such a taxon, and Round (1984) has supported this kindgom concept but suggested the name be changed to Chromoplantae. Acceptance of this idea still leaves the problem of divisions (phyla), and I believe that the concept of Heterokontophyta is too inclusive of diverse phylogenetic lines. While this name is a useful adjectival (heterokont) term, it is subsumed in the taxon Chromoplantae. I am convinced that the divergent lines (and divergence is still only hypothetical) form divisions when one takes an overall view of the organismal world. Hence, I prefer to retain the divisions Chrysophyta, Tribophyta, Bacillariophyta, and the recently separated Eustigmatophyta and Prymnesiophyta. The latter are not regarded as chrysophytes by Hibberd (1976), who earlier

(Hibberd 1972) clearly stated that these eustigmatophytes and prymnesiophytes could not be accommodated in a heterokont division; this raises a problem in the system of Cavalier-Smith (1981) who includes them. The title of the symposium contains the term "chrysophyte," and I am sure that this in most contributors' minds refers to the Pascherian group of algae.

The systematic scheme of Bourrelly (1968) is one of the most detailed based on his comprehensive (Bourrelly 1957) work, but when one compares it with that generally adopted for the chlorophytes at an ordinal level, one is struck immediately by the fusion of flagellate, colonial flagellate, coccoidal, palmelloid, rhizopodial, even filamentous, and parenchymatous forms within the same order. The basis for this scheme is the form of the motile cell that is used to group orders into subclasses, and its roots lie in the variation or absence of flagellation in the genera. The only other pigmented group where such flagellar variation is prominent is the euglenophytes. Such a scheme may represent best the phylogeny – only further studies will decide this – but it is certainly a confusing system, especially in the light of new ultrastructural data. An alternative is to have a greater number of orders within the subclasses, e.g., raise some of the subordinal groups to orders. The general reluctance to do this may be related to the small number of genera in the nonflagellate clusters, although, of course, such a systematic arrangement was proposed by Pascher (1914) and has been adopted by Fritsch (1935, 1951) and Bold & Wynne (1978) with their 11 orders (10 orders if the Dictyochales are discounted and they do appear an aberrant group or a side line off the general chrysophyte evolutionary trend). A reinvestigation of the Bold and Wynne orders is required to see if some fusion of the varied series is warranted, especially when it seems that uniflagellate forms are extremely rare [the only well authenticated one seems to be the *Rhizochromulina* of Hibberd & Chretiennot-Dinet (1979)] and the nonflagellate forms are surely merely life cycle variants of the various morphological series. In fact, many genera have been removed from the nonflagellate subclass (Acontochrysophycideae) as their motile stages were discovered. On the other hand, it could be argued that this approach would overemphasize gross morphology (cf. the chlorophyte series). The recent tendency in many groups has been to upgrade the importance of ultrastructural studies at the expense of the gross morphology. Ultrastructure is undoubtedly of immense value in defining a whole class (or division), as Hibberd (1976) showed for the Chrysophyceae and Prymnesiophyceae, but gross morphology/life cycle may be more important at the ordinal and familial level. Norris (1977) has, in fact, argued for the greater importance of morphology rather than flagellation in the classification of chrysophytes. Good examples of this are the loricate series, *Dinobryon* (Karim & Round 1967), *Epipyxis, Chrysolykos* (Kristiansen 1969), *Pseudokephyrion* (Belcher 1968), *Stenocalyx, Bitrichia, Chrysopyxis* (Kristiansen 1972), and *Poterioochromonas* (Peterfi,

1969) with their microfibrillar organization and the Synuraceae with their distinctive scales. Belcher and Swale (1972a) even recorded some microfibrils in the thecate genus *Chrysococcus*, which perhaps indicates an underlying biochemical uniformity.

The relationship of the chrysophytes to the Phaeophyceae is one that requires much further study; again, such a relationship was hinted at as early as 1914 by Pascher. On the surface, the phaeophytes seem to be very different biochemically and cytologically, and only some aspects of the flagellate cells form the main evidence for this connection – although there is no typical transitional helix (*spiralkörper*) at least in the advanced genus *Fucus*. It is clearly desirable to investigate simpler genera (Hibberd 1978). There is a general lack of simple coccoid/palmelloid genera in the Phaeophyceae and no really authenticated free-living motile forms. Several genera previously thought to be Phaeophyceae have been transferred to the Chrysophyceae [e.g., *Phaeosaccion* (Chen, McLachlan, & Craigie 1974)]. However perhaps there is still some doubt here; the sections of the flagellar base do not show a clear transitional helix that is a characteristic of the Chrysophyta. This may be due to fixation because such a structure is not clear in freshwater chrysophyte motile cells sectioned by Dop (1980), although he does refer to it. The presence of the transitional helix in 19 chrysophyte species (Hibberd 1978) and also in a number of eustigmatophyte and xanthophyte species (but not in phaeophyte or prymnesiophyte species) is clearly an important comparative feature, and it should be sought in diatom sperm and in any motile cell of doubtful affinity. Hibberd (1978) suggests it is a feature indicative of monophyly, but if this is so it will require considerable rethinking of the systematics of the heterokont series.

Gayral and Billard (1977) created a new chrysophycean order, Sarcinochrysidales, for the marine benthic chrysophytes (see also Gayral, 1972), and these certainly have many features allying them to the Phaeophyceae (see also comment in Dop 1980) or in some instances to the prymnesiophytes. There is a real need to investigate the relationships of these and simple ectocarpalean forms at the biochemical systems level, for it is indeed remarkable if no simple phaeophytes have survived from the early period of serial endosymbiosis even in such a modified form as to make recognition difficult (see comments on ecology). It is also necessary to compare Sarcinochrysidales with Prymnesiophyceae because in my view, from studying the papers of Gayral and Haas (1969), Loiseaux (1967), and Gayral and Lepailleur (1971), they may possess some haptophycean features. One cannot deny that the whole system both of Bourrelly (1957, 1968) and Bold and Wynne (1978) will need extensive revision as more genera are examined in detail. Bold and Wynne separate the Phaeophyta into a separate division and leave the remainder of the "chromophyte" series in another, as do many authors, but such disparate treatment has little justification. Either there is a single chromophyte division (or better, a kingdom, see section on classification) or there are several.

Bold and Wynne (1978) and Pienaar (1980) use a classification based partly on Bourrelly (1968). Christensen (1980) and Kristiansen (1982) use a morphological basis, since flagellar number is a difficult criterion unless, and even if, the details are checked by electron microscopy. The various systematic treatments have been drawn together in Table 1.1.

Morphology

The majority of Chrysophytes are unicellular, but, whereas other divisions tend to have a rather conservative form for their flagellate orders, the range in external form of chrysophytes is great. Internally, as Hibberd (1976) points out, there is greater consistency. There are, however, a number of anomalous genera included in the Chrysophyceae [e.g., *Sphaleromantis, Olisthodiscus* – the latter sometimes placed in Tribophyceae but recently removed to the Raphidophyceae (Chloromonadophyceae) by Loeblich and Fine (1977) – and *Dictyocha*] and all these need reassessment. If they prove to be true chrysophytes, then they should probably be allocated to new taxa at the ordinal or higher level (as indeed *Dictyocha* has in the Dictyochales) or as a subclass – Dictyochophycidae – by Kristiansen (1982). The most consistent group is the one possessing siliceous scales [*Mallomonas, Mallomonopsis* [only a variant on flagellar form (?)] *Synura,* possibly *Catenochrysis*], and they probably form a distinctive evolutionary line (but see comments on cysts). *Chrysosphaerella* cannot be included in this line and is now separated from *Synura* and *Mallomonas* in the Paraphysomonadaceae (Preisig & Hibberd 1983). The scales do have a distinct ordered heteromorphy, with the most complex occurring at the flagella pole, a main mass of body scales, and a number of simplified antapical scales (cf. the numerous publications on these genera). Only very rarely have additional structures been found on the scales, e.g., the flask-shaped structures on *Mallomonas cyathellata* (Wujek & Asmund 1979). It is only members of this group (=Synuraceae) Mallomonadaceae that have siliceous structures that might conceivably relate the whole series to the Bacillariophyceae. Scales of a different type are recorded in *Syncrypta.* These seem to be hollow discs, the outer part of which is irregularly microfibrillar (Clarke & Pennick 1975). The origin of scales varies in genera commonly placed in the Chrysophyceae [e.g., those of the Synuraceae are associated with the chloroplast membrane, but those of *Paraphysomonas* and *Sphaleromantis* are possibly Golgi–endoplasmic reticulum (ER) associated].

The existence of rhizopodial forms in the Chrysophyceae is a striking feature as Pascher originally pointed out, and it is a feature almost confined to this group of algae. Pascher also pointed out the parallels in morphology between this group and the green algae, but such a feature should be placed more in context since in the Chrysophyceae there is a vast base of variability at the monad level declining to a few filaments at the more complex level whereas the reverse is true of the Chlorophyta.

The so-called benthic chrysophytes mostly have a coccoid/palmelloid form (especially those living in fresh water) and only rarely are truly filamentous, e.g., *Phaeothamnion* (see Dop 1980). The marine forms, on the other hand, seem to require further study; the growth of some species from a basal cup seems rather distinctive.

Cysts

Cysts (or stomatocysts, statospores, or endocysts) are generally considered a feature of the Chrysophyta, and indeed in genera such as *Ochromonas, Dinobryon, Mallomonas, Synura,* and *Uroglena* they are conspicuous. The cyst wall is very variable from smooth in some *Dinobryon* and *Synura* species to ornamented with stellate spines in *Mallomonas torquata* (Asmund & Cronberg 1979). In *Hydrurus,* the cyst has a winglike structure (Bourrelly 1957). In spite of their occasional occurrence or even absence in many chrysophycean genera, it was the cyst above all that Pascher (1914) considered the characteristic feature of the group. In fact, if one checks back into the early literature, many species were described together with their cysts (e.g., by Doflein 1923 and Conrad 1926). Has its significance and the attempt to fit cysts of other groups into the evolutionary framework been overdone? Although Pascher (1914 and 1932) wrote at length concerning cysts, it was Scherffel in 1911 who first quite clearly stated that the endogenous origin and the presence of a "plug" were the characteristic features. Both Cienkowsky (1870) and Prowazek (1903) had investigated the cysts before Pascher. Are the various cyst forms homologous structures? This topic is not new, and the argument was aired 60 years ago between Scherffel (1924) and Pascher (1924). In xanthophyte genera, cysts are less obvious and consist of two equal (subequal?) halves, although there is very little detail of these in the literature apart from the scattered illustrations in Pascher's early publications and a series of figures in his introductory remarks to the Heterokontae (Pascher 1939). The superficial resemblance of the diatom resting spore (common only in *Chaetoceros*) is only coincidence in that the two halves are a result of the bipartite mode of cell growth of this group. Diatom resting spores can resemble heavily silicified vegetative cells or depart considerably from this (e.g., *Leptocylindrus*), but they are still composed of two valves (Hargraves 1976). Pascher comments on at least one cyst with equal halves in the Chrysophyceae, but I have been unable to locate this. The cysts in the filaments of *Giraudyopsis* (Loiseaux 1967; Gayral & Haas 1969) really do not resemble the "plugged" chryso-

Table 1.1. *The major schemes of classification of the Chrysophyta* sensu stricto. *Bracketed taxa are those which clearly do not belong in the division*

Pascher (1914) Chrysophyta Chrysophyceae	Fritsch (1935)[a] Chrysophyceae
1. Chrysomonadales	1. Chrysomonadales
Chrysomonadineae	Chromulinaceae
Rhizochrysidineae	Oicomonadaceae
	Mallomonadaceae
	Cyrtophoraceae
	Isochrysidaceae
	[Coccolithophoridaceae]
	Synuraceae
	Ochromonadaceae
	Monadaceae
	Lepochromonadaceae
	[Prymnesiaceae]
	Rhizochrysidaceae
	Lagyniaceae
2. Chrysocapsales	Chrysocapsaceae
Chrysocapsaceae	Naegelliellaceae
Hydruraceae	Hydruraceae
3. Chrysosphaerales	2. Chrysosphaerales
Chrysosphaeraceae	Chrysosphaeraceae
	Chrysostomataceae
	[Pterospermaceae]
4. Chrysotrichales	3. Chrysotrichales
Thallochrysidaceae	Nematochrysidaceae
Chrysotrichaceae	Thallochrysidaceae

phyte cysts. Those described by Gayral and Lepailleur (1971) in *Nematochrysopsis* and *Chrysowaerniella* are described as pseudocysts in order to distinguish them clearly from typical chrysophycean "plugged" cysts. The pseudocysts in the filamentous Sarcinochrysidales are calcareous and are purely exogenous tubular structures that surround one or more cells in the filaments. I am a little skeptical of the idea that there exists a range of cyst form from those having two equal halves to those with a "plug." Cysts without plugs are also reported for *Epipyxis* (Hilliard & Asmund 1963). The ability to form cysts, especially in the Chrysophyceae/Xanthophyceae, may well be derived from a common (?) ancestral form.

How much importance should one attach to the production of cysts? They are rather uncommon in all three groups when an overall view of the

Table 1.1. *(cont.)*

Bourelly (1968)[b] Chromophyta Chrysophycées	Fott (1971) (in part only) Chromophyta Chrysophyceae
1. Heterochrysidiophycidae	1. Chrysomonadales
Ochromonadales	Ochromonadaceae
Ochromonadaceae	Dinobryaceae
Dinobryaceae	Synuraceae
Synuraceae	Isochrysidaceae
Naegeliellaceae	[Prymnesiaceae]
Phaeothamniaceae	[Pedinellaceae]
Chrysapionaceae	
Chromulinales	2. Rhizochrysidales
Thallochrysidaceae	Rhizochrysidaceae
Chrysomeridiaceae	Lagyniaceae
Chrysosphaeraceae	Myxochrysidaceae
Chromulinaceae	
Chrysococcaceae	3. Chrysocapsales
Chrysocapsaceae	Chrysocapsaceae
Chrysochaetaceae	Naegelliellaceae
Hydruraceae	Hydruraceae
Chrysamoebaceae	4. Chrysosphaerales
Kybotionaceae	Chrysosphaeraceae
Myxochrysidaceae	Chrysapionaceae
	Stichogloeaceae
2. Acontochrysophycidae	Chrysotilaceae
Chrysosaccales	[Pterospermaceae]
Chrysosaccaceae	
Rhizochrysidales	
Rhizochrysidaceae	
Stylococcaceae	
Stichogloeales	
Chrysotilaceae	
Stichogloeaceae	
Phaeoplacales	5. Phaeothamniales
Sphaeridiotrichaceae	Sphaeridiotrichaceae
Phaeoplacaceae	Phaeoplacaceae
	Phaeothamniaceae
	Phaeodermatiaceae

(continued)

genera is made. Are the genera evolving cysts, are they being lost, or have a small number of genera always possessed them? How widespread are they in the colorless genera; Scherffel (1924) records them in *Monas* but without good illustration.

Table 1.1. (*cont.*)

Bold & Wynne (1978) Chrysophycophyta (excluding Phaeophycophyta) Chrysophyceae	Christensen (1980) Chromophyta Chrysophyceae
1. Ochromonadales	1. Ochromonadales
	Synuraceae
	Monadaceae
	Chromulinaceae
	Bicosoecaceae
	Lepochromulinaceae
	Stylococcaceae
	Chrysamoebaceae
	Rhizochrysidaceae
	Myxochrysidaceae
	Chrysocapsaceae
	Chrysosaccaceae
	Naegelliaceae
	[Pedinellaceae]
	Stylochromonadaceae
2. Chrysapiales	2. Chrysapiales
	Chrysapiaceae
	Stichogloeaceae
	Sarcinochrysidaceae
3. Phaeothamniales	3. Phaeothamniales
	Phaeothamniaceae
	Chrysomeridaceae
4. Chromulinales	4. Chrysosphaerales
	Aurosphaeraceae
	Chrysosphaeraceae
5. [Craspedomonadales]	5. Thallochrysidales
	Thallochrysidaceae
6. [Dictyochales]	6. Hydrurales
	Hydruraceae
7. Chrysosphaerales	7. [Dictyochales]
	Vallecertaceae
	Dictyochaceae
	Cornuaceae
8. Thallochrysidales	
9. Chrysococcales	
10. Rhizochrysidales	
11. Stichogloeales	
12. Phaeoplacales	

Table 1.1. *(cont.)*

Ettl (1980) (in part only) Chrysophyta Chrysophyceae	Kristiansen (1982) Chromophycota Chrysophyceae
	A. Chrysophycidae
1. Chrysomonadales	1. Ochromonadales
Ochromonadaceae	Ochromonadaceae
Dinobryaceae	Dinobryaceae
Synuraceae	Synuraceae
[Dictyochaceae]	
2. Rhizochrysidales	2. Chrysamoebidales
Rhizochrysidaceae	Chrysamoebidaceae
Lagyniaceae	Stylococcaceae
Myxochrysidaceae	Myxochrysidaceae
3. Chrysocapsales	3. Chrysocapsales
Chrysocapsaceae	Chrysocapsaceae
Naegeliellaceae	Naegeliellaceae
Hydruraceae	Hydruraceae
4. Chrysosphaerales	4. Chrysosphaerales
Chrysosphaeraceae	Chrysosphaeraceae
Stichogloeaceae	Chrysapionaceae
[Pterospermaceae]	Stichogloeaceae
	[Pterospermaceae]
5. Phaeothamniales	5. Phaeothamniales
Phaeoplacaceae	Sphaeridiotrichaceae
Sphaeridiothriaceae	Phaeoplacaceae
Phaeothamniaceae	Phaeothamniaceae
	Phaeodermatiaceae
	6. Sarcinochrysidales[c]
	Sarcinochrysidaceae
	Chrysomeridaceae
	Phaeosaccionaceae
	Nematochrysopsideae
	7. [Pedinellales]
	Pedinellaceae
	B. Dictyochophycidae
	1. Dictyochales

[a] By 1951 *(Manual of Phycology)* he recognized Rhizochrysidales and Chrysocapsales

[b] Excluding some families shown by later workers to belong elsewhere.

[c] NB: A group with certain chrysophycean and certain phaeophycean features and therefore difficult to contain within the Chrysophyceae.

Phylogeny

There is a tendency to construct phylogenetic schemes based on present-day genera, but as long ago as 1914, Pascher warned of the danger.

> Es sei hier ausdrucklich betont, dass in keiner Flagellatenreihe sich ursprüngliche Formen finden . . . . Es wäre aber verfehlt und darauf wurde auch von verschiedenen Seiten hingewiesen, die rezenten Flagellaten als die tatsächlichen phylogenetischen Ausgangspunkte für die Algen zu betrachten, denn auch die Flagellaten haben eine Entwicklung als solche hinter sich und solche während der Zeit, in der die Entwicklung zu Algen erfolgte, nicht station[d.ar geblieben.

He also considered the Phaeophyceae to be regarded as within the heterokont systems but did not consider that they were derived from the Chrysophyceae, and he did not think it worthwhile to try to derive them from organisms of lower organization. Care has to be taken when freshwater genera, such as *Porterinema,* have been shown to be phaeophytes rather than chrysophytes (Dop 1979).

Such schemes are only valid if one assumes that the basic organism [*Ochromonas* is usually chosen (e.g., Hibberd 1976)] has retained many of the original features, i.e., it contains conservative (primitive) features and not derived features. This may be so of course, but we have no proof, although there is experimental proof for other plants (e.g., Plane trees), and it is now generally thought that there has been great constancy over millions of years (Stebbins 1982). If it is so, it strengthens the hand of those who support punctuated evolution. Hibberd (1976) does, however, maintain that, "the *Ochromonas* type of organization is basic within the majority of the group."

The occurrence of siliceous scales and cysts may indeed indicate some affinity between groups (Takahashi 1964), although I suspect it is rather remote. The assumed primitive nature of the auxospore scales of centric diatoms (Round & Crawford 1981) may indicate a distant connection with simple chrysophyte scales, but the scales of modern Synuraceae are formed in a different position to those of diatoms. In addition, they do not have a central initiation zone and their growth and relationship to microtubules is quite different (McGrory & Leadbeater 1981). The cyst also differs in that the whole silica deposition vesicle (SDV) is laid down before any deposition of silica occurs, and this is then patchy (Hibberd 1977). Thus, it does not grow like the SDV of the diatom valve. On the other hand, the heterogenicity of the synuracean scales is somewhat paralleled by that of girdle band series (scales) of some diatoms. The recently described *Mallomonas plumosa* (Croome & Tyler 1983) has "clawed" spines that are extremely similar to the spines on the marine diatom *Corethron* (Fryxell &

Hasle 1971). This "convergence" of form in freshwater and marine genera is very striking indeed, but can anything more than convergence be read into it?

Much has been made of the similarity of the general organization of the motile cells in the "heterokont" series, and this coupled with certain cytological features of the vegetative phases undoubtedly supports a degree of relatedness; however, it is generally conceded that the structure of motile vegetative zoospores or gametes tends to be conservative.

Flagellar hairs are apparently very characteristic for the Chrysophyceae (tripartite – the shaft consisting of two helical fibers and three spreading terminal fibers). Such a type does not occur in Prymnesiophyceae; does this reflect a different colorless host in the early evolution? If this is so, then the grouping into a heterokont series is a consequence of the secondary symbiotic event – a probable event in the origin of many (all?) algal lines and one that raises interesting problems for the evolutionist. The "chrysophyte" flagellar hair type seems to occur in *Chrysomeris,* but a different type occurs in *Giraudyopsis* (Gayral & Haas 1969) and *Nematochrysopsis* (Gayral & Lepailleur 1971). The presence of a third appendage on the motile cells of some of the Sarcinochrysidales is an artifact as pointed out by Gayral and Haas (1969).

What is of greater consequence is the variation in the lines of evolution followed by the "heterokont" series – the play on unicellular/colonial habit in the Chrysophyceae, the restriction of evolution to a coccoid form in the Bacillariophyceae [for possible explanation, see Round (1981)], the more algal development of the Xanthophyceae, the impressive radiation into thalloid/alternation of generations in the Phaeophyceae. These are the diverging aspects, recognized by many workers who consider each cluster as an independent line of evolution and, I believe, rightly so. The evolutionist, on the other hand, is equally correct in taking a convergent view; the convergence being backward into a "Precambrian" cluster of forms.

The view from ecology is often ignored, but it is important [as shown for the Chlorophytes (Mattox & Stewart 1984; Round 1984)]. It must be significant that the Chrysophyceae and Xanthophyceae are virtually freshwater clusters, Phaeophyceae and Prymnesiophyceae are marine, and Bacillariophyceae divide on generic lines (with only a few exceptions) into marine or freshwater (Round & Sims 1980). The true habitat of diatoms collected in coastal regions requires careful study as many could be washed in, and it is essential to study unfixed material. Gene pools of each class have been separated over most of geological time, and this, coupled with inherent genetic factors, determines the vast differences between the groups. The similarities are surely the remnants of the original organisms that came together in the symbiotic system – the primitive features as opposed to the derived.

Protozoal forms or apochlorotic chrysophytes?

Can colorless flagellates really be part of the systematic entity Chrysophyceae? (NB: colorless coccoid/filamentous forms have not been recorded, but of course they may be difficult to recognize or may have been placed in the heterokont fungi.) Colorless flagellates with apparent chrysophycean features may be apochlorotic forms and thus truly "belong" in the group, [cf. arguments for apochlorotic green flagellates (Round, 1980)]. Round (1980) pointed out that apochlorotic forms tend to be in the minority (in chlorophytes, diatoms, angiosperms), and hence one assumes that loss of the pigmentation system has been a rare phenomenon. On the other hand, some colorless flagellates may have continued their evolution without entertaining a symbiosis with a pigmented unit. Should these then be classified with the pigmented series or in the Protozoa? There is no doubt that the overall ultrastructural features of some colorless "chrysophytes" do resemble the autotrophic series when their flagellation, parabasal position of the nucleus and Golgi body are compared [e.g., of *Paraphysomonas* (Preisig & Hibberd 1982a,b) and *Anthophysa* (Belcher & Swale 1972b)]. It is interesting that Scherffel (1911), commenting on the query in Klebs (1982) (*"ob nun die Monaden aus Chrysomonaden oder diese aus ersteren entstanden sind"*) comments that forms with a stigma are derived from pigmented species, *"Denn das Stigma ist eigentlich eine weitgehende Differenzierung am Chromatophor, welche das Vorhandsein eines solches zur notwendigen Voraussetzung hat."*

References

Allorge, P. 1930. Heterocontées ou Xanthophycées. *Rev. Alg.* 5: 230.
Asmund, B. & Cronberg, G. 1979. Two new taxa of *Mallomonas* (Chrysophyceae). *Bot. Notiser* 132: 409–18.
Belcher, J.H. 1968. Lorica construction in *Pseudokephyrion pseudospirale* Bourrelly. *Br. phycol. Bull.* 3(3): 495–9.
Belcher, J.H. & Swale, E.M.F. 1972a. Some features of the microanatomy of *Chrysococcus cordiformis* Naumann. *Br. Phycol. J.* 7: 53–9.
– 1972b. The morphology and fine structure of the colourless colonial flagellate *Anthophysa vegetans* (O.F. Müller) Stein. *Br. Phycol. J.* 7: 335–46.
Bold, H.C. & Wynne, M.J. 1978. *Introduction to the Algae.* Prentice-Hall, Englewood Cliffs, N.J.
Bourrelly, P. 1957. Recherches sur les Chrysophycées: Morphologie, phylogénie, systématique. *Rev. Algol., Mém. Hors-Sér.* 1: 1–412.
– 1968. *Les Algues d'Eau Douce. Initiation à la Systématique. Tome II: Les Algues Jaunes et Brunes.* Boubée, Paris.
Cavalier-Smith, T. 1981. Eukaryote kingdoms: seven or nine? *BioSystems* 14: 461–81.
Chadefaud, M. 1960. *Traité de Botanique Systématique. Tome I: Les végétaux non Vasculaires. Cryptogamie.* Masson, Paris.
Chen, L.C.M., McLachlan, J. & Craigie, J.S. 1974. The fine structure of the

marine chrysophycean alga *Phaeosaccion collinsii*. *Can. J. Bot.* 52: 1621-4.
Christensen, T. 1962. Alger. *In:* Böcher, T.W., Lange, M.C., Sørensen, T. [Eds.] *Botanik,* Bd. 2, *Systematisk Botanik Nr 2*. Munksgaard, Copenhagen.
- 1979. Annotation to a text book of phycology. *Bot. Tidskr.* 73: 65-70.
- 1980. *Algae. A Taxonomic Survey*. AiO Tryk as, Odense.
Cienkowski, L. 1870. Über Palmellaceen und einige Flagellaten. *Arch. f. mikrosk. Anat.* 6: 421.
Clarke, K.J. & Pennick, N.C. 1975. *Syncrypta glomerifera* sp. nov., a marine member of the Chrysophyceae bearing a new form of scale. *Br. Phycol. J.* 10: 363-70.
Conrad, W. 1926. Recherches sur les flagellätes de nos eaux saumâtres. 2[e] Parte. Chrysomonadines. *Arch. Protist.* 56: 167-231.
Croome, R.L. & Tyler, P.A. 1983. *Mallomonas plumosa* (Chrysophyceae) a new species from Australia. *Br. Phycol. J.* 18: 151-8.
Doflein, F. (1923). Untersuchungen über Chrysomonadinen. 3. Arten von *Chromulina* und *Ochromonas* aus dem bädischen Schwarzwald und ihre Cystenbildung. *Arch. Protist.* 46: 267-344.
Dop, A.J. 1979. *Porterinema fluviatile* (Porter) Waern (Phaeophyceae) in The Netherlands. *Acta Bot. Neerl.* 28: 449-58.
- 1980. Benthic Chrysophyceae from The Netherlands. Thesis. University of Amsterdam.
Ettl, H. 1980. *Grundriss der allgemeinen Algologie*. Fischer, Jena.
Fott, B, 1971. *Algenkunde* 2nd ed. Fischer, Jena.
Fritsch, F.E. 1935. *The Structure and Reproduction of the Algae,* Vol. I. Cambridge University Press.
- 1951. Chrysophyta. *In* Smith, G.M. [Ed.] *Manual of Phycology - an Introduction to the Algae and Their Biology*. Chronica Botanica, Waltham, Mass.
Fryxell, G.A. & Hasle, G.R. 1971. *Corethron criophilum* Castracane: its distribution and structure. *Ant. Res. Ser.* 17: 335-46.
Gayral, P. 1972. Sur les Chrysophycées à zoides phéophycéens notamment *Sarcinochrysis marina* Geitler. *Bull. Soc. Phycol. Fr.* 17: 40-5.
Gayral, P. & Billard, C. 1977. Synopsis du nouvel ordre des Sarcinochrysidales (Chrysophyceae). *Taxon* 26: 241-5.
Gayral, P. & Haas, C. 1969. Etude comparédes genres *Chrysomeris* Carter et *Giraudyopsis* P. Dangeard. Position systématique des Chrysomeridaceae (Chrysophyceae). *Rev. Gén. Bot.* 76: 659-66.
Gayral, P. & Lepailleur, H. 1971. Etude de deux Chrysophycées filamenteuses *Nematochrysopsis roscoffensis* Chadefaud, *Nematochrysis hieroglyphica* Waern. *Rev. Gén. Bot.* 78: 61-74.
Hargraves, P.E. 1976. Studies on marine plankton diatoms. II. Resting spore morphology. *J. Phycol.* 12: 118-28.
Hibberd, D.J. 1972. Chrysophyta: definition and interpretation. *Br. Phycol. J.* 7: 281.
- 1976. The ultrastructure and taxonomy of the Chrysophyceae and Prymnesiophyceae (Haptophyceae): a survey with some new observations on the ultrastructure of the Chrysophyceae. *Bot. J. Linn. Soc.* 72: 55-80.
- 1977. Ultrastructure of the cyst formation in *Ochromonas tuberculata* (Chrysophyceae). *J. Phycol.* 13: 209-320.
- 1978. Possible phylogenetic value of the transitional helix in some chromophyte algal classes and some colourless protists. *BioSystems* 10: 115-16.
- 1980. Prymnesiophytes (= Haptophytes). *In* Cox, E.R. [Ed.] *Phytoflagellates*. Elsevier Science (North Holland), Amsterdam, pp. 223-317.

– 1981. Notes on the taxonomy and nomenclature of the algal classes Eustigmatophyceae and Tribophyceae (synonym Xanthophyceae). *Bot. J. Linn. Soc.* 82: 93–119.
Hibberd, D.J. & Chretiennot-Dinet, M-J. 1979. The ultrastructure and taxonomy of *Rhizochromulina marina* gen. et sp. nov., an amoeboid marine chrysophyte. *J. Mar. Biol. Ass. U.K.* 59: 179–93.
Hibberd, D.J. & Leedale, G.F. 1972. Observations on the cytology and ultrastructure of the new algal class, Eustigmatophyceae. *Ann. Bot.* 36: 49–71.
Hilliard, D.K. & Asmund, B. 1963. Studies on Chrysophyceae from some ponds and lakes in Alaska. 2. Notes on the genera *Dinobryon, Hyalobryon* and *Epipyxis* with descriptions of new species. *Hydrobiol.* 22: 331–400.
Karim, A.G.A. & Round, F.E. 1967. Microfibrils in the lorica of the freshwater alga *Dinobryon. The New Phytologist,* 66: 409–12.
Klebs, G. 1892. Flagellatenstudien. II. *Z. Wiss. Zool.* 55: 353–445.
Kristiansen, J. 1969. Lorica structure in *Chrysolykos* (Chrysophyceae). *Bot. Tidskr.* 64: 162–8.
– 1972. Studies on the lorica structure in Chrysophyceae. *Svensk. Bot. Tidskr.* 66: 184–90.
– 1982. Chromophycota–Chrysophyceae. *In:* Parker, S.P. [Ed.] *Synopsis and Classification of Living Organisms.* McGraw-Hill, New York, pp. 81–6.
Leedale, G.F. 1974. How many are the kingdoms of organisms? *Taxon,* 23: 261–70.
Loeblich, A.R. III. & Fine, F.E. 1977. Marine chloromonads: More widely distributed in neritic environments than previously thought. *Proc. Biol. Soc. Wash.* 90(2): 388–99.
Loiseaux, S. 1967. Sur la position systematique du genre *Giraudyopsis* P. Dangeard. *Rev. Gen. Bot.* 74: 389–97.
Luther, A. 1899. Ueber *Chlorosaccus* eine neue Gattung der Susswasseralgen nebst Bemerkungen zur Systematik verwandter Algen. *Bih K. Svenska Vetensk. Akad. Handl. 24,* Afd. III, No. 13: 1–22.
Maekawa, F. 1960. A new attempt in phylogenetic classification of plant kingdom. *J. Fac. Sci. Univ. Tokyo. Bot.* 7: 543–69.
Manton, I. 1955. Observations with the electron microscope on *Synura caroliniana* Whitford. *Proc. Leeds Phil. Lit. Soc.* 6: 306–16.
Manton, I. & Harris, K. 1966. Observations on the microanatomy of the brown flagellate *Sphaleromantis tetragona* Skuja with special reference to the flagellar apparatus and scales. *J. Linn. Soc. (Bot.)* 59: 397–403.
Mattox, K.R. & Stewart, K.D. 1984. Classification of the green algae: A concept based on comparative cytology. Chap 2. *In:* Irvine, D.E.G. & John, D.M. [Eds.] *Systematics of the Green Algae.* Academic Press, London, pp. 29–72.
McGrory, C.B. and Leadbeater, B.S.C. 1981. Ultrastructure and deposition of silica in the Chrysophyceae. *In:* Simpson, T.L. and Volcani, B.E. [Eds.] *Silicon and Siliceous Structures in Biological Systems.* Springer-Verlag, New York, pp. 201–30.
Moestrup, Ø 1982. Flagellar structure in algae: a review with new observations particularly on the Chrysophyceae, Phaeophyceae (Fucophyceae), Euglenophyceae and *Reckertia. Phycologia,* 21: 427–528.
Norris, R.E. 1977. Flagellate cells in the life-history of *Stichogloea* (Chrysophyceae). *Phycologia,* 16(1): 75–8.
Parke, M. 1961. Some remarks concerning the class Chrysophyceae. *Br. Phycol. Bull.* 2: 47–55.
Pascher, A. 1914. Über Flagellaten und Algen. *Ber. Deutschen Bot. Ges.* 32: 136–60.

– 1924. Zur Homologisierung der Chrysomonadencysten mit den Endosporen den Diatomeen. *Arch. Protist.* 48: 196–203.
– 1925. Die braune Algenreihe der Chrysophyceae. *Arch. Protist.* 52: 489–563.
– 1931. Systematische Übersicht über die mit Flagellaten in Zusammenhang stehenden Algenreihen und Versuch einer Einreihung dieser Algenstämme in die Stämme des Pflanzenreiches. *Beih. Bot. Zbl.* 48: 317–32.
– 1932. Über die Verbreitung endogener bzw. endoplasmatisch gebildeter Sporen bei den Algen. *Beih. Bot. Centralbl.* 49: 293–308.
– 1939. Heterokonten. *Rabenhorst's Kryptogamen-Flora von Deutschland, Österreich und der Schweiz,* Vol. II. Akademische Verlagsgesellshaft, Leipzig.
Peterfi, L.S. 1969. The fine structure of *Poterioochromonas malhamensis* Pringsheim comb. nov. with special reference to the lorica. *Nov. Hedwig.* 17: 93–103.
Pienaar, R.N. 1980. Chrysophytes. *In:* Cox, E.R. [Ed.] *Phytoflagellates.* Elsevier Science (North-Holland), Amsterdam, pp. 213–42.
Preisig, H.R. & Hibberd, D.J. 1982a. Ultrastructure and taxonomy of *Paraphysomonas* (Chrysophyceae) and related genera. 1. *Nord. J. Bot.* 2: 397–420.
– 1982b. Ultrastructure and taxonomy of *Paraphysomonas* (Chrysophyceae) and related genera. 2. *Nord. J. Bot.* 2: 601–38.
– 1983. Ultrastructure and taxonomy of *Paraphysomonas* (Chrysophyceae) and related genera. 3. *Nord. J. Bot.* 3: 695–723.
Prowazek, S. 1903. Flagellatenstudien. *Arch. f. Protistenk.* 2: 195–212.
Round, F.E. 1980. The evolution of pigmented and unpigmented unicells—A reconsideration of the Protistia. *BioSystems* 12: 61–9.
– 1981. Some aspects of the origin of diatoms and their subsequent evolution. *BioSystems* 14: 483–6.
– 1984. The systematics of the Chlorophyta: An historical review leading to some modern concepts [taxonomy of the Chlorophyta III]. Chap. 1. *In:* Irvine, D.E.G. & John, D.M. [Eds.] *Systematics of the Green Algae.* Academic Press, London, pp. 1–27.
Round, F.E. & Crawford, R.M. 1981. The lines of evolution of the Bacillariophyta. I. Origin. *Proc. R. Soc. Lond. B.* 211: 237–60.
Round, F.E. & Sims, P.A. 1980. The distribution of diatom genera in marine and freshwater environments and some evolutionary considerations. *6th Int. Diat. Symp. Budapest,* pp. 301–20.
Sandgren, C.D. 1980a. An ultrastructural investigation of resting cyst formation in *Dinobryon cylindricum* Imhof (Chrysophyceae, Chrysophycota). *Protistologica* 16: 259–76.
– 1980b. Resting cyst formation in selected chrysophyte flagellates: An ultrastructural survey including a proposal for the phylogenetic significance of interspecific variations in the encystment process. *Protistologica* 16: 289–303.
Scherffel, A. 1911. Beitrag zur Kenntnis der Chrysomonadineen. *Arch. Protist.* 22: 299–344.
– 1924. Über die Cyste von *Monas. Arch. Protist.* 48: 187–95.
Sheath, R.G., Hellebust, J.A. & Sawa, T. 1975. The statospore of *Dinobryon divergens* Imhof: formation and germination in a subarctic lake. *J. Phycol.* 11: 131–8.
Stebbins, G.L. 1982. *Darwin to DNA, Molecules to Humanity.* Freeman, San Francisco.
Takahashi, E. 1964. Studies on genera *Mallomonas, Synura* and other plankton in freshwater with the electron microscope. (5) On the similarity of the

fine structure between scale of *Mallomonas* and frustule of diatom. *Bull. Yamagata Univ. Agr.Sci.* 4: 137–45.

Thomsen, H.A., Zimmerman, B., Moestrup, Ø., & Kristiansen, J. 1981. Some new freshwater species of *Paraphysomonas* (Chrysophyceae). *Nord. J. Bot.* 1: 559–81.

Wujek, D.E. & Asmund, B.C. 1979. *Mallomonas cyathellata* sp. nov. and *Mallomonas cyathellata* var. *kenyana* var. nov. (Chrysophyceae) studied by means of scanning and transmission electron microscopy. *Phycologia* 18: 115–19.

2

Ultrastructure of the Chrysophyceae – phylogenetic implications and taxonomy

DAVID J. HIBBERD[1]
Institute of Terrestrial Ecology, The Culture Centre of Algae and Protozoa
36, Storey's Way, Cambridge CB3 ODT, England

As originally defined (Pascher 1914), the class Chrysophyceae comprised golden-brown flagellates together with their presumed nonmotile and colorless relatives; it was not distinguished by the possession of some easily recognized characters, as are, for example, the Cryptophyceae and Dinophyceae. Data from electron microscopy (EM) have subsequently shown that the Chrysophyceae *sensu lato* is an unnatural assemblage, containing a wide range of mutually unrelated taxa. I propose that it should now be possible to define the Chrysophyceae *sensu stricto* as a firmly circumscribed class and to distribute the remaining, excluded taxa between a number of new or existing algal classes and protozoan orders.

Data on the Chrysophyceae *sensu stricto* have been surveyed in a previous paper (Hibberd 1976a), and the ultrastructure of the group has also recently been reviewed by Pienaar (1980). The cytological features of the class are recapitulated here, and against this background are examined, in turn, all of the taxa previously classified within the Chrysophyceae whose characters do not conform to this basic pattern.

Chrysophyceae *sensu stricto*

The main conclusion from my earlier survey (Hibberd 1976a) was that the EM data then available supported the original idea of Bourrelly (1957) that the *Ochromonas* type of organization was the basic form from which a large part of the Chrysophyceae could be derived by a variety of modifications and reductions. This type of organization is summarized diagrammatically in Hibberd (1976a, Fig. 1) and redrawn by Pienaar (1980, Fig. 2). Its basic features include (a) naked bilaterally symmetrical cells with a pair of anteriorly directed, unequal length flagella, inserted subterminally into the cell at an oblique angle to each other; (b) the presence of tripartite

[1] Present address: Axle Tree Cottage, Starvecrow Lane, Peasmarsh, Rye, East Sussex TN31 6XL, England.
I am grateful to Dr. F.K. Kupicha for her help in preparing this paper.

flagellar hairs on the long flagellum only; (c) a swelling, usually with electron-dense contents, at the proximal end of the short flagellum, this swelling lying in a shallow depression in the cell surface directly over the eyespot (stigma); (d) a flagellar transition region containing a transitional helix (*sensu* Hibberd 1979) and a transverse partition at the level of the cell surface with a central axosomal thickening; (e) two categories of flagellar root, one consisting of relatively small numbers of microtubules that run superficially and the other a larger cross-banded root extending deeply into the cell, its branches ramifying over the surface of the nucleus; (f) a single pyriform nucleus with its narrow end extended toward the flagellar basal bodies, the outer membrane of the nuclear envelope confluent with that of the chloroplast ER; (g) one or two chloroplasts per cell with regularly arranged three-thylakoid lamellae and a peripheral girdle lamella; (h) a single-layered eyespot at the edge of the chloroplast, this region being pressed closely against the flagellar depression mentioned above; (i) a single, large Golgi body, relatively easily visible in the light microscope, lying against the nucleus in the anterior end of the cell, often in a concavity in the nuclear envelope, and with one edge directed toward the flagella with the few cisternae proximal to the nucleus being structurally modified and extended as far as the basal bodies; (j) chrysolaminaran vesicles usually filling the posterior part of the cell; and (k) a contractile vacuole system typically occupying the extreme anterior end of the cell.

Chrysophycean cysts, termed stomatocysts or statospores, are also highly characteristic in both their structure and formation. Thus, whereas the general type of cell organization outlined above is also characteristic of motile cells of the Tribophyceae (Xanthophyceae) and Fucophyceae (Phaeophyceae) (see later in this section), the presence of stomatocysts is the sole feature by which a member of the Chrysophyceae may unequivocally be recognized. The cyst wall is composed of silica and may be either smooth or variously ornamented, is formed endogenously, and has a small pore, mostly surrounded by a collar, which is closed at maturity by a plug. The extracystic cytoplasm is either withdrawn into the cyst before plug formation or is lost. The ultrastructure of cyst formation has been described by Hibberd (1977a) and Sandgren (1980a, b), and while details vary, particularly those relating to nuclear events, the cyst wall in all cases is laid down between the membranes of a large silica deposition vesicle, the plug being formed later from Golgi-derived material.

The majority of EM data since the 1976 survey support the view that the *Ochromonas* type of organization is basic for the Chrysophyceae (see, for example, Schnepf et al. 1977, Mesquita & Santos 1976, Pienaar 1976, Hibberd 1977b, 1978, Kristiansen & Walne 1977, Mignot 1977, Dop 1978, Couté & Preisig 1981, Andersen 1982, Preisig & Hibberd 1983). It must be pointed out, however, that a chloroplast girdle lamella has been shown to be absent in several species from a range of genera (Hibberd 1971, Hibberd & Chretiennot-Dinet 1979, Couté & Preisig 1981, Preisig & Hibberd 1983). In addition, three basic types of microtubular flagellar root are now recognized: one running around the flagellar depression; the second running beneath the plasmalemma into the anterior end of the cell; and the third running posteriorly (Schnepf et al. 1977, Mignot 1978). Modification from this basic pattern occurs by loss of one or two of these roots.

Against this background will now be examined the cell structure of the other forms hitherto included within the Chrysophyceae, dealing first with pigmented and then with colorless forms. In order to assess the phylogenetic implications, and therefore the possible taxonomic conclusions resulting from these comparisons, it is necessary to mention briefly the phylogenetic relationships of the Chrysophyceae *sensu stricto.* As recently discussed (Hibberd 1979), similarities in motile cell structure show that the Chrysophyceae, Fucophyceae, and Tribophyceae must have a common ancestry, and the Raphidophyceae (chloromonads) and Diatomophyceae (Bacillariophyceae) also appear closely related to this line of evolution. The relationship between the Chrysophyceae and brown algae is so close that the Fucophyceae could be said to consist of the clearly differentiated multicellular representatives of a group comprising both classes (Christensen 1980). The brown algae are maintained as a separate class because of their differentiation into vegetative and reproductive cells, the existence of cytoplasmic connections between the cells in advanced forms, and the mostly laterally inserted flagella in the motile cells. If one aims to construct a natural classification, then any group within such a classification should contain only organisms that are more closely related to one another than to organisms outside that group. There is no minimal size requirement for any taxonomic category, and the individuality of a taxon should not be obscured by tucking it into another group for the sake of neatness (Leedale 1974). In the present context, the Chrysophyceae *sensu stricto* is monophyletic, as far as can be judged, and can be included with the Tribophyceae, Fucophyceae, and Raphidophyceae and other possibly related groups, within a variously conceived and arguably natural division: Chrysophyta or Heterokontophyta (Hibberd 1979). The taxonomic position of any group excluded from the Chrysophyceae must be considered not only in relation to the Chrysophyceae *sensu stricto* but also to this larger group.

Prymnesiophyceae

The most obvious change in the classification of the Chrysophyceae influenced by the results of EM has been the separation of most forms possessing a pair of more or less equal length flagella into a separate class known as the Prymnesiophyceae or Haptophyceae. Algae with this type of flagellation have always formed a separate group within the Chrysophyceae, originally being placed in the order Isochrysidales by Pascher (1910) and forming the subclass Isochrysophycidae in the classifications of Bourrelly (1968, 1981). The separation of the Haptophyceae from the Chrysophyceae by Christensen in 1962 has been strongly supported by all the information from EM produced since then. The recognition of the new class is now almost universal, but its acceptance was slow probably because of the lack of comparative data on the Chrysophyceae *sensu stricto.* My detailed comparison of the two groups (Hibberd 1976a) remains essentially unmodified by subsequent information, the only significant new data concerning the structure and formation of prymnesiophycean cysts. These have now been studied by EM in a single genus (*Prymnesium*) and found to be fundamentally different from those of the Chrysophyceae, being produced exogenously by the deposition and accretion of siliceous material on the distal face of the outermost layer of scales (Green, Hibberd, & Pienaar 1982, R.N. Pienaar personal communication).

Dictyochophyceae

The silicoflagellates were not included in my 1976 review, being considered too remote from the Chrysophyceae to warrant comparison. They have been assigned by various authors to a separate class, for which the typified name Dictyochophyceae is now available (Silva 1980). However, several contemporary workers place them within the Chrysophyceae, and this appears to be the consensus of taxonomic opinion. Thus Kristiansen (1982) classifies silicoflagellates in the subclass Dictyochophycidae of Chrysophyceae; Christensen (1980) gives them only ordinal status as Dictyochales; and Van Valkenburg (1980) states that "silicoflagellates are obviously chrysophytes." The characters on which these judgments are based are apparently the type of pigmentation and storage product, the structure of chloroplasts and mitochondria, and the deposition of biological silica. Nevertheless, the work of Van Valkenburg on *Dictyocha fibula* Ehr. and data on the ultrastructure of *Distephanos speculum* (Ehr.) Haeckel (Ø. Moestrup & H.A. Thomsen personal communication) show clearly that silicoflagellates differ from chrysophytes in virtually every aspect of cell structure. Typical forms also possess an "internal" silica skeleton composed of tubular elements, a feature unknown in the Chrysophyceae, and stomatocysts are never produced. Silicoflagellates are certainly less closely related to the Chrysophyceae than are the heterokont classes, and they should be segregated in a separate class.

Pedinellaceae

The family Pedinellaceae has always been a well-defined group within the Chrysophyceae, having originally been erected by Pascher in 1910 on the basis of light microscope characteristics alone. The EM studies of *Pedin-*

ella by Swale (1969), of *Apedinella* by Throndsen (1971), and of *Pseudopedinella* by Ostroff and Van Valkenburg (1978) demonstrate that these taxa show little or no resemblance to the *Ochromonas* pattern of organization. Thus, the cells are radially symmetrical, more or less apple shaped, and contain several regularly arranged chloroplasts and a large central nucleus. There is only a single emergent flagellum, with a second basal body parallel to the flagellar base (a feature known in only one other chrysophyte, *Rhizochromulina marina*: see section on aberrant Chrysophyceae). The flagellum bears unilateral flagellar hairs, and its membrane is expanded into a fin supported by a paraxial rod. There are apparently no microtubular flagellar roots, and a rhizoplast also appears to be absent, the flagellar basal body and supernumerary basal body lying very close to the nuclear envelope. Finally *Pedinella* and *Pseudopedinella* possess a peculiar and apparently unique filiform but contractile peduncle at their posterior end.

These considerations led me in 1976 to propose that the family Pedinellaceae should be given a more isolated position in the Chrysophyceae, and in the absence of any species showing an organization intermediate between the pedinellids and the main chrysophycean line, I suggested that it may be separated from the Chrysophyceae by a phyletic distance at least as great as that between the classes of heterokont algae (i.e., it could form the basis of a new algal class).

The one feature that would unequivocally link pedinellids with the Chrysophyceae would be the production of stomatocysts. These, however, have never been found, and preliminary work on an isolate of *Apedinella* (D.J. Hibberd, personal observations) demonstrated the formation of cysts that were not of the chrysophycean type and were apparently unsilicified.

Both Davidson (1982) and Patterson (Chapter 4) have recently demonstrated some remarkable similarities between pedinellids and helioflagellates (working on *Ciliophrys marina* Caullery and *Pteridomonas* sp., respectively), although some differences in flagellar structure were also found. Both authors propose a phylogenetic link between pedinellids and the actinophryd heliozoa. Patterson derives all the actinophryd heliozoa from the Chrysophyceae via pedinellids and *Pteridomonas,* thus envisaging the heliozoa as "aberrant" chrysophytes.

Bearing in mind current ideas concerning the evolution of chloroplasts in chromophyte algae by symbiosis with eukaryotes (Whatley & Whatley 1981, Cavalier-Smith 1982), the present evidence makes it most likely that pedinellids arose as an offshoot from a primarily colorless flagellate line leading to the heliozoa, a line remote from that of the Chrysophyceae and other heterokont classes. As more data become available, it seems probable that the pedinellids will be segregated into a separate algal class. Their undoubted relationship with the heliozoans may not be very close, and in zoological nomenclature it would seem most sensible to classify helioflagellates only within the Zoomastigophorea, the pedinellids forming a separate order in the Phytomastigophorea.

Olisthodiscus luteus

When first described by Carter in 1937, *Olisthodiscus luteus* was doubtfully assigned to the Xanthophyceae, but it has since more often been treated as a chrysophyte. This was the conclusion of Leadbeater (1969) on the basis of an early EM study, but Loeblich & Fine (1977) subsequently transferred *O. luteus* to the Raphidophyceae. In spite of the arguments presented by Gibbs, Chu, and Magnussen (1980) in favor of regarding this organism as a member of the Chrysophyceae, I feel that its taxonomic position is still uncertain and will need to be reconsidered in future. The reasons for this are outlined briefly as follows.

The photosynthetic pigments of *O. luteus* include a large proportion of fucoxanthin, which suggests a position in the Chrysophyceae rather than the Tribophyceae (=Xanthophyceae) or Raphidophyceae. On the other hand, the cells have multiple chloroplasts, a ring of Golgi bodies surrounding the anterior end of the nucleus, and no trace of a photoreceptor apparatus. In addition, the short flagellum is relatively longer than in typical chrysophytes. The key to the natural affinities of *Olisthodiscus* probably lies in the structure of the flagellar root system. This is still very poorly illustrated in the literature, but Leadbeater has shown a most unusual flagellar root, interpreted by Moestrup (1982) as consisting of about 22 microtubules, associated on one side with an electron-dense plate and on the other with projections extending to a multilayered structure, the whole arrangement being constant and precise. A closely similar root is present in typical members of the Raphidophyceae, but is unknown in other groups (Moestrup 1982). The multiple chloroplasts, Golgi system, and absence of a photoreceptor apparatus all are consistent with a position in the Raphidophyceae. However, the cells of *Olisthodiscus* are smaller than those of most raphidophytes, and the cytoplasm is not divided into endoplasmic and ectoplasmic regions. Returning to the subject of photosynthetic pigment composition, raphidophytes apparently vary in this respect: the best-known genera, *Vacuolaria* and *Gonyostomum,* lack fucoxanthin, but this has been found in isolates identified as species of *Chattonella* (Loeblich & Fine 1977), and *Chattonella subsalsa* Biecheler is undoubtedly a member of the Raphidophyceae (Mignot 1976). This evidence suggests that the Raphidophyceae may have acquired chloroplasts by endosymbiosis with more than one group of eukaryotes relatively late in its evolutionary history.

"Aberrant" Chrysophyceae

Exclusion from the Chrysophyceae of the taxa mentioned above and the colorless forms dealt with below leaves a much more unified group. However, there are a few more taxa that differ from the basic pattern of chrysophycean organization and yet are placed within the Chrysophyceae, because links cannot be found elsewhere. It is not yet clear whether these forms are misplaced in the Chrysophyceae or whether the accepted range of variation of this class should be widened to encompass them.

Rhizochromulina marina *Hibberd & Chretiennot-Dinet*

This amebo-flagellate possesses fucoxanthin as a photosynthetic pigment but is so far unique among the Chrysophyceae in having lageniform zoospores with no emergent second flagellum, the latter being reduced to a very short supernumerary basal body. Stomatocysts are unknown. As when it was first described (Hibberd & Chretiennot-Dinet 1979), there still seems no alternative to placing this monospecific genus in the Chrysophyceae, and it will be interesting to see whether any other chrysophytes lack an emergent second flagellum.

Pelagococcus subviridis *Norris*

The very small, pale green coccoid cells of this oceanic species, recently described by Lewin et al. (1977), have a simple cell structure not at variance with chrysophycean organization, but motile cells and stomatocysts are unknown. Photosynthetic pigment composition is unusual, although fucoxanthinlike carotenoids are present. At present, there is no alternative but to place this species in the Chrysophyceae. It is possible that *P. subviridis* represents only one of a number of oceanic chrysophytes that have formed a separate line of morphological development; this idea is supported by the parallel existence of some coccoid marine eustigmatophytes (Hibberd 1981), and one coccoid marine prasinophyte (Foss et al. 1984), all of which have very small cells and a much simpler structure than their more typical relatives.

Colorless forms

Bicosoeca

The genus *Bicosoeca* consists wholly of colorless, loricate flagellates, with the cells attached to the base of their lorica by the tip of the shorter of the two flagella. Such an arrangement is unknown in the Chrysophyceae, but

Bicosoeca does show some points of resemblance with the *Ochromonas* type of organization. It has been classified in a variety of positions: as Chrysophyceae order Bicosoecales (Bourrelly 1968, 1981) or family Bicosoecaceae (Christensen 1980); as the algal class Bicosoecophyceae (Casper 1974); and as the protozoan order Bicosoecida in the class Zoomastigophorea (Honigberg et al. 1964). For reasons unknown, this order is excluded from the most recent classification of the Society of Protozoologists (Levine et al. 1980).

Details of the internal ultrastructure of several species of *Bicosoeca* have been given by Mignot (1974a), Belcher (1975), and Moestrup & Thomsen (1976), and from these studies and personal observations on other species it can be seen that *Bicosoeca* resembles the Chrysophyceae in possessing tripartite tubular flagellar hairs (although only in a single row, in most species) and in having a similar spatial relationship between Golgi body, nucleus, and flagellar basal bodies. Also, the fibrillar structure of the lorica of *B. maris* Picken resembles that of some undoubted chrysophytes, e.g., *Dinobryon.* The main features of *Bicosoeca* that are unusual for Chrysophyceae include the peristome; a paraxial inclusion in the hind flagellum in some species; probable absence of a transitional helix and descending microfibrillar root; and a more complex flagellar root system, which, however, cannot be fully compared with that of the chrysophytes, which is still insufficiently known. All authors who have made ultrastructural studies on *Bicosoeca* agree that this genus is related to the Chrysophyceae, and this argument is phylogenetically feasible. However, since all members are colorless, and their cell structure differs from typical chrysophycean organization more than that of the heterokont algal classes, it is unreasonable to classify *Bicosoeca* within the Chrysophyceae *sensu stricto.* The relationship with the Chrysophyceae while plausible does not appear to be very close, and therefore the most logical taxonomic decision is to classify *Bicosoeca* only in the zooflagellate order Bicosoecida.

Rhipidodendron *and* Spongomonas

These genera were erected by Stein (1878) for colorless flagellates forming colonies with a granular matrix. In botanical classification they have been referred to the Chrysophyceae, but data from EM (Hibberd 1976b) has shown that their fine structure is inconsistent with this placing; they differ from typical chrysophytes in every feature of cell organization, including the flagellar apparatus. Moreover, their structure also differs from that of all other classes of algae and they do not appear to be related to any of the existing groups of protozoa. In a recent paper including fresh information on *Spongomonas,* I have referred these genera to the new order Spongomonadida in the protozoan class Zoomastigophorea (Hibberd 1983).

Pseudodendromonas *and* Cyathobodo

The genus *Pseudodendromonas* was erected by Bourrelly (1953) for colorless chrysophytes having flattened cells borne at the ends of branches of dichotomous corymbose colonies; it was placed in the subclass Isochrysophycidae (Bourrelly 1968) because each cell has two subequal homodynamic flagella. *Cyathobodo* was described to include individually stalked cells with their two subequal flagella inserted at one edge of the flat anterior end of the cell (Petersen & Hansen 1961); its authors referred *Cyathobodo* to an undefined position in the protozoa.

The subsequent discovery of more species in both genera, and further studies on their ultrastructure (Mignot 1974b, Swale & Belcher 1975, Hibberd 1976c, 1985) have shown that they are closely related and that their organization differs fundamentally from the chrysophycean and prymnesiophycean types. Thus both have cells with a characteristic shape and symmetry, two smooth flagella, a cytopharynx functioning in the phagotrophy of bacteria, and a basal pad to which the stalk is attached. They show no trace of a haptonema, and although they produce unmineralized scales in Golgi cisternae, these are of a different type from those prymnesiophytes. This small group appears to be of isolated taxonomic position within the protozoa, and I have recently described the new order Pseudodendromonadida to accommodate them within the class Zoomastigophorea (Hibberd 1985).

Choanoflagellates

These were included by Bourrelly (1968) as the subclass Craspedomonadophycidae of the Chrysophyceae, and have also been treated as the separate class Craspedophyceae by Chadefaud (1960) and Christensen (1962). In the protozoa, they form the order Choanoflagellida in the class Zoomastigophorea (Levine et al. 1980). It is known from EM studies that they have no structural similarities with any group of algae (Leadbeater & Morton 1974; Hibberd 1975), and it is now generally accepted that they should be classified only as protozoa. A few forms have been reported to contain chloroplasts: *Stylochromonas minuta* Lackey (Lackey 1940), *Microsportella fiordensis* Scagel & Stein, and *Pseudomicrosportella ornata* Scagel & Stein (Scagel & Stein 1961). None has yet been studied by EM, and one suspects that they are almost certainly unrelated to either chrysophytes or choanoflagellates; it would be interesting to bring them into culture and clarify the situation.

Phalansterium

This small genus of colonial colorless flagellates forms dichotomously branched colonies and has hitherto been classified with the choanoflagellates (e.g. by Bourrelly 1957, 1968, Skuja 1964) because the cells are uniflagellate and have a long thin collar surrounding the proximal end of the flagellum. A recent EM study (Hibberd 1983) shows that *Phalansterium* differs in cell structure from both the Chrysophyceae and the choanoflagellates in virtually every feature of cell organization (Hibberd 1983). Thus the cells are truly uniflagellate, with no supernumerary basal body as in choanoflagellates; the collar is a continuous structure rather than being composed of separate tentacles; and the mitochondria have tubular, not flattened, cristae. This latter feature in particular emphasizes the wide phylogenetic distance between *Phalansterium* and choanoflagellates. *Phalansterium* is known to produce cysts, but these are not of the chrysophycean type, being unsilicified and produced exogenously. It is clear that this isolated group does not belong with the algae, and after comparison with other groups of protozoa, it has been classified in the new order Phalansteriida of the class Zoomastigophorea (Hibberd 1983)

Conclusion

Removal from the Chrysophyceae of prymnesiophytes, silicoflagellates, pedinellids, choanoflagellates, and the genera *Olisthodiscus, Bicosoeca, Rhipidodendron, Spongomonas, Pseudodendromonas, Cyathobodo,* and *Phalansterium* leaves a more or less well circumscribed, natural group. The alternative to accepting a narrow definition of the Chrysophyceae is to include one or more of these segregate taxa. This must depend on individual judgement of phyletic distances and the concept of the Chrysophyceae will undoubtedly change with time. My own preference for a narrowly circumscribed class is influenced by the belief that any other course brings with it the practical difficulty of submerging or understating important differences between taxa. This in turn discourages further investigation by nontaxonomists who may be unaware of the areas of special interest associated with segregate taxa.

Electron microscopic data will, of course, be invaluable in establishing a natural taxonomic system within the Chrysophyceae *sensu stricto,* and the segregation of certain silica-scale-bearing genera formerly classified in the Mallomonadaceae into the separate family Paraphysomonadaceae is a recent example (Preisig & Hibberd 1983). The phylogenetic, and hence taxonomic, significance of the extensive range of cytological and biochemical differences between the Mallomonadaceae *sensu stricto* and the rest of the Chrysophyceae pointed out by Andersen & Mulkey (1983) remains enigmatic, especially in view of the fact that members of this group possess Golgi bodies with the unique chrysophycean structure and also produce stomatocysts. The phyletic distance between the marine benthic Chrysophyceae producing zoospores with laterally inserted flagella (order Sarcinochrysidales *sensu* Gayral & Billard 1977) and the main chrysophycean line may also be greater than formerly realized (Gayral & Billard, Chapter

3) and this group may occupy a truly intermediate position between Chrysophyceae and Fucophyceae.

References

Andersen, R.A. 1982. A light and electron microscopical investigation of *Ochromonas sphaerocystis* Matvienko (Chrysophyceae): the statospore, vegetative cell and its peripheral vesicles. *Phycologia* 21: 390-8.

Andersen, R.A. & Mulkey, T.J. 1983. The occurrence of chlorophylls c_1 and c_2 in the Chrysophyceae. *J. Phycol.* 19: 289-94.

Belcher, J.H. 1975. The fine structure of the loricate colourless flagellate *Bicoeca planctonica* Kisselew. *Arch. Protistenk.* 117: 78-84.

Bourrelly, P. 1953. Flagellés incolores rares ou nouveaux. *Öst. Bot. Z.* 100: 533-9.

- 1957. Recherches sur les Chrysophycées. *Revue Algol., Mém. Hors. Sér.*, 1: 1-412.

- 1968. *Les Algues d'Eau Douce, 2: Algues Jaunes et Brunes.* Boubée, Paris.

- 1981. *Les Algues d'Eau Douce, 2: Algues Jaunes et Brunes,* 2nd ed. Boubée, Paris.

Carter, N. 1937. New or interesting algae from brackish water. *Arch. Protistenk.* 90: 1-68.

Casper, T. 1974. Abteilung Goldalgen, Chrysophyta. *Urania Pflanzenreich* 1: 145-58.

Cavalier-Smith, T. 1982. The origins of plastids. *Biol. J. Linn. Soc.* 17: 289-306.

Chadefaud, M. 1960. Les végétaux non vasculaires (Cryptogamie). *In:* Chadefaud, M. & Emberger, L. [Eds.] *Traité de Botanique, Systématique,* Vol. 3. Massot, Paris, pp. 1-1018.

Christensen, T. 1962. Alger. *In:* Böcher, T.W., Lange, M. & Sørensen, T. [Eds.] *Botanik* Bd. 2, *Systematisk Botanik,* Nr. 2. Munksgaard, Copenhagen.

- 1980. *Algae, a Taxonomic Survey.* Fasc. 1. AiO Tryk, Odense, Denmark.

Couté, A. & Preisig, H.R. 1981. Sur l'ultrastructure de *Microglena butcheri* Belcher (Chrysophyceae, Ochromonadales, Synuraceae) et sur sa position systematique. *Protistologica* 17: 465-77.

Davidson, L.A. 1982. Ultrastructure, behaviour, and algal flagellate affinites of the helioflagellate *Ciliophrys marina,* and the classification of the helioflagellates (Protista, Actinopoda, Heliozoea). *J. Protozool.* 29: 19-29.

Dop, A.J. 1978. Systematics and morphology of *Chrysochaete brittanica* (Godward) Rosenberg and *Phaeoplaca thallosa* Chodat (Chrysophyceae). *Acta Bot. Neerl.* 27: 35-60.

Foss, P., Guillard, R.R.L. & Liaaen-Jensen, S. 1984. Prasinoxanthin - a chemosystemic marker for algae. *Phytochemistry* 23: 1629-33.

Gayral, P. & Billard, C. 1977. Synopsis du nouvel ordre des Sarcinochrysidales (Chrysophyceae). *Taxon* 26: 241-5.

Gibbs, S.P., Chu, L.L. & Magnussen, C. 1980. Evidence that *Olisthodiscus luteus* is a member of the Chrysophyceae. *Phycologia* 19: 173-7.

Green, J.C., Hibberd, D.J. & Pienaar, R.N. 1982. The taxonomy of *Prymnesium* (Prymnesiophyceae) including a description of a new cosmopolitan species, *P. patellifera* sp.nov., and further observations on *P. parvum* N. Carter. *Br. Phycol. J.* 17: 363-82.

Hibberd, D.J. 1971. Observations on the cytology and ultrastructure of *Chrysamoeba radians* Klebs (Chrysophyceae). *Br. Phycol. J.* 6: 207-23.

– 1975. Observations on the ultrastructure of the choanoflagellate *Codosiga botrytis* (Ehr.) Saville-Kent with special reference to the flagellar apparatus. *J. Cell Sci.* 17: 191–219.
– 1976a. The ultrastructure and taxonomy of the Chrysophyceae and Prymnesiophyceae (Haptophyceae): a survey with some new observations on the ultrastructure of the Chrysophyceae. *Bot. J. Linn. Soc.* 72: 55–80.
– 1976b. The fine structure of the colonial colorless flagellates *Rhipidodendron splendidum* Stein and *Spongomonas uvella* Stein with special reference to the flagellar apparatus. *J. Protozool.* 23: 374–85.
– 1976c. Observations on the ultrastructure of three new species of *Cyathobodo* Petersen & Hansen (*C. salpinx, C. intricatus* and *C. simplex*) and on the external morphology of *Pseudodendromonas vlkii* Bourrelly. *Protistologica* 12: 249–61.
– 1977a. Ultrastructure of cyst formation in *Ochromonas tuberculata* (Chrysophyceae). *J. Phycol.* 13: 309–20.
– 1977b. The cytology and ultrastructure of *Chrysonebula holmesii* Lund (Chrysophyceae), with special reference to the flagellar apparatus. *Br. Phycol. J.* 12: 369–83.
– 1978. The fine structure of *Synura sphagnicola* (Korsh.) Korsh. (Chrysophyceae). *Br. Phycol. J.* 13: 403–12.
– 1979. The structure and phylogenetic significance of the flagellar transition region in the chlorophyll *c*-containing algae. *BioSystems* 11: 243–61.
– 1981. Notes on the taxonomy and nomenclature of the algal classes Eustigmatophyceae and Tribophyceae (synonym Xanthophyceae). *Bot. J. Linn. Soc.* 82: 93–119.
– 1983. Ultrastructure of the colonial colourless zooflagellates *Phalansterium digitatum* Stein (Phalansteriida ord. nov.) and *Spongomonas uvella* Stein (Spongomonadida ord. nov.). *Protistologica* 19: 523–35.
– 1985. Observations on the ultrastructure of new species of *Pseudodendromonas* Bourreily (*P. operculifera* and *P. insignis*) and *Cyathobodo* Petersen & Hansen (*C. peltatus* and *C. gemmatus*), Pseudodendromonadida ord. nov *Arch. Protistenk.* 129: 3–11.
Hibberd, D.J. & Chretiennot-Dinet, M.-J. 1979. The ultrastructure and taxonomy of *Rhizochromulina marina* gen. et sp. nov., an amoeboid marine chrysophyte. *J. Mar. Biol. Ass. U.K.* 59: 179–93.
Honigberg, B.M., Balamuth, W., Bruce, E.C., Corliss, J.O., Gojdics, M., Hall, R.P., Kudo, R.R., Levine, N.D., Loeblich, A.R. Jr., Weiser, J., & Wenrich, D.H. 1964. A revised classification of the phylum Protozoa. *J. Protozool.* 11: 7–20.
Kristiansen, J. 1982. Chrysophyceae. *In:* Parker, S.P. [Ed.] *Synopsis and Classification of Living Organisms.* McGraw-Hill, New York, pp. 81–6.
Kristiansen, J. & Walne, P.L. 1977. Fine structure of photokinetic systems in *Dinobryon cylindricum* var. *alpinum* (Chrysophyceae). *Br. Phycol. J.* 12: 329–41.
Lackey, J.B. 1940. Some new flagellates from the Woods Hole area. *Am. Midl. Nat.* 23: 463–71.
Leadbeater, B.S.C. 1969. A fine structural study of *Olisthodiscus luteus* Carter. *Br. Phycol. J.* 4: 3–17.
Leadbeater, B.S.C. & Morton, C. 1974. A microscopical study of a marine species of *Codosiga* James-Clark (Choanoflagellata) with special reference to the ingestion of bacteria. *Biol. J. Linn. Soc.* 6: 337–47.
Leedale, G.F. 1974. How many are the kingdoms of organisms? *Taxon* 23: 261–70.

Levine, N.D., Corliss, J.O., Cox, F.E.G., Deroux, G., Grain, J., Honigberg, B.M., Leedale, G.F., Loeblich, A.R.III, Lom, J., Lynn, D., Merinfield, E.G., Page, F.C., Poljansky, G., Sprague, V., Vavra, J. & Wallace, F.G. 1980. A newly revised classification of the Protozoa. *J. Protozool.* 27: 37–58.

Lewin, J., Norris, R.E., Jeffrey, S.W. & Pearson, B.E. 1977. An aberrant chrysophycean alga, *Pelagococcus subviridis* gen. nov. et sp. nov. from the North Pacific Ocean. *J. Phycol.* 13: 259–66.

Loeblich, A.R., III & Fine, K.E. 1977. Marine chloromonads: more widely distributed in neritic environments than previously thought. *Proc. Biol. Soc. Wash.* 90: 388–99.

Mesquita, J.F. & Santos M.F. 1976. Études cytologiques sur les algues jaunes (Chrysophyceae). I. Ultrastructure de *Chrysocapsa epiphytica* Lund. *Bol. Soc. Brot.* 50: 63–98.

Mignot, J.-P. 1974a. Étude ultrastructurale des *Bicoeca,* protistes flagellés. *Protistologica* 10: 543–65.

– 1974b. Étude ultrastructurale d'un protiste flagellé incolore; *Pseudodendromonas vlkii* Bourrelly. *Protistologica* 10: 397–412.

– 1976. Compléments à l'étude des Chloromonadines. Ultrastructure de *Chattonella subsala* Biecheler. Flagellé d'eaux saumâtre. *Protistologica* 12: 279–93.

– 1977. Étude ultrastructurale d'un flagellé du genre *Spumella* Cienk. Chrysomonadine leucoplastidiée. *Protistologica* 13: 219–31.

– 1978. Sur l'organisation ultrastructurale de la cinétide chez quelques Chrysomonadida et son intérêt dans l'étude phylétique de ces Protistes. *In* Hutner, S.H. [Ed.] *Abstracts of papers read at the Fifth International Congress on Protozoology, 1977.*

Moestrup, Ø. 1982. Flagellar structure in algae: a review, with new observations particularly on the Chrysophyceae, Phaeophyceae (Fucophyceae), Euglenophyceae, and *Reckertia.* (Phycological Reviews 7) *Phycologia* 21: 427–528.

Moestrup, Ø. & Thomsen, H.A. 1976. Fine structural studies on the flagellate genus *Bicoeca.* I. *Bicoeca maris,* with particular emphasis on the flagellar apparatus. *Protistologica* 12: 101–20.

Ostroff, C.R. & Van Valkenburg, S.D. 1978. The fine structure of *Pseudopedinella pyriforme* Carter (Chrysophyceae). *Br. Phycol. J.* 13: 35–49.

Pascher, A. 1910. Chrysomonaden aus dem Hirschberger Grossteiche. *Monogr. Abh. Intern. Rev. Gesamt. Hydrobiol. Hydrogr.,* 1: 1–66.

– 1914. Über Flagellaten und Algen. *Ber. Dt. Bot. Ges.* 32: 136–60.

Petersen, J.B. & Hansen, J.B. 1961. On some neuston organisms. III. *Bot. Tidsskr.* 57: 293–305.

Pienaar, R.N. 1976. The microanatomy of *Sphaleromantis marina* sp. nov. (Chrysophyceae). *Br. Phycol. J..* 11: 83–92.

– 1980. Chrysophytes. *In* Cox, E.R. [Ed.] *Phytoflagellates (Developments in Marine Biology,* 2). Elsevier North-Holland, New York, pp. 213–42.

Preisig, H.R. & Hibberd, D.J. 1983. Ultrastructure and taxonomy of *Paraphysomonas* (Chrysophyceae) and related genera 3. *Nord. J. Bot.* 3: 695–723.

Sandgren, C. 1980a. An ultrastructural investigation of resting cyst formation in *Dinobryon cylindricum* Imhof (Chrysophyceae, Chrysosphyta). *Protistologica* 16: 259–76.

– 1980b. Resting cyst formation in selected chrysophyte flagellates. *Protistologica* 16: 289–303.

Scagel, R.F. & Stein, J.R. 1961. Marine nannoplankton from a British Columbia fjord. *Can. J. Bot.* 39: 1205–13.

Schnepf, E., Deichgräber, G., Röderer, G. & Herth, W. 1977. The flagellar root apparatus, the microtubular system and associated organelles in the chrysophycean flagellate *Poterioochromonas malhamensis* Peterfi (syn. *Poteriochromonas stipitata* Scherffel and *Ochromonas malhamensis* Pringsheim). *Protoplasma* 92: 87-107.

Silva, P.S. 1980. Names of classes and families of living algae. *Regnum Veg.* 103: 1-156.

Skuja, H. 1964. Grundzuge der Algenflora und Algenvegetation der Fjeldgegenden um Abisko in Schwedisch-Lappland. *Nova Acta R. Soc. Scient. Upsal., Ser. IV* 18(3): 1-404.

Stein, F.R. 1878. *Der Organismus der Infusionsthiere,* Vol. 3. W. Engelmann, Leipzig, pp. 1-154.

Swale, E.M.F. 1969. A study of the nannoplankton flagellate *Pedinella hexacostata* Vўsotskiĭ by light and electron microscopy. *Br. Phycol. J.* 4: 65-86.

Swale, E.M.F. & Belcher, J.H. 1975. A study of three new species of the colourless scaly flagellate *Cyathobodo* Petersen et Hansen: *C. reticulatus, C. crucifer* and *C. umbraculum. Arch. Protistenk.* 117: 269-75.

Throndsen, J. 1971. *Apedinella* gen. nov. and the fine structure of *A. spinifera* (Throndsen) comb. nov. *Norw. J. Bot.* 18: 47-64.

Van Valkenburg, S.D. 1980. Silicoflagellates. *In:* Cox, E.R. [Ed.] *Phytoflagellates* (*Developments in Marine Biology,* 2). Elsevier North-Holland, New York, pp. 335-50.

Whatley, J.M. & Whatley, F.R. 1981. Chloroplast evolution. *New Phytol.* 87: 233-47.

3

A survey of the marine Chrysophyceae with special reference to the Sarcinochrysidales

PAULETTE GAYRAL AND CHANTAL BILLARD
Laboratoire d'Algologie fondamentale et appliquée
Université de Caen, 39 rue Desmoueux, 14000-CAEN, France

In marine environments true Chrysophyceae are represented by relatively few taxa. As a result of the creation of the class Haptophyceae by Christensen (1962), an important number of marine genera, previously placed in the Chrysophyceae, were transferred into the Haptophyceae. The Haptophyceae were subsequently renamed Prymnesiophyceae for nomenclatural reasons (Hibberd 1976). The following marine taxa now belong to that class: *Phaeocystis* Lagerheim, *Prymnesium* Conrad, *Platychrysis* Geitler, *Chrysotila* Anand, *Chrysochromulina* Lackey, *Isochrysis* Parke, *Dicrateria* Parke, *Pavlova* Butcher, *Diacronema* Prauser, and all the coccolithophorid genera. The marine organisms that remain in the Chrysophyceae *sensu* Christensen are rarely observed and generally neglected by most specialists of the class.

Two major works historically contributed to our knowledge of the marine Chrysophyceae *sensu stricto:* the study of Conrad (1926) on flagellated representatives in brackish areas of Belgium, and Carter's (1937) investigation on the algae of brackish waters of Great Britain. To a lesser extent, Pascher (1915, 1925, 1940), Geitler (1930), and Schussnig (1940) also contributed to the study of the European marine chrysophycean flora. In 1937, Anand described the "Chrysophyceae belt," a remarkable supralittoral phytocoenosis on the British chalk cliffs. This same belt was subsequently discovered, symmetrically, on the cliffs of France (Bourrelly & Denizot 1967, Billard & Fresnel, 1980a), but today ought to be more properly designated as the "Prymnesiophyceae belt" because we showed that the algae in that zone are in fact members of this class (Billard & Fresnel 1980a). Anand's pioneering study stimulated interest among phycologists (e.g., Parke in Britain and Chadefaud, Magne, and Bourrelly in France), which led to the discovery of new marine taxa and also fostered a better understanding of their morphology and cytology. Ultimately a number of these microalgae were transferred to the Prymnesiophyceae.

True marine Chrysophyceae are usually modest contributors in the different biotopes where they are found; in spite of this, our group at the University of Caen has prompted a new interest in marine Chrysophyceae by demonstrating the existence of a distinct order Sarcinochrysidales and its phylogenetic affinities.

After this brief historical survey, we propose:

1. To show the necessity of a biosystematic method for the identification and the study of the marine Chrysophyceae
2. To review the marine representatives of large groups of the class and the diversity of their biotopes and to bring forth their originality compared to the freshwater ones
3. To emphasize finally the features of the Sarcinochrysidales, an order restricted to marine habitats and remarkable by its phaeophycean affinities.

Problems with identifying Chrysophyceae in different habitats: necessity of a biosystematic method

In the motile flagellated stage, Chrysophyceae are relatively easy to recognize by the color of their plastids, their chrysolaminarin vacuoles, their heterokont flagella, and the muciferous bodies present at the cell surface. Such planktonic organisms can be collected by the usual methods and concentrated by filtration or centrifugation. These organisms are rare and restricted to a small number of genera that will be reviewed; up to now none are pelagic and most of them are collected from tidal pools or oyster ponds.

The great majority of the marine Chrysophyceae has been obtained indirectly when various samples are maintained in plain seawater or enriched media. These samples are very diverse in type and distribution (rock or plant fragments, sandy–silty friable pieces) from the intertidal zone, the supralittoral zone, or salt marshes in estuarine areas. The mucilaginous pellicles present on the inner sides of marine aquariums where organisms are concentrated by continuous circulation of seawater also make favorable samples.

Such subcultures enable observation of flagellate or nonflagellate Chrysophyceae, whose identification will be estimated by examination of morphological and cytological characters and confirmed after isolation, culturing, and obtaining reproductive cells. Frequently, poorly defined organisms will develop consisting of yellow cells dispersed in more or less structured mucilaginous investments. Isolation will then be necessary and require subsequent subculturing before obtaining unialgal cultures that in time produce swarmers. We insist here on the priority that must be given to the flagellate cells of benthic marine chromophytes. Morphological similarities between benthic forms of Chrysophyceae and Prymnesiophyceae are such that only examination of the zoids permits the correct identification and classification. Demonstration of this is given by the notorious "Chrysophyceae belt" described by Anand. Its components grown in unialgal cultures were shown to be Prymnesiophyceae: *Chrysotila lamellosa* Anand and *C. stipitata* Anand with benthic filaments of coccolithophorids; the only authentic chrysophycean alga is *Pulvinaria feldmannii* (Bourrelly & Magne) Billard & Fresnel (Figure 3.3), a minor species in the belt (Billard & Fresnel 1980a).

Groups of Chrysophyceae present both in fresh water and in marine waters

Aside from the Dictyochales that will not be discussed here, the marine Chrysophyceae *sensu stricto* can be grouped into four series according to their flagellar organization: (a) a nonflagellate series, (b) a heterokont series with two subapical flagella (the second eventually missing by regression), (c) a uniflagellate series with a special ribbonlike flagellum, and (d) a heterokont series with lateral flagella. The first three series have representatives both in fresh water and in marine waters, the fourth, which will be dealt with separately, is a strictly marine group with phaeophycean affinities for which we erected the order Sarcinochrysidales.

Series with no flagella

The vegetative stage or the dispersion cells may be ameboid: the rare species *Leukapsis vorax* Pascher (Pascher 1940) and *Heliapsis achromatica* Norris (Norris 1964) are colorless, but *Chrysothylakion vorax* Pascher (Pascher 1915) and *Chrysopodocystis socialis* Billard (Billard 1978) are pigmented and both rhizopodial, enclosed in a lorica; the second species liberates ameboid cells. The coccoid forms belong to the Tetrapionales (Bourrelly 1981): *Podochrysis roscoffensis* Magne lives attached to various substrates, and *Polypodochrysis teissieri* Magne is distinguished by its open tubelike expansions (Magne 1975). The pelagic *Pelagococcus subviridis* Norris has an aberrant pigment composition, and its systematic position is uncertain (Lewin et al. 1977).

Heterokont series with two (or only one) subapical flagella

Typical freshwater species of this series are sometimes mentioned in coastal areas, probably brought by river flow (i.e., species of *Dinobryon, Pseudokephyrion, Mallomonas, Microglena, Synura, Chrysococcus, Kephyrion, Calycomonas*). Much more significant are representatives of the genera *Ochromonas, Chrysosphaerella* [*C. salina* Birch-Andersen (Birch-Andersen 1973), *C. tripus* Takahashi & Hara, and *C. patelliformis* Takahashi & Hara (Takahashi & Hara 1984)], and *Sphaleromantis* [*S. subsalsa* Conrad (Conrad 1926), *S. marina* Pienaar (Pienaar 1976), the latter with remarkable organic superficial scales], *Syncrypta glomifera* Clarke & Pennick

(Clarke & Pennick 1975), and a colonial form of the genus *Chrysobotriella* [*C. spondylomorum* (Conr.) Strand ex-Bourrelly] known from brackish waters (Conrad 1926). The nonpigmented genus *Paraphysomonas* with siliceous scales is well represented, with so far nine marine species.

Besides these biflagellate species, a few uniflagellate organisms with a second vestigial flagellum have been described in marine waters: various species of *Chromulina, Chrysapis yserensis* Conrad (Conrad 1926), and ameboid organisms such as the ameboflagellate *Chrysamoeba nana* Scagel & Stein (Scagel & Stein 1961) and *Rhizochromulina marina* Hibberd & Chrétiennot-Dinet (Hibberd & Chrétiennot-Dinet 1979), an ameboid genus with uniflagellate zoospores.

The Aurosphaeraceae, a family of unsufficiently known planktonic organisms can be included here, although they are sometimes considered related to the protozoa (see Leadbeater 1974).

Among benthic organisms, the following species are marine: *Pascherella yserensis* described by Conrad (1926) as palmelloid colonies on salt marsh halophytes and *Nematochrysis sessilis* Pascher, a pseudofilamentous species with biflagellate swarmers (Pascher, 1925). *Thallochrysis pascheri* Conrad, a pseudoparenchymatous form, produces uniflagellate zoospores (Conrad 1926). It must be reminded here that marine creeping filamentous organisms attributed to the genus *Apistonema* Pascher are in fact Prymnesiophyceae: those with a coccolith-bearing diploid generation must be included in the genus *Pleurochrysis* Pringsheim, those whose sexuality has been lost in the genus *Gloeothamnion* Cienkowsky. Therefore, the genus *Apistonema* does not stand, at least for marine forms.

Series with a single ribbonlike flagellum

This assemblage, including pigmented or nonpigmented marine organisms belonging to the genera *Pedinella, Pseudopedinella, Apedinella,* is distinguished on the basis of radially symmetrical cells with regularly arranged chloroplasts and a single apical mastigoneme-bearing flagellum often surrounded by a ring of tentacles. This group stands well apart from the two others.

Conclusions

If we consider the marine taxa of the preceeding series, the following facts may be pointed out:

1. In each series, the number of either generic or specific taxa is reduced in marine environments as opposed to the number of freshwater taxa (over 100 genera and 700 species).
2. Very few marine Chrysophyceae utilize silica to form scales or statospores. Only the colorless genus *Paraphysomonas* and the pigmented

one *Chrysosphaerella* are covered with siliceous scales. As for siliceous cysts, there are only few species for which a well-defined type of cyst has been demonstrated in culture, and the cysts mentioned in habitats yielding various Chrysophyceae probably belong to the genus *Ochromonas*.

3. Loricate species are rare.
4. Sexuality is unknown in these marine Chrysophyceae.
5. Few genera are present both in freshwater and in marine habitats. In the series we have reviewed, it appears as if increasing salinities had limited the species' range of adaptation and gone against a diversification of the group.

Heterokont Chrysophyceae restricted to marine habitats: the Sarcinochrysidales

The opinion put forward by Gayral (1972) that certain Chrysophyceae with heterokont swarmers are clearly different from all others, including the Ochromonadales, is based on successive studies (Gayral & Haas 1969, Gayral & Morange 1970, Gayral & Lepailleur 1971). The investigative method we developed enabled us some years back to isolate benthic filamentous species and study their reproduction, which was unknown at the time. In 1969, Gayral and Haas in *Chrysomeris ramosa* Carter (Figure 3.7) and *Giraudyopsis stellifer* Dangeard (Figure 3.5) and in 1971, Gayral and Lepailleur in *Nematochrysopsis roscoffensis* Chadefaud (Figure 3.1) and *Chrysowaernella hieroglyphica* (Waern) Gayral & Lepailleur (=*Nematochrysis hieroglyphica* Waern) (Figure 3.2) demonstrated that reproduction occurs by means of heterokont swarmers with lateral flagella. Each cell liberates a single swarmer that germinates to give a new filament. The ultrastructural study of flagellated cells of *Nematochrysopsis* and *Ochromonas* by Gayral and Morange in 1970 had shown for the first time the morphology of the tubular mastigonemes present on the long flagellum of *Ochromonas*. It also demonstrated that they are different in *Nematochrysopsis* and that these dissimilarities confirm the more evident difference between both types of monads, i.e., the position of the flagella.

Finally it was the observation of the planktonic species *Sarcinochrysis marina* (Figure 3.6), for which Geitler (1930) in his description had indicated the phaeophycean organization of the zoids, that enabled Gayral (1972) to propose for the Chrysophyceae with this type of swarmer an order Sarcinochrysidales with two subdivisions: Sarcinochrysidineae and Chrysomeridineae. In the hope of new information or discovery of new organisms that might confirm this view, and in order to achieve a better understanding of the families in the suborders, validation of the new taxa was postponed until 1977 (Gayral & Billard 1977). At this time a synopsis was presented that comprised the species just mentioned but also a num-

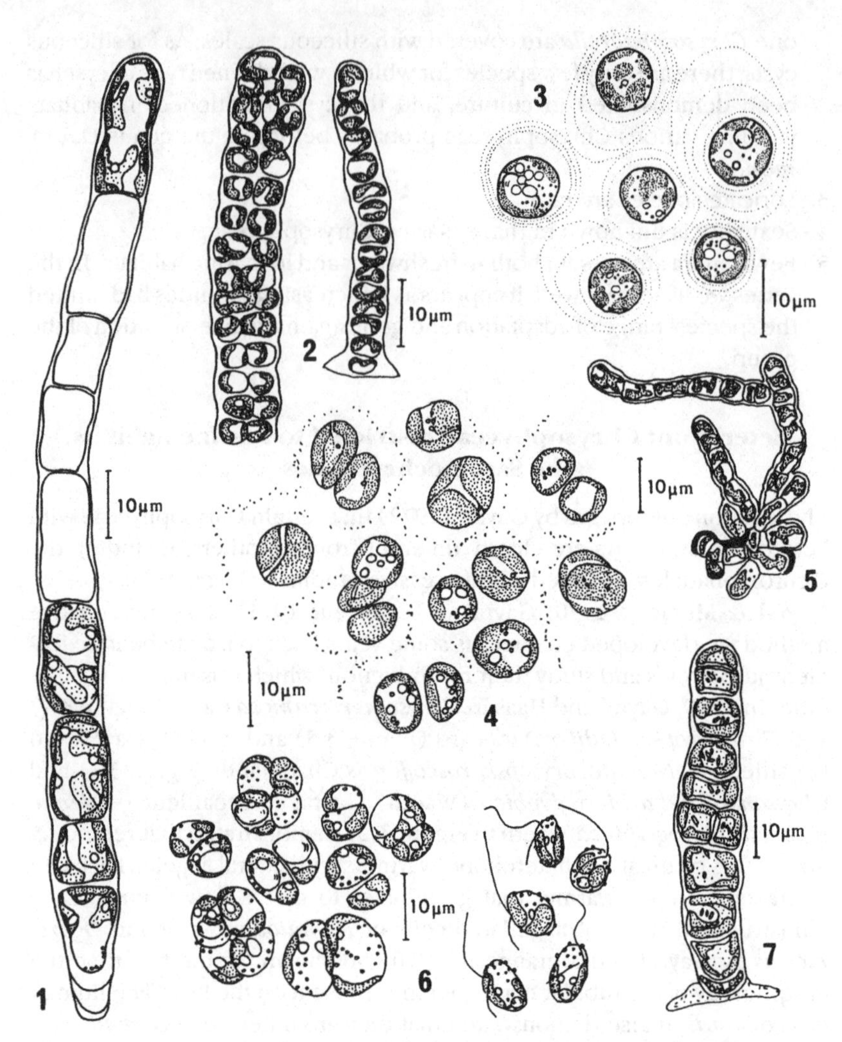

Figs. 3.1–3.7. Morphological aspects of Sarcinochrysidales. ***Fig. 3.1.*** ***Nematochrysopsis roscoffensis.*** Typical uniseriate filament showing thick-walled basal cell. Note prominent pyrenoids in the cells which carry four plastids. ***Fig. 3.2.*** ***Chrysowaernella hieroglyphica.*** Young uniseriate filament (right) with differentiated basal cell, and distal pluriseriate segment of adult filament (left). Note irregular arrangement of the plastids. ***Fig. 3.3.*** ***Pulvinaria feldmannii.*** Typical palmelloid organization showing cells surrounded with stratified mucilage. Note prominent pyrenoids. ***Fig. 3.4.*** ***Chrysoderma mucosa.*** Palmelloid organization with cells embedded in thick non stratified mucilage. Note large overlaping plastids. ***Fig. 3.5.*** ***Giraudyopsis stellifer.*** Young uniseriate erect axes with disclike basal attachment. Note filament branching. ***Fig. 3.6.*** ***Sarcinochrysis marina.*** Palmelloid cells (left) without apparent cell walls; monadoid cells (right) laterally biflagellate. Both types of cells carry prominent pyrenoids. ***Fig. 3.7.*** ***Chrysomeris ramosa.*** Young filament showing basal cell surrounded with mucilage. Note begining of longitudinal division.

ber of other organisms: *Rhamnochrysis* described from Massachusetts (Wilce & Markey 1974), the subarctic genus *Phaeosaccion* (McLachlan et al. 1971) and its equivalent in the southern hemisphere *Antarctosaccion* (Delepine et al. 1970), and perhaps *Chrysonephos* known from Florida and Bermuda (Taylor 1951, 1952).

On the whole, the order includes unicellular or multicellular organisms, either planktonic or benthic, producing heterokont motile cells with the following characteristics: unequal lateral flagella; the longer front flagellum is blunt tipped; a typical long hair point is present on the shorter hind flagellum (acronema; Figures 3.9 and 3.10). The front flagellum carries two equally balanced rows of tripartite smooth mastigonemes terminated by three filaments, two short and the middle one longer (Figure 3.11).

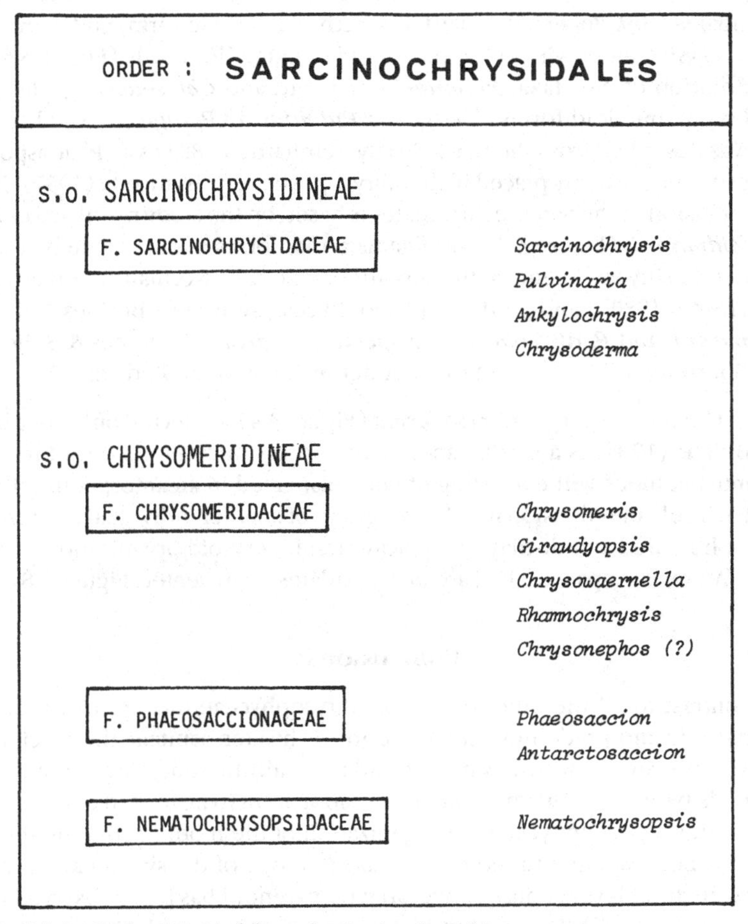

Fig. 3.8. Revised synopsis of the order Sarcinochrysidales.

Three sets of modifications are here brought to the 1977 synopsis:

1. Deletion of the genus *Olisthodiscus* with *O. luteus* Carter (Carter 1937) that had been treated as chrysophycean and tentatively placed in the Sarcinochrysidales. We feel today that *Olisthodiscus* should be included in the Raphidophyceae because of its multiple plastids, its numerous perinuclear dictyosomes, and the presence of a peculiar flagellar root similar to the one described in the raphidophycean genus *Chattonella*. Loeblich and Fine (1977) and Moestrup (1982) are of the same opinion.
2. Transfer from the Xanthophyceae to the Sarcinochysidales of the organism Van der Veer described as *Ankylonoton luteum* (Van der Veer, 1970). We were able to isolate and reexamine this organism, which is in fact sarcinochrysidalean. It is included in the order under the new combination *Ankylochrysis lutea* (Van der Veer) Billard (Billard 1984). Two genera of planktonic organisms are now recognized in the Sarcinochrysidales: *Ankylochrysis* Billard, a strictly planktonic form, and *Sarcinochrysis* Geitler with either monadoid or palmelloid cells (Figure 3.6).
3. Addition of new taxa: *Pulvinaria* Reinhard and *Chrysoderma* Billard, both palmelloid forms. The genus *Pulvinaria* (*P. algicola* Reinhard) was described from the Black Sea by Reinhard in 1885 as a "Phaeosporeae" and correctly placed in the Chrysophyceae by Bourrelly (1957). Its inclusion in the Sarcinochrysidales is based on the study by Billard of *P. feldmannii* (Figure 3.3), an organism that forms a remarkable belt on friable silty sediments in the supralittoral zone in Normandy (Billard & Fresnel 1980b) and produces phaeophycean swarmers. Besides *P. feldmannii* and *P. algicola*, a third species, *P. giraudii* (Derbès & Solier) Bourrelly, is known from the Mediterranean and the Red Sea.

The second genus, *Chrysoderma* (Figure 3.4), collected only once by Billard (1984) is a marine species that forms in culture a thick brown mucilaginous veil consisting of cells dispersed in an amorphous jelly. Each cell after division may form a group of smaller individuals that will be liberated and display the typical sarcinochrysoid organization.

A new synopsis of the Sarcinochrysidales is presented Figure 3.8.

Conclusions

In contrast to all the other orders of Chrysophyceae that are essentially planktonic and predominantly found in freshwater habitats, the Sarcinochrysidales are so far strictly marine and generally benthic. The nonmotile phase is palmelloid, filamentous, or pseudoparenchymatous in contrast to the uniform phaeophyceanlike organization of the motile cell. This similarity is not restricted to the general morphology of the swarmers. At the ultrastructural level, a phaeophycean type pyrenoid has been observed in some (but not all) Sarcinochrysidales, viz., *Sarcinochrysis, Nematochry-*

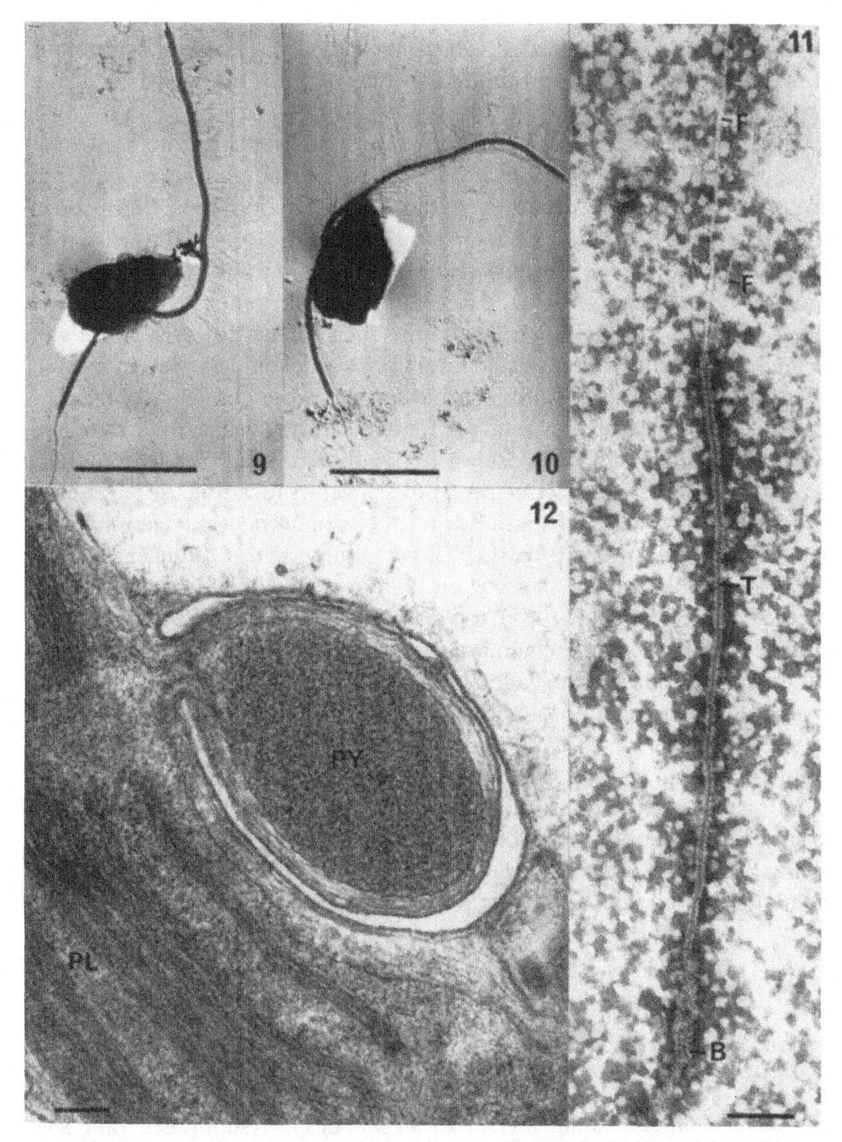

Figs. 3.9–3.12. Structural features of the Sarcinochrysidales. ***Fig. 3.9. Giraudyopsis stellifer.*** Shadow-cast whole mount of the swarmer showing lateral flagella with heterokont organization; note the long tapering hair point terminating the short hind flagellum. Scale = 5 μm. ***Fig. 3.10. Chrysoderma mucosa.*** Shadow-cast whole mount of the swarmer. Organization of the laterally biflagellate swarmer in this palmelloid organism is similar to that of the filamentous *Giraudyopsis.* Scale = 5 μm. ***Fig. 3.11.*** Negative staining of a single isolated tripartite mastigoneme of *Giraudyopsis stellifer* showing the tapering basal region (B), the smooth tubular shaft (T) and three terminal filaments (F); the two short lateral filaments are thinner than the middle one which extends further than shown in the illustration. Scale = 0.1 μm. ***Fig. 3.12.*** Transverse section through the phaeophycean type pyrenoid (PY) of the plastid (PL) of *Sarcinochrysis marina.* Note electron transparent cap. Scale = 0.1 μm.

sopsis, and *Pulvinaria* (Billard 1984). This projecting, stalked pyrenoid is devoid of thylakoids but surrounded by a cap of electron-transparent material (Figure 3.12).

At the biochemical level, two features are significant. Mannitol is present (Craigie et al. 1971) in *Phaeosaccion* Farlow, the most advanced genus in the group, and is generally absent in the other Chrysophyceae. Cellulose in the cell wall of *Phaeosaccion* is also indicative of a close relationship with the Phaeophyceae. The ecological characteristics of the Sarcinochrysidales are: marine organisms restricted to the neritic zone, generally benthic, and often growing in the higher tidal zone thus subject to long periods of drought. The following question might be asked: could the Sarcinochrysidales be primitive Phaeophyceae rather than advanced Chrysophyceae? Indeed, rarely have such natural phylogenetic relationships been detected and such a direct affiliation established without the help of hypothetical ancestors. To this question we prefer to answer by recalling the features of the Sarcinochrysidales that are not yet those of authentic Phaeophyceae: lack of plasmodesmata (Billard 1974) and alginic acid (Craigie et al. 1971) in the cell wall; absence of fucosterol but presence of two rare C_{30} sterols (Raederstorff & Rohmer 1984), the main one with the same position of the double bond in the side chain as in fucosterol; and, finally, presence of a "transitional helix" in the flagellar transition region (Billard 1984). Moreover, the swarmers are produced individually in vegetative cells and consequently there is not even a primitive type of sporocyst, and, finally, sexuality is also lacking. The differences may be regarded as subtle, and wanting to settle the question might lead to a rigid attitude. Certainly, this would be regrettable in one of the few cases where we are before actually living organisms establishing undoubtedly one of these bridges so often vainly sought by the taxonomist tracing evolutionary pathways.

References

Anand, P.L. 1937. A taxonomic study of the algae of the British chalk-cliffs. *J. Bot. Lond.* 75, Suppl. II: 1–51.

Billard, C. 1974. Microanatomie comparée de deux Chrysoméridacées. Individualisation des Sarcinochrysidales au sein des Chrysophycées. *Bull. Soc. Phycol. Fr.* 19: 17–30.

– 1978. *Chrysopodocystis socialis* gen. et sp. nov. (Chrysophyceae), une nouvelle Rhizochrysidale marine loriquée. *Bull. Soc. Bot. Fr.* 125: 307–12.

– 1984. Recherches sur les Chrysophyceae marines de l'ordre des Sarcinochrysidales. Biologie, Systématique, Phylogénie. Thèse de Doctorat d'Etat, Université de Caen.

Billard. C. & Fresnel, J. 1980a. Nouvelles observations sur les ceintures supralittorales de Chrysophycophytes des côtes de la Manche. *105ème Congrès National des Sociétés Savantes, Caen, 1980, Sciences, fasc.* III: 213–24.

– 1980b. Nouvelles observations sur le *Pulvinaria feldmannii* (Bourrelly &

Magne) comb. nov. (Chrysophycée, Sarcinochrysidale) formant une ceinture sur substrat meuble. *Cryptogamie: Algologie,* I: 281–92.
Birch-Andersen, P. 1973. *Chrysosphaerella salina,* a new species of the Chrysophyceae from salt marsh pools. *Bot. Tidsskrift,* 68: 140–4.
Bourrelly, P. 1957. Recherches sur les Chrysophycées: morphologie, phylogénie, systématique. *Rev. algol., Mém. Hors Sér.,* 1: 1–412.
– 1981. *Les Algues d'Eau Douce.* II. *Les Algues Jaunes et Brunes,* 2nd ed. Boubée, Paris.
Bourrelly, P. & Denizot, M. 1967. Quelques algues marines des niveaux élevés de la falaise d'Etretat (Seine-Maritime).Extrait des Travaux de Biologie Végétale dédiés au Professeur P. Dangeard. *Le Botaniste, Sér. L,* pp. 43–56.
Carter, N. 1937. New or interesting algae from brackish water. *Arch. Protistenk.,* 90: 1–68.
Christensen, T. 1962. Alger. *In:* Böcher, T.W., Lange, M., & Sørensen, T. [Eds.] *Botanik* Bd 2, *Systematisk Botanik,* Nr. 2 Munksgaard, Copenhagen.
Clarke, K. & Pennick, N. 1975. *Syncrypta glomifera* sp. nov., a marine member of the Chrysophyceae bearing a new form of scale. *Br. Phycol. J.,* 10: 363–70.
Conrad, W. 1926. Recherches sur les Flagellates de nos eaux saumâtres, 2ème partie; Chrysomonadines. *Arch. Protistenk.,* 56: 167–231.
Craigie, J.S., Leigh, C., Chen, L.C-M. & McLachlan, J. 1971. Pigments, polysaccharides and photosynthetic products of *Phaeosaccion collinsii. Can. J. Bot.* 49: 1067–74.
Delepine, R., Mackenzie Lamb, I. & Zimmermann, M. 1970. Sur les algues marines antarctiques rapportées au genre *Monostroma* Thuret. *C.R. Acad. Sci. Paris,* 270: 1973–6.
Gayral, P. 1972. Sur les Chrysophycées à zoïdes phéophycéens notamment *Sarcinochrysis marina* Geitler. *Bull. Soc. Phycol. Fr.,* 17: 40–5.
Gayral, P. & Billard, C. 1977. Synopsis du nouvel ordre des Sarcinochrysidales (Chrysophyceae). *Taxon,* 26: 241–5.
Gayral, P. & Haas, C. 1969. Etude comparée des genres *Chrysomeris* Carter et *Giraudyopsis* P. Dangeard. Position systématique des Chrysomeridaceae (Chrysophyceae). *Rev. Gén. Bot.* 76: 659–66.
Gayral, P. & Lepailleur, H. 1971. Etude de deux Chrysophycées filamenteuses: *Nematochrysopsis roscoffensis* Chadefaud, *Nematochrysis hieroglyphica* Waern. *Rev. Gén. Bot.,* 78: 61–74.
Gayral, P. & Morange, J. 1970. Structure fine du fouet pleuronématé chez quelques Chrysophycées. *Septième Congr. Int. Micr. électr., Grenoble, 1970,* 3: 387–8.
Geitler, L. 1930. Ein grünes Filarplasmodium und andere neue Protisten. *Arch. Protistenk.* 69: 615–36.
Hibberd, D.J. 1976. The ultrastructure and taxonomy of the Chrysophyceae and Prymnesiophyceae (Haptophyceae): a survey with some new observations on the ultrastructure of the Chrysophyceae. *Bot. J. Linn. Soc.* 72: 55–80.
Hibberd, D.J. & Chrétiennot-Dinet, M.J. 1979. The ultrastructure and taxonomy of *Rhizochromulina marina* gen. et sp. nov., an amoeboid marine Chrysophyte. *J. Mar. Biol. Assoc. U.K.* 59: 179–83.
Leadbeater, B.S.C. 1974. Ultrastructural observations on nanoplankton collected from the coast of Jugoslavia and the bay of Algiers. *J. Mar. Biol. Assoc. U.K.* 54: 179–96.
Lewin, J., Norris, R.E., Jeffrey, S.W. & Pearson, B.E. 1977. An aberrant chryso-

phycean alga *Pelagococcus subviridis* gen. nov. et sp. nov. from the North Pacific Ocean. *J. Phycol.* 13: 259-66.
Loeblich, A.R. III & Fine, K.E. 1977. Marine chloromonads: more widely distributed in neritic environments than previously thought. *Proc. Biol. Soc. Wash.* 90(2): 388-99.
Magne, F. 1975. Contribution à la connaissance des Stichogloéacées (Chrysophycées, Stichogloéales). *Cah. Biol. Mar.* 16: 531-9.
McLachlan, J., Chen, L.C.-M., Edelstein, R. & Craigie, J.S. 1971. Observations on *Phaeosaccion collinsii* in culture. *Can. J. Bot.* 49: 563-6.
Moestrup, Ø. 1982. Flagellar structure in algae: a review, with new observations particularly on the Chrysophyceae, Phaeophyceae (Fucophyceae), Euglenophyceae, and *Reckertia*. *Phycologia* 21: 427-528.
Norris, R.E. 1964. Studies on phytoplankton in Wellington harbour. *N. Z. J. Bot.* 2: 258-78.
Pascher, A. 1915. Studien über die rhizopodiale Entwicklung der Flagellaten, I. Uber einige rhizopodiale Chromatophoren führende Organismen aus der Flagellatenreihe der Chrysomonaden. *Archiv. Protistenk.* 36: 92-117.
- 1925. Die braune Algenreihe der Chrysophyceen. *Archiv. Protistenk.* 52: 489-564.
- 1940. Filaraplasmodiale Ausbildungen bei Algen. *Archiv. Protistenk.* 94: 295-309.
Pienaar, R.N. 1976. The microanatomy of *Sphaleromantis marina* sp. nov. (Chrysophyceae). *Br. Phycol. J.* 11: 83-92.
Raederstorff, D. & Rohmer, M. 1984. Sterols of the unicellular algae *Nematochrysopsis roscoffensis* and *Chrysotila lamellosa*: isolation of (24*E*)-24-*n*-propylidenecholesterol and 24-*n*-propylcholesterol. *Phytochem.* 23: 2835-8.
Reinhard, L. 1885. Recherches algologiques. I. Materiaux sur la morphologie et la systématique des Algues de la Mer Noire. *Notes de la Société de Chercheurs en Sciences Naturelles de Novorosisk* 9(2): 201-512 (in Russian).
Scagel, R. & Stein, J.R. 1961. Marine nannoplankton from a British Columbia fjord. *Can. J. Bot.*, 39: 1205-13.
Schussnig, B. 1940. Uber einige neue Protophyten aus der Adria. *Archiv. Protistenk.* 93: 317-30.
Takahashi, E. & Hara, S. 1984. Two new marine species of *Chrysosphaerella* (Chrysophyceae) with a reinvestigation of *C. salina*. *Phycologia* 23: 103-9.
Taylor, W.R. 1951. Structure and reproduction of *Chrysophaeum lewisii*. *Hydrobiologica,* 3: 122-30.
- 1952. The algal genus *Chrysophaeum*. *Bull. Torrey Bot. Club.* 79: 79.
Van der Veer, J. 1970. *Ankylonoton luteum* (Chrysophyta), a new species from the Tamar estuary, Cornwall. *Acta Bot. Neerl.* 19: 616-36.
Wilce, R. & Markey, D. 1974. *Rhamnochrysis aestuarinae,* a new monotypic genus of benthic marine chrysophytes. *J. Phycol.* 10: 82-8.

4

The actinophryid heliozoa (Sarcodina, Actinopoda) as chromophytes

DAVID J. PATTERSON

Department of Zoology
University of Bristol
Bristol BS8 1UG, England

Introduction

The heliozoa are predatory protozoa that are characterized by having stiff pseudopodia that radiate out from a central body mass. The heliozoa are classified as amebae, forming a class within the Actinopoda (Levine et al. 1980). The remaining actinopods make up the Radiolaria and are unlike the heliozoa in having elaborate inorganic skeletons. There are four orders of heliozoa: Desmothoracida, Taxopodida, Centrohelida, and Actinophryida. This chapter concerns itself primarily with the phylogenetic affinities of the Actinophryida, an order that contains three well-studied genera: *Actinophrys, Actinosphaerium,* and *Echinosphaerium.*

There have been two proposals concerning the evolutionary source of the actinophryid heliozoa (Figure 4.1). Trégouboff (1953) suggested that they are derived from filose amebae that have thin but unsupported pseudopodia. This, or a similar relationship, is implicit in the classification of heliozoa as a kind of ameba. The second suggestion is that the actinophryids are derived from chrysophyte or chromophyte algae. Specifically, this lineage would involve the pedinellid algae and some colorless helioflagellates (these groups are described below) (Pascher 1917, Gadea 1973, Davidson 1975, 1982).

The helioflagellates are colorless flagellates that, like heliozoa, have stiffened radiating pseudopodia. These organisms have had an unsettled taxonomic history, being placed in ill-defined taxa such as the Rhizomastigida (Bütschli 1889, Honigberg et al. 1964, Krylov et al. 1980), the pseudoheliozoa, (Penard 1904) or the Protomyxidea (Doflein 1916). In the most recent classification of protozoa endorsed by the Society of Protozoologists (Levine et al. 1980), only one genus of helioflagellates is mentioned

I am very grateful to Claire Wyatt for technical assistance. I thank T. Fenchel for providing cultures of *Pteridomonas* sp., the Royal Society of London for provision of light microscopic facilities, and the National Science Foundation for travel funds.

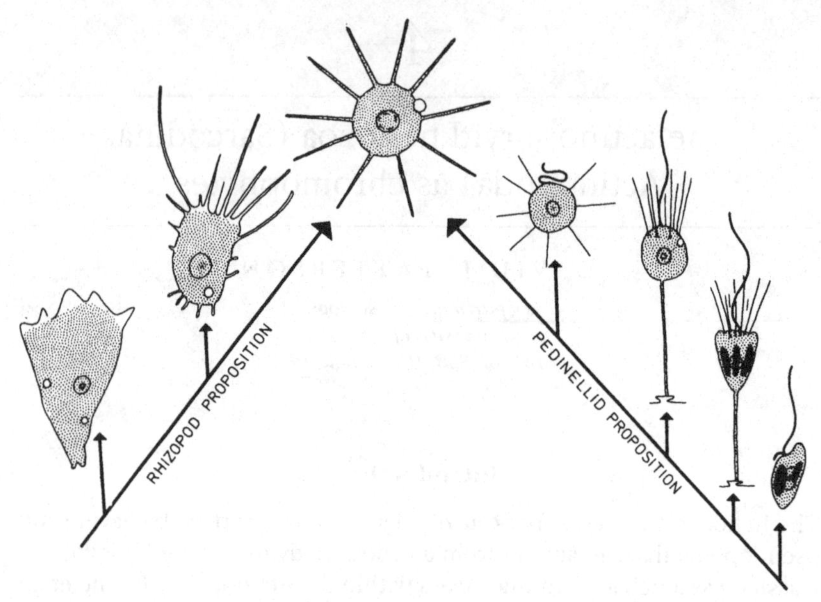

Fig. 4.1. Alternative propositions for the origins of the actinophryid heliozoa. For details see text.

(*Ciliophrys*), and this is included among the heliozoa. Elsewhere the same organisms have been treated as "algae" (Bourrelly 1968).

The helioflagellates are divided into two groups, depending upon whether the axonemes that support the pseudopodia do or do not terminate on the surface of the nucleus. There are three reasonably well-studied genera that have the axonemes terminating on the nucleus (the "actinomonad" helioflagellates). All actinomonad helioflagellates have a single apical flagellum. The first genus, *Ciliophrys,* was designated by Cienkowski (1876) for spherical organisms with pseudopodia emanating from the entire body surface. Shortly after, Kent (1882) added a second genus, *Actinomonas,* for organisms that also have arms radiating from all parts of the body, but that have a stalk. In 1890, Penard added the genus *Pteridomonas* for stalked helioflagellates in which the arms are restricted to a wreath around the flagellum. Penard misinterpreted the nature of developing flagella (cf. Griessmann 1914) and also cast doubt on the nature of the arms and of the flagellum (Penard 1921). This created sufficient confusion to lead some (Davidson 1975, 1982) to doubt the validity of the genus. In addition to these three genera, Bourrelly (1968) mentions a further poorly characterized genus, *Stylomonas,* which may also be an actinomonad helioflagellate.

There are several additional genera of helioflagellates in which the axonemes do not terminate on the nucleus, such as *Dimorpha, Dimorphella* and *Tetradimorpha* (Gruber 1882, Schouteden 1907, Bělař 1926, Valkanov 1928, Trégouboff 1953, Bovee 1960). These have been implicated in the evolution of the centrohelidian heliozoa (Trégouboff 1953, Honigberg et al. 1964, Davidson 1976, 1982, Brugerolle and Mignot 1983, 1984).

In view of the success with which ultrastructural features of unicellular organisms have been employed to define taxa and to clarify phylogenetic relationships (Hibberd & Leedale 1972, Dodge 1973, Taylor 1976, Hibberd 1976, 1979, Corliss, 1979, Cox 1980, Moestrup 1982, Patterson 1983), the present chapter will consider the two suggestions of actinophryid evolution in the light of available ultrastructural data and will indicate some possible taxonomic consequences.

The fine structure of actinophryid heliozoa is well documented (Tilney & Porter 1965, Patterson 1979, Mignot 1979, 1980a, Suzaki et al. 1980, Patterson and Thompson 1981, Patterson & Hausmann 1981, 1982, Hausmann & Patterson 1982, Lennebach, Hausmann & Patterson, 1983 *inter alia*).

The centrohelidian heliozoa are ultrastructurally more diverse than the preceding group. In some species, the axonemes terminate on a central trilamellate structure (the centroplast), and the fine structure of this type of centrohelid is described by Bardele (1975, 1977), Davidson (1976) and Rieder (1979). Jones (1974) and Febvre-Chevalier (1973, 1975, 1980) have provided ultrastructural accounts of several other organisms currently classified as centrohelid heliozoa but which lack the centroplast. In regard to the remaining two orders of heliozoa, Bardele (1972) has described the fine structure of the desmothoracid heliozoon, *Clathrulina,* and Cachon and Cachon (1978) have summarized the ultrastructure of the only known taxopodid, *Sticholonche zanclea.*

Available ultrastructural data on the actinomonad helioflagellates is limited to Davidson's accounts of *Ciliophrys* (Davidson, 1975, 1976, 1982) and a preliminary account of the structure of *Pteridomonas* (under the name *Actinomonas mirabilis*) by Fenchel (1982). Further data are presented in this chapter and a more extensive description will follow (D.J. Patterson, in preparation). Brugerolle and Mignot (1983, 1984) have given an account of some aspects of the fine structure of *Dimorpha* and *Tetradimorpha.* The fine structure of three pedinellids has been documented by Swale (1969a), Throndsen (1971), and Ostroff and Van Valkenburg (1978). The ultrastructural organization of the chrysophytes is well known and has been ably reviewed by Hibberd (1976, and Chapter 2 of this treatise) and by Pienaar (1980). The filose amebae have not been extensively studied, but our current understanding of their organization and affinities is discussed by Patterson (1983, 1984).

Materials and methods

The sources of *Actinophrys sol* Ehrenberg and of *Nuclearia* spp., methods of their culture, and preparation for electron microscopy are given elsewhere (Patterson 1979, 1983, 1984). *Pteridomonas* was obtained from Dr. T. Fenchel (University of Aarhus, Denmark). The identity of this organism is discussed in Patterson and Fenchel (1985). It is maintained in artificial sea water with *Klebsiella aerogenes* added weekly as food. Cells were fixed in 2.5% (w/v) glutaraldehyde with 1% (w/v) osmium tetroxide in 50 mM/liter sodium cacodylate buffer at pH 7.4 in calcium-free sea water with 5 mM/liter added magnesium chloride. For some fixes, the cells were placed in 1% (w/v) Taxol for several hours prior to fixation. The in vivo stabilizing effect of this drug on heliozoan axonemes has been described elsewhere (Hausmann, Linnenbach, & Patterson 1983), but its effect on *Pteridomonas* was negligible. After fixation, cells were washed, embedded in 2%(w/v) agar, dehydrated, and embedded in Araldite resin. Negative staining was carried out with 2% (w/v) aqueous uranyl acetate. Light microscopy was as described elsewhere (Patterson 1982).

Results

The bulk of the original observations presented here are restricted to light and electron microscopic observations of *Pteridomonas* sp. (Figure 4.2). *Ciliophrys marina* and a representative of the filose amebae, both of which have been implicated in the evolution of the actinophryid heliozoa (Figures 4.5 and 4.6) are illustrated in Figures 4.3 and 4.5

The general organization of *Actinophrys sol* is illustrated by Figures 4.7–4.12. The body measures about 50 μm in diameter and gives rise to about 40 stiff arms (axopodia). The arms are supported internally by microtubular axonemes in which the microtubules are arrayed in a double polygonal spiral (Figures 4.7B and 4.12). The axonemes terminate on the surface of the nucleus. Nucleoli are located at the periphery of the nucleus and give the impression that the nucleus has a thick wall when it is viewed with the light microscope. The nuclear envelope contains two electron-dense laminae (Figures 4.7, and 4.10). Mitochondria have short bleblike cristae without prominent intracristal material (Figure 4.11). Extrusomes lie adjacent to the plasma membrane (Figure 4.8). These structures are involved in food capture. There are two types of extrusomes: the first measures about 0.1 μm in diameter and has granular contents; the second has a diameter of about 0.5 μm and has varied appearance. Some appear dense and homogeneous, while others appear to be fibrillar. These differences are a result of their series of transformations associated with the development, maturation, and use (D.J. Patterson, unpublished data). Ingested food is initially enclosed within a single ingestion vacuole, but shortly after ingestion numerous digestion vacuoles appear in the outer regions of the cell. The

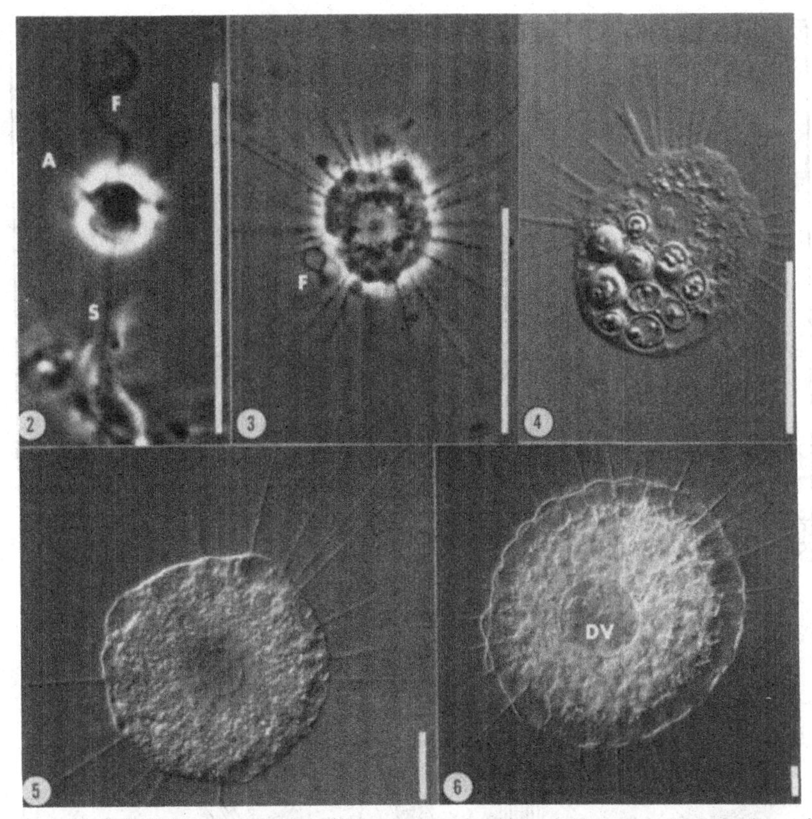

Figs. 4.2–4.6. Light micrographs of living actinophryids and other organisms implicated in their evolution. All scale bars represent 20 μm. ***Fig. 4.2.*** *Pteridomonas* sp., a helioflagellate that attaches to the subsrate by means of a stalk (S), beating of the flagellum (F) create currents of water from which particles are caught using the arms (A). ***Fig. 4.3.*** *Ciliophrys* sp., a floating or swimming helioflagellate with a flagellum (F) and with axopodia radiating from the organism and which bear extrusomes. The center of the cell is occupied by the nucleus. ***Fig. 4.4.*** ***Nuclearia moebiusi.*** An ameba with thin unsupported pseudopodia. The posterior region of the cell is occupied by ingested yeast cells lying in food vacuoles. ***Fig. 4.5.*** ***Actinophrys sol.*** The single nucleus lies in the centre of the cell. Stiff axopodia radiate from the surface of the body. ***Fig. 4.6.*** ***Echinosphaerium nucleofilum,*** a multinucleated actinophryid with a peripheral layer of transparent vacuoles and with a central vacuole (DV) containing ingested food.

vacuoles shrink and get denser as digestion progresses (Linnenbach, Hausmann, & Patterson 1983). During nuclear division, the spindle microtubules arise in the cytoplasm and penetrate the nucleus at polar fenestrae. Actinophryid heliozoa produce cysts with complex walls (Figure 4.7C) within which autogamy occurs.

The general organization of *Pteridomonas* sp. is shown by Figures 4.13–4.23. This study is not yet complete, and the information presented here represents an interim understanding of the cytology of this species. The single apical flagellum draws a current of water down toward the cell and

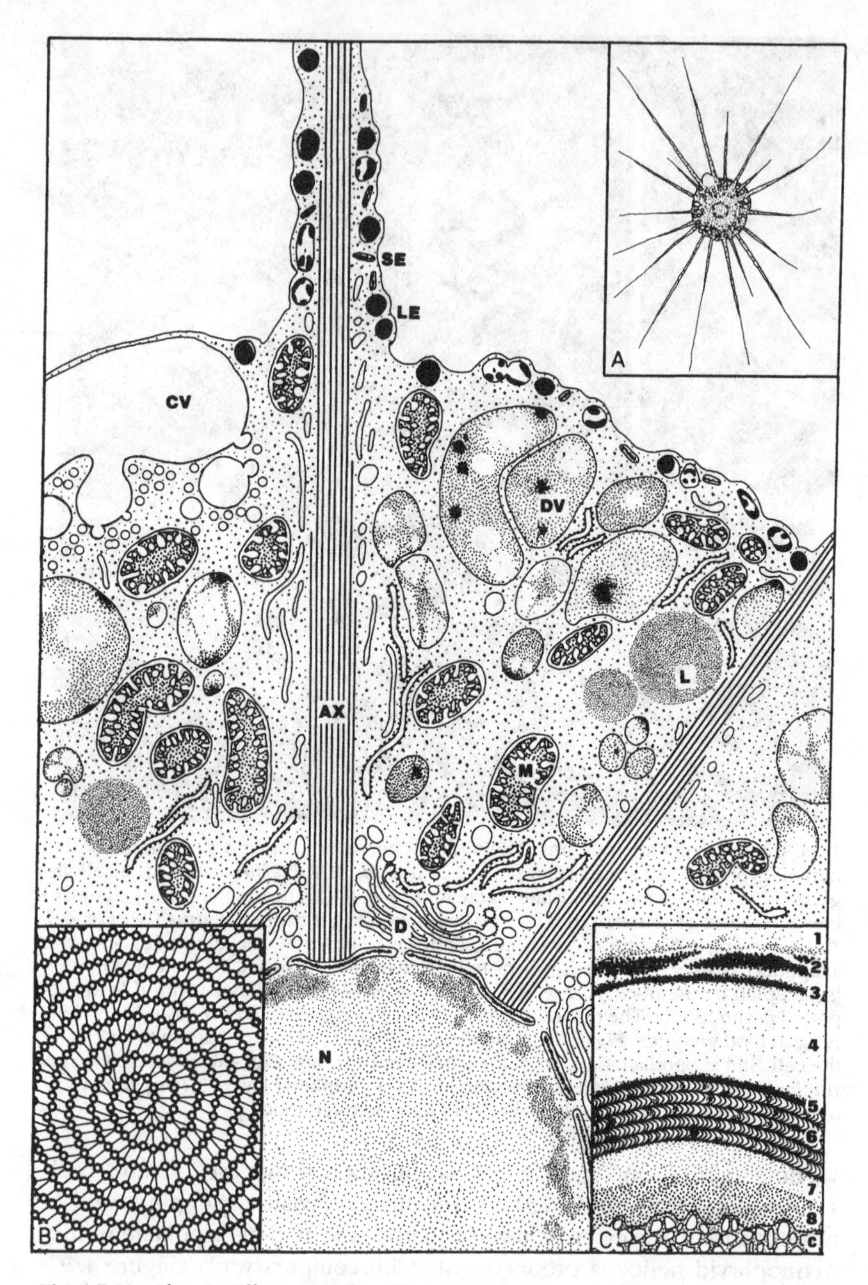

Fig. 4.7. Line drawing illustrating the ultrastructural organization of *Actinophrys sol.* For source references see text. The nucleus (N) lies in the center of the cell (see inset A, whole cell), and from it extend the microtubular axonemes (AX) that support the arms. The microtubules of the axonemes are packed in a double polyogonal spiral array (inset B). The nuclear envelope contains electron-dense laminae. Dictyosomes (D) lie in a zone around the nucleus. The cytoplasm contains "lipid" droplets (L). Mitochondria (M)

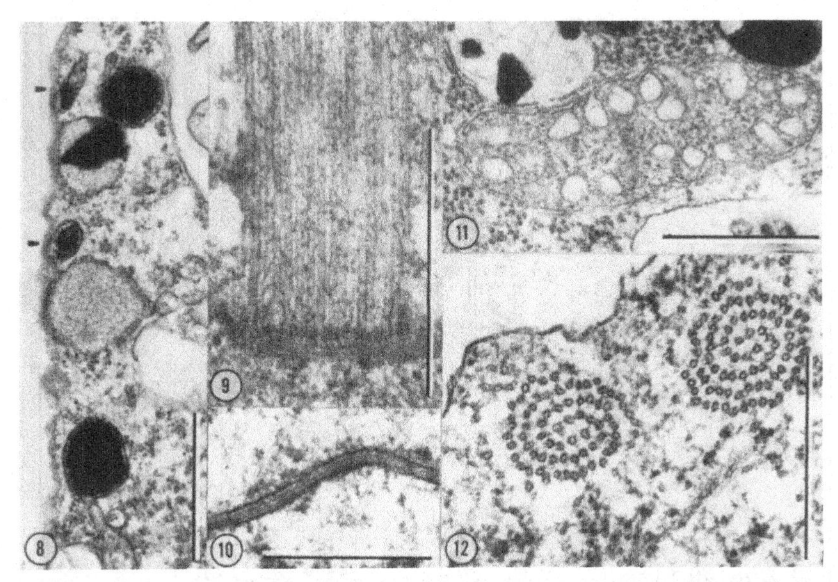

Figs. 4.8–4.12. Transmission electron micrographs of *Actinophrys sol.* All scale bars represent 0.5 μm. ***Fig. 4.8.*** The large and small (arrowheads) extrusomes underlying the plasma membrane. ***Fig. 4.9.*** Termination of the axoneme on the nuclear envelope. ***Fig. 4.10.*** Electron-dense laminae within the nuclear envelope. ***Fig. 4.11.*** Typical mitochondrion with bleblike cristae having narrow bases and without prominent intracristal material. ***Fig. 4.12.*** Axonemes in cross section.

through a wreath of about 12 stiff arms that arise at the anterior lateral shoulders of the cell. Each arm is supported internally by a triad of microtubules (Figures 4.13B and 4.14–4.16). The microtubules terminate on the surface of the nucleus where they are supported by short struts (Figure 4.14). Near the surface of the body, the triads are interconnected by cytoskeletal material (Figure 4.15).

The single flagellum arises from one of two kinetosomes that are interconnected and joined to the nucleus by electron-dense material. The kinetosome that bears the flagellum has a transitional plate located slightly distal to the surface of the body. The plate has a central thickening below which appear to be two rings or a two-gyred helix (Figures 4.17 and 4.18). The kinetosomes contain electron-dense material. Small crescents lie between the bases of the kinetosomes and the nuclear envelope (Figures

---

have bleblike cristae. The peripheral region of the cytoplasm contains many digestion vacuoles (DV) associated with the digestion of food. The cell contains one or more contractile vacuoles (CV). Under the plasma membrane lie numerous small (SE) and large (LE) extrusomes that are employed in food capture and digestion. Inset C shows the complex wall of the cyst: 1, mucus; 2, siliceous plates; 3, electron-dense layer; 4, electron-lucent layer; 5, electron-dense layer; 6, helicoidal layer; 7 and 8, granular layers; c, cytoplasm.

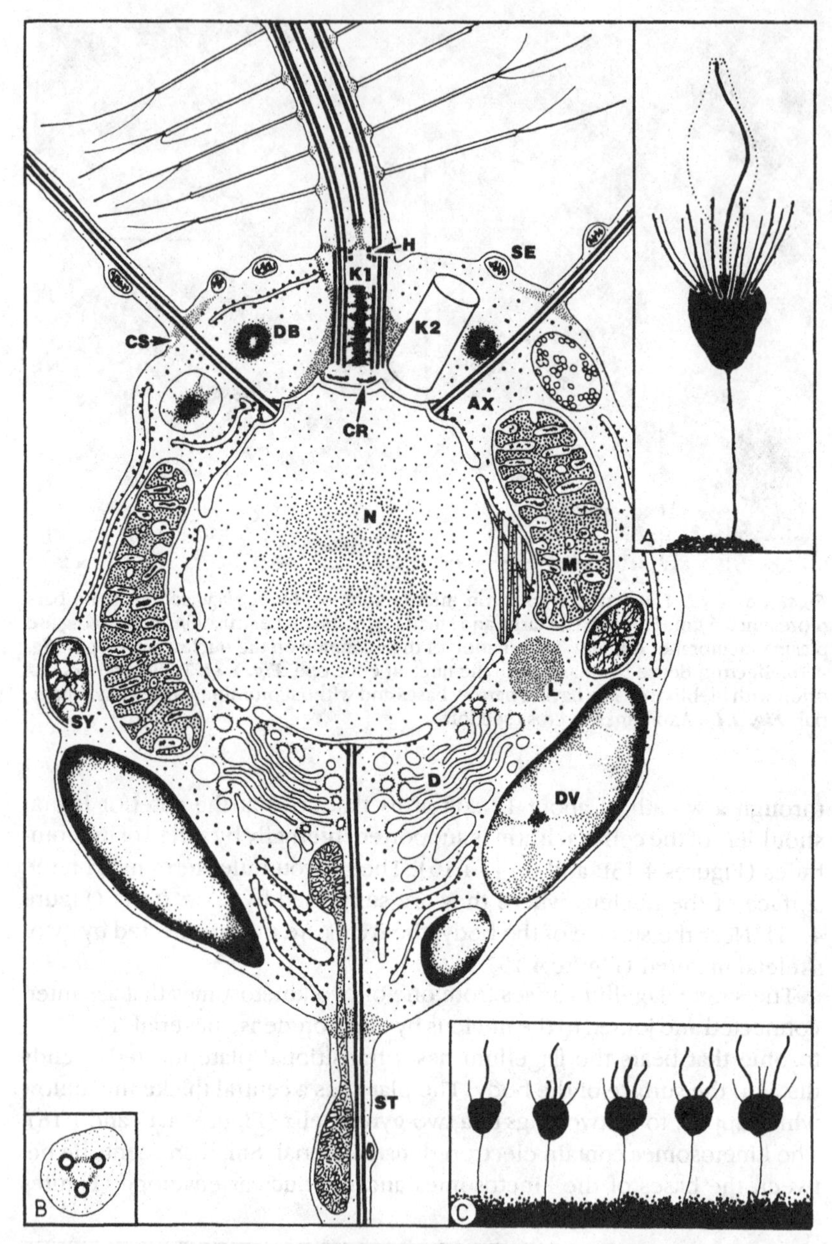

Fig. 4.13. Line drawing illustrating some of the characteristics of *Pteridomonas* sp. The cell (inset A) attaches to the substrate by a contractile stalk. The anterior flagellum (dotted line shows beat envelope, the beat being planar) is surrounded by 12 contractile arms. The arms are supported by three microtubules (inset B) that terminate on the nuclear envelope (main drawing). Near the surface of the cell, they are embedded in a cytoskeletal material (CS) as are the contractile element and microtubules of the stalk. There are two kinetosomes, only one (K1) bears a flagellum that carries tubular mastigonemes that in turn arise in the nuclear envelope. This kinetosome has a helix or rings

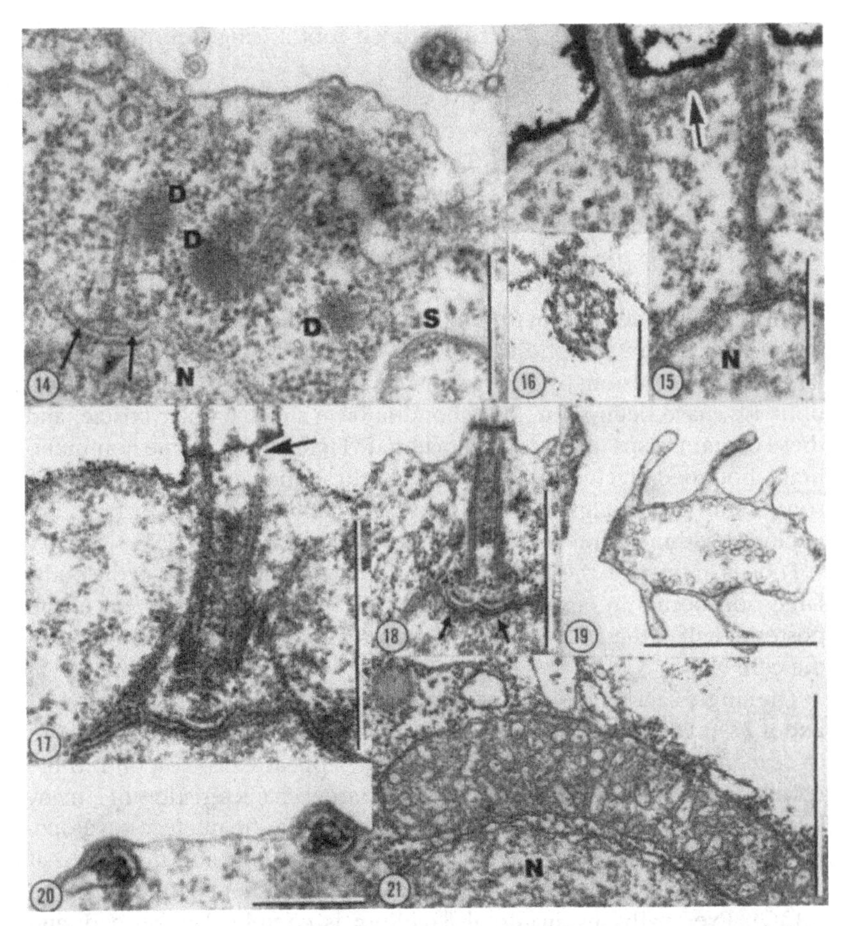

Figs. 4.14–4.21. Transmission electron micrographs of *Pteridomonas* sp. Scale bars represent 0.5 μm, except in Fig. 4.20 where it represents 0.2 μm. ***Fig. 4.14.*** Nuclear (N) termination of microtubular triads. At the level of the nuclear envelope, the microtubules are supported by short struts (arrow). The cytoplasm contains dense bodies (D) and a symbiotic bacterium (S). ***Fig. 4.15.*** The axonemes are interconnected at the periphery of the cell by cytoskeletal material (arrow). ***Fig. 4.16.*** An arm in cross section revealing a triad of microtubules. ***Fig. 4.17.*** The flagellate kinetosome, with electron-dense contents, helix, or rings (arrow) and rootlets connecting it to the nucleus. One crescent may be seen under the kinetosome, and two are evident in ***Fig. 4.18. Fig. 4.19.*** A contracted stalk, the contractile element has a globular appearance. ***Fig. 4.20.*** Extrusomes. ***Fig. 4.21.*** Mitochondrion, the tubular cristae contain obvious intracristal filaments.

(H) lying below the transitional plate. Crescents (CR) lie between the kinetosomes and the nucleus. The two kinetosomes are joined to each other and to the nuclear envelope by electron-dense rootlet structures. Adjacent to the kinetosomes and arms are electron-dense and non-membrane-bounded bodies (DB). Small extrusomes (SE) lie over the body, arms, and stalk. Mitochondria (M) have tubular cristae with intracristal material. Dictyosomes (D) are located in the posterior part of the cell. The cytoplasm contains "lipid" droplets (L), symbiotic bacteria (SY) and, peripherally, digestion vacuoles (DV). Prior to cell division, two new daughter flagella develop alongside the maternal flagellum (inset C).

4.17 and 4.18). There is no cross-striated root arising from the kinetosomes.

The posterior end of the cell gives rise to a contractile stalk, with which the organism attaches to the substrate. When contracted, the stalk appears flaccid and irregularly swollen. The stalk contains a membrane-bound organelle with vesicular contents (Figure 4.19), and this is believed to be the contractile element. This and several microtubules pass through a cytoskeletal sheath at the surface of the cell and terminate against the nuclear envelope.

Small extrusomes (about 0.1 μm in diameter) may be found lying under the plasma membrane of the body and of the arms (Figure 4.20). Some larger dark inclusions, reminiscent of the larger extrusomes of actinophryids, have been seen. Mitochondria have short, tubular cristae, and these contain wisps of intracristal material (Figure 4.21). Some non-membrane-bounded electron-dense bodies occur in the vicinity of the kinetosomes and triads (Figure 4.14). The cells contain bacteria that are not enclosed within vacuoles.

Food is trapped on the arms and ingested in the anterior part of the body. Digestion occurs in large vacuoles that are particularly common in the posterior half of the cell. Dictyosomes are restricted to the posterior half of the cell.

The single flagellum bears two rows of stiff mastigonemes (Figures 4.22 and 4.23). These are tubular when seen in cross section and give rise to several fine filaments at their tips. They appear to correspond to the tubular triparite mastigonemes that are characteristic of many heterokont organisms. The mastigonemes arise in the nuclear envelope. Prior to cell division, two new flagella develop alongside the maternal flagellum. This maternal flagellum shortens as the other two grow (Figure 4.13C). Eventually the maternal flagellum is completely resorbed, and each daughter cell receives one of the new flagella. In actively growing and feeding cultures, multinucleated and multiflagellated individuals are not infrequently encountered. Nuclear division has not been observed.

Discussion

Electron microscopic studies have greatly increased the amount of information that is available for use in assessing the relationships of the actinophryid heliozoa. The analysis that is summarized in Figure 4.24 employs data relating to flagella and associated structures, kinetosomes and associated structures, nuclei, arms, mitochondria, stalks, extrusomes, and a variety of other cytoplasmic inclusions in addition to information relating to the morphology, behavior (Patterson 1980, 1981), and life cycle derived from light microscopic studies.

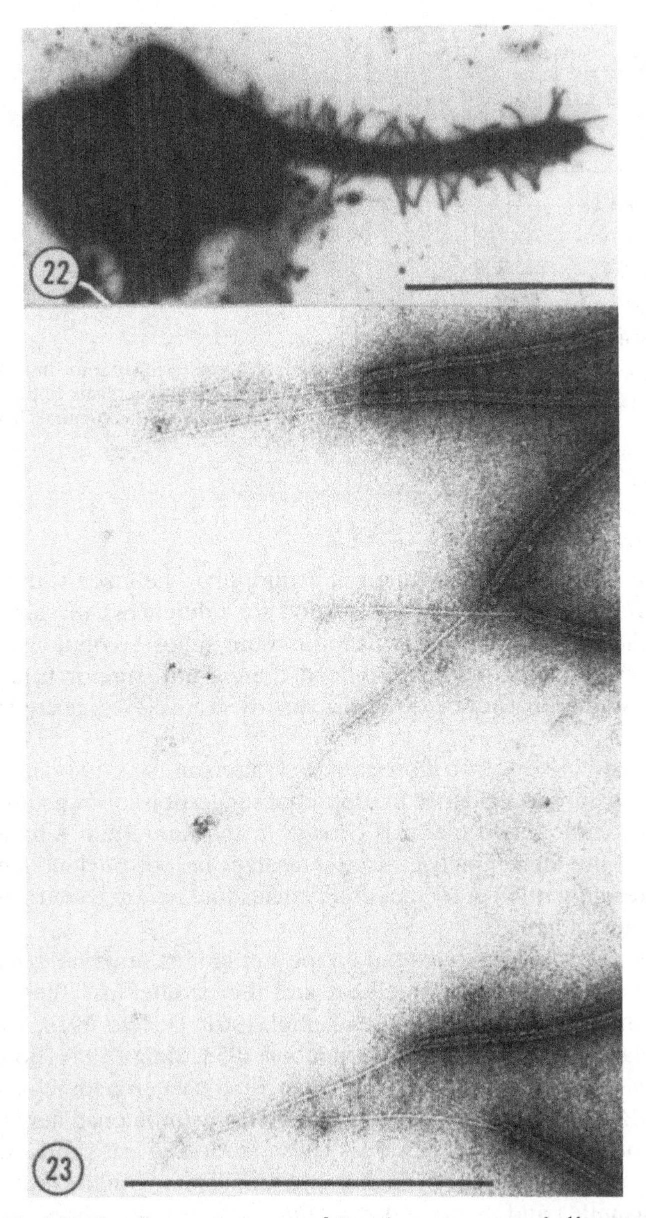

Figs. 4.22–4.23. Flagellar mastigonemes of *Pteridomonas* sp. revealed by negative staining. ***Fig. 4.22.*** Whole cell. Scale bar represent 5 μm. ***Fig. 4.23.*** Detail showing terminal filaments arising from the tubular mastigonemes. Scale bar represents 0.5 μm.

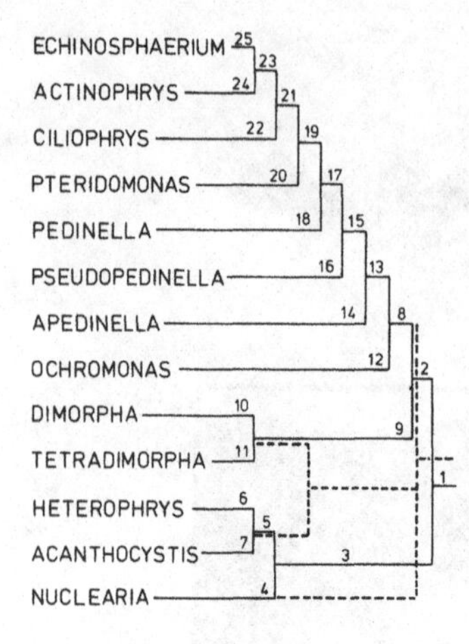

Fig. 4.24. Possible clustering of some heliozoa, helioflagellates, and other organisms implicated in their evolution, based on ultrastructural characters. For details see text.

Comparison of the organization of actinophryid heliozoa with that of other heliozoa (see Introduction section for references) reveals differences sufficiently great to allow us to discount a close evolutionary relationship with the other heliozoa. Indeed, there is little support from ultrastructural studies that any of the heliozoan orders are closely related to any others.

Fine structural studies on filose amebae (Patterson 1983, 1984) have also failed to provide any evidence in support of suggestions of ameboid affinities of the actinophryid heliozoa. However, the same studies have suggested that the filose amebae may themselves be polyphyletic, and we cannot presently rule out the possibility that some heliozoa are related to some of these amebae.

Many workers have commented on the similarities, and possible affinities, of the colorless helioflagellates and the pedinellids (Cienkowski 1876, Kent 1882, Bütschli 1889, Scherffel 1901, Doflein 1916, Pascher 1917, Penard 1921, Skuja, 1948, Trégouboff 1953, Sleigh 1964, Bourrelly 1968, Davidson 1975, 1982). The ultrastructural data on pedinellids (see Introduction section for references) and on the actinomonad helioflagellates do support these suggestions. These studies reveal similarities in extrusomes, in the number of flagella and kinetosomes, in the location of these organelles and associated structures, in the mitochondria, in the arms and supporting axonemal triads, in the location of dictyosomes, and in the nature of the stalk. Dimorphid helioflagellates have similar mitochondria, but differ in respect to most remaining features (Brugerolle & Mignot 1983, 1984).

Pteridomonas sp. described here very closely resembles *Pedinella hexacostata* Vysotskii in the nature and location of its organelles and differs primarily in the lack of plastids that are characteristic of all pedinellids. The flagellum of *P. hexacostata* carries a swollen wing and contains a paraxial rod that appears reduced or absent from *Pteridomonas* sp. Swale (1969a) refers to the flagellar mastigonemes being arranged in a single row, and this also distinguishes it from the colorless organism under investigation here. Other pedinellids may have a more usual double row (Moestrup 1982).

Ciliophrys marina, the only other actinomonad previously studied, also resembles *Pteridomonas* sp. in the mode of food capture, arm contractility, structure and nucleation of the axonemes that support the ams, and presence of a single flagellum bearing tubular mastigonemes (Davidson 1982). According to Davidson (1982), the mastigonemes of *Ciliophrys* do not arise in the nuclear envelope as in *Pteridomonas,* the pedinellids, and other chrysophytes. *Ciliophrys* differs from *Pteridomonas* in having arms emanating from all parts of the body, lacking a stalk, and having large extrusomes. Davidson's account of the kinetid structure is not detailed, and further information here would be helpful in corroborating proposed affinities. However, there seems little doubt that the actinomonad flagellates and the pedinellids are closely related.

As well as having similarities with the helioflagellates, *Ciliophrys* also resembles the actinophryid heliozoa in three respects: (1) the axonemes that support the arms terminate on the nucleus, (2) the cristae of the mitochondria are circular in cross section, and (3) the cells contain similar extrusomes. The value of the last character is uncertain as similar electron-dense bodies have also been observed in other, presumably unrelated, taxa (Swale 1969b, Hausmann 1978, Anderson & Hoeffler 1979, Anderson 1980, Old & Darbyshire 1980). In view of the more striking similarities with other actinomonad helioflagellates, *Ciliophrys* should be placed in a sister group to the actinophryids and not as a subgroup of them as has been done by Levine et al. (1980). On balance, the ultrastructural data are consistent with helioflagellate affinities of the actinophryid heliozoa. However, the data are not sufficient to state confidently that *Ciliophrys* is the closest helioflagellate relative about which we have ultrastructural data.

Using the data that have been di: ussed above, we can attempt to cluster the taxa into nested sets and attempt to identify the evolutionary novelties (synapomorphies *sensu* Hennig 1966) that characterize each cluster. Such an attempt is presented as Figure 4.24. It must be emphasized that while alternative interpretations of the data are possible, such exercises are thought to be a valuable way of clarifying the ancestor–descendent relationships and in guiding us when revising the taxonomy of the organisms involved (see Smith & Patterson 1986).

Features held in common by all organisms included within Figure 4.24 are the only universal eukaryotic features (nuclei, endoplasmic reticulum, dictyosomes, mitochondria, etc.). Organisms with these features (step 1) represent the common ancestor to all of the organisms under consideration. The first division segregates organisms with tubular (2) and flattened mitochondrial cristae (3). Within the organisms with flattened cristae, *Nuclearia*, the filose amoeba (4), most closely represents the ancestral state, having few features not encountered elsewhere. By comparison, the centrohelid heliozoa (5) share a number of distinguishing features (synapomorphies). These include the centroplast, ball-and-cone extrusomes, siliceous scales, and axonemes. The two genera may be distinguished by subsequent diversification of the structure of the siliceous scales (6 and 7).

Within the second lineage, step 8 corresponds to appearance of tubular mastigonemes and associated kinetid features of the heterokont flagellates. The dimorphid flagellates (step 9) may be distinguished by the appearance of concentric extrusomes, elongated kinetosomes, and elaborate microtubular and non-microtubular flagellar roots (Brugerolle and Mignot 1983, 1984). The two species can be distinguished by the number of flagella and by a number of differences in ultrastructural organization (steps 10 and 11). Within the heterokont lineage, *Ochromonas* (step 12) is taken as representing the symplesiomorphic state. At step 13, some of the features that distinguish the pedinellids and the actinomonad flagellates appeared. These include the appearance of microtubular triads, the multiplication of chloroplasts, the loss of the eyespot, flagellar swelling, presence of one flagellum, simplification of the flagellar roots, posterior movement of the dictyosomes, etc. The diagram suggests that the stalkless *Apedinella* (step 14) most closely resembles the symplesiomorphic state of this group, being a parsimonious but not biologically satisfying interpretation. The stalk, which characterizes most actinomonad flagellates and many pedinellids, would characterize step 15. *Pseudopedinella* (step 16) resembles this state, whereas the remaining species (step 17) contain extrusomes. The transition from the remaining pedinellids (step 18) to the actinomonad flagellates simply involved the loss of chloroplasts and a reduction in the paraxial rod. This amply describes *Pteridomonas* (step 20). At step 21, the stalk has been lost, the arms have become more numerous, and larger extrusomes are present. *Ciliophrys* (22) is the only genus currently known that satisfies these conditions. The actinophryid heliozoa (23) are associated with the elaboration of the axonemes that support the arms into spiral arrays, by the occurrence of two kinds of extrusomes, and by the development of a cyst stage in the life cycle in which autogamy occurs. The two genera (24 and 25) may be distinguished by the occurrence of multiple nuclei in *Echinosphaerium*, some associated cytological changes, and the appearance of the cyst.

Clearly, this is a cursory account of what might have happened based on a limited selection of organisms and characters. As stated above, the data are open to alternative suggestions. If one accepts the homologies suggested by Brugerolle and Mignot (1983, 1984) for certain cytological features of the dimorphid helioflagellates and the centrohelid heliozoa, then the clustering indicated by the dashed line in Figure 4.24 ensues. This places the two groups mentioned closely together, but in placing the dimorphids that have tubular mitochondrial cristae in the lower cluster, it leaves the remainder without any synapomorphies.

Another aspect of the scenario presented in Figure 4.24 that is open to debate is whether the pedinellids are ancestral to the colorless actinomonads. In having chloroplasts similar to those of chrysophytes, the pedinellids have been presented as being close to *Ochromonas,* taken to represent the ancestral heterokont situation. Hibberd (1976) has argued that the pedinellids are cytologically very dissimilar to the ochromonads, and in respect to aspects of the flagellar structure (two rows of mastigonemes, reduced or absent paraxial rod) *Pteridomonas* and *Ciliophrys* could be argued to be closer to the ancestor. Evidently, we require more information before we are in a position to provide a convincing account of the interrelationships among all of the taxa under discussion.

While a full understanding of the interrelationships among these taxa is still elusive, it seems probable that the suggestion of a chrysophyte link for the actinophryids will be corroborated by future data. This would create a situation in which organisms initially described in accordance with the zoological code of nomenclature are placed in a taxon with some organisms that are plants by nomenclature. There are areas of nomenclatural conflict between the two codes, and this has led some authors to consider how those conflicts might be resolved or minimized (Jeffrey 1982, Ride 1982, Corliss 1983).

Four options of resolving areas of conflict have been discussed. The first would be a unified code, which seems impractical in view of the amount of revisions that would be required and because the majority of taxonomists, working with bona fide plants or animals, could see little cause for such a major change in tradition. A second possibility is that each new kingdom should have its own code. However, there is little likelihood of agreement over how many kingdoms there should be (see Leedale 1974, Cavalier-Smith 1981, Jeffrey 1982, Corliss 1983, for some conflicting opinions). Indeed, in view of the nomenclaturally casual fashion in which many of these high level taxa are proposed, there is a good case for the introduction of a code of conduct for the formal proposition of kingdoms and phyla. The third option is that each case of disharmony be resolved independently. This approach has been adopted by Silva (1980). This is the only proper approach and is one that is to some extent accommodated by the codes because they permit transfer of organisms across kingdoms. It does require a working familiarity with both codes and does tend to perpetuate the misleading tradition of envisaging the eukaryotic world as being comprised only of plants and animals. The final option is to have a code for protists. This would not be satisfactory because it would encounter the problems encountered with other paraphyletic taxa. The boundaries of the taxon could not be defined and, therefore, the organisms to which the new code should apply could not be unambiguously identified.

A flexible option needs to be found. This option should not be in conflict with tradition so that the existing codes may still be employed. The answer may lie in the use of some clearly defined, but "colloquial" taxa that permit the coclassification of some organisms that are, from the nomenclatural point of view, animals, and some that are plants. If further information corroborates the view expressed in this chapter that pedinellids, actinomonad helioflagellates, and actinophryid heliozoa are related, then some resolution of this problem will need to be encountered. It can also be predicted that this will not be the only case in which "plants" and "animals" will need to be grouped together. It may well provide a test case, revealing if the codes of nomenclature have sufficient flexibility not to inhibit the development and acceptance of a natural classification of lower eukaryotes.

References

Anderson, O.R. 1980. Radiolaria. *In:* Levandowsky, M. & Hutner, S.H. [Eds.] ***Biochemistry and Physiology of Protozoa,*** 2nd Ed. Vol. 3. Academic Press, New York, pp. 1–42.

Anderson, O.R. & Hoeffler, W.K. 1979. Fine structure of a marine proteomyxid and cytochemical changes during encystment. *J. Ultrastruct. Res.* 66: 276–87.

Bardele, C.F. 1972. Cell cycle, morphogenesis, and ultrastructure in the pseudoheliozoan *Clathrulina elegans. Z. Zellforsch.* 130: 219–42.

– 1975. The fine structure of the centrohelidian heliozoan *Heterophrys marina. Cell Tiss. Res.* 161: 85–102.

– 1977. Comparative study of axopodial microtubule patterns and possible mechanisms of pattern control in the centrohelidian heliozoa *Acanthocystis, Raphidiophrys* and *Heterophrys. J. Cell Sci.* 25: 205–32.

Bělař, K. 1926. *Der Formwechsel der Protistenkerne.* Fischer, Jena.

Bourrelly, P. 1968. *Les Algues d'Eau Douce,* Tome II, *Les Algues jaunes et brunes. Chrysophycées, Pheophycées, Xanthophycées et Diatomées.* Boubée, Paris.

Bovee, E.C. 1960. Studies on the helioflagellates. I. The morphology and fission of *Dimorpha floridanis* n. sp. *Arch. Protistenk.* 104: 503–14.

Brugerolle, G. & Mignot, J-P. 1983. Charactéristiques ultrastructurales de l'helioflagelle *Tetradimorpha* Hsiung et leur interet pour l'etude phyletique des heliozoaires. *J. Protozool.* 30: 473–80.

– 1984. The cell characters of the two centrohelidian helioflagellates related to the centrohelidion lineage: *Dimorpha* and *Tetradimorpha. Origins of Life* 13: 305–14.

Bütschli, O. 1889. Protozoa. *In:* Bronn, H.G. [Ed.] *Klassen und Ordnungen des Thier-Reichs.* Winter'sche, Leipzig.

Cachon, J. & Cachon, M. 1978. *Sticholonche zanclea* Hertwig: A reinterpretation of its phylogenetic position based upon new observations on its ultrastructure. *Arch. Protistenk.* 120: 148–68.

– 1982. Actinopoda. *In:* Parker, S.P. [Ed.] *Synopsis and Classification of Living Organisms* Vol. 2, McGraw-Hill, New York, pp. 553–68.

Cavalier-Smith, T. 1981. Eukaryote kingdoms: seven or nine? *BioSystems* 14: 461–81.

Cienkowski, L. 1876. Ueber einige Rhizopoden und verwandte Organismen. *Arch. Mikr. Anat.* 12: 15–50.

Corliss, J.O. 1979. The impact of electron microscopy in ciliate systematics. *Amer. Zool.* 19: 573–87.

– 1983. Consequences of creating new kingdoms of organisms. *BioScience* 33: 314–18.

Cox, E.R. [Ed.] 1980. *Phytoflagellates.* Elsevier North-Holland, New York.

Davidson, L.A. 1975. Studies of the actinopods *Heterophrys marina* and *Ciliophrys marina:* Energetic and structural analysis of their contractile axopodia, general ultrastructure, and phylogenetic relationships. Ph.D. thesis, University of California, Berkeley (Xerox University Microfilms 76-15,155).

– 1976. Ultrastructure of the membrane attachment sites of the extrusomes of *Ciliophrys marina* and *Heterophrys marina* (Actinopoda). *Cell Tiss. Res.* 170: 353–65.

– 1982. Ultrastructure, behavior, and algal flagellate affinities of the helioflagellate *Ciliophrys marina,* and the classification of the helioflagellates (Protista, Actinopoda, Heliozoea). *J. Protozool.* 29: 19–29.

Dodge, J.D. 1973. *The Fine Structure of Algal Cells.* Academic Press, London.

Doflein, F. 1916. *Lehrbuch der Protozoenkunde.* Fischer, Jena.

Febvre-Chevalier, C. 1973. *Hedraiophrys hovassei,* morphologie, biologie et cytologie. *Protistologica* 9: 503–20.

– 1975. Etude cytologique de *Gymnosphaera albida* Sassaki, 1894 (Heliozoaire Centrohelidié). *Protistologica* 11: 331–44.

– 1980. Behaviour and cytology of *Actinocoryne contractilis,* nov. gen., nov. sp., a new stalked heliozoan (centrohelidia): comparison with other related genera. *J. Mar. Biol. Ass., U.K.* 60: 909–28.

Fenchel, T. 1982. Ecology of some heterotrophic microflagellates. I. Some important forms and their functional morphology. *Mar. Ecol. Progr. Ser.* 8: 211–23.

Gadea, E. 1973. La implacacion filetico-sistematico de los Rizopodos. *Publ. Inst. Biol. Apl. Barcelona* 54: 93–8.

Griessmann, K. 1914. Uber marine Flagellaten. *Arch. Protistenk.* 32: 1–78.

Gruber, A. 1882. *Dimorpha mutans.* Eine Mischform von Flagellaten und Heliozoen. *Zeitschr. f. wiss. Zool.* 36: 445–58.

Hausmann, K. 1978. Bakterien und Virusähnliche Partikel im Cytoplasma des Rhizopoden *Vampyrella lateritia. Ann. St. Biol. Besse-en-Chandesse* 11: 102–7.

Hausmann, K. & Patterson, D.J. 1982. Pseudopod formation and membrane production during prey capture by a heliozoon (Feeding by *Actinophrys,* II). *Cell Motility* 2: 9–24.

Hausmann, K., Linnenbach, M. & Patterson, D.J. 1983. The effects of Taxol on microtubular arrays: in vivo effects on heliozoan axonemes. *J. Ultrastruct. Res.* 82: 212–20.

Hibberd, D.J. 1976. The ultrastructure and taxonomy of the Chrysophyceae and Prymnesiophyceae (Haptophyceae): a survey with some new observations on the ultrastructure of the Chrysophyceae. *Bot. J. Linn. Soc.* 72: 55–80.

– 1979. The structure and phylogenetic significance of the flagellar transition region in the chlorophyll c-containing algae. *BioSystems* 11: 243–61.

Hibberd, D.J. & Leedale, G.F. 1972. Observations on the cytology and ultrastructure of the new algal class, Eustigmatophyceae. *Ann. Bot.* 36: 49-71.
Honigberg, B.M., Blamauth, W., Bovee, E.C., Corliss, J. O., Gojdics, M., Hall, R.P., Kudo, R.R., Levine, N.D., Loeblich, A.R., Weiser, J. & Wenrich, D.H. 1964. A revised classification of the phylum Protozoa. *J. Protozool.* 11: 7-20.
Jeffrey, C. 1982. Kingdoms, codes and classification. *Kew Bull.* 37: 403-16.
Jones, W.C. 1976. The ultrastructure of *Gymnosphaera albida* Sassaki, a marine axopodiate protozoon. *Phil. Trans. Roy. Soc. Lond. B* 275: 349-84.
Kent, W.S. 1882. *A Manual of the Infusoria*, Vols. 1-3. Bogue, London.
Krylov, M.V., Dobrovolsky, A.A., Issi, I.V., Mikhalevich, V.I., Podlipaev, S.A., Reschetnyak, V.V., Seravin, L.N., Starabogatov, Y.I., Shulman, S.S. & Jankowski, A.V. 1980. New proposals regarding the system of unicellular organisms. *Trudy. Zool. Inst. Akad. Nauk. S.S.S.R.*, 94: 122-32 (in Russian).
Leedale, G.F. 1974. How many are the kingdoms of organisms? *Taxon* 23: 261-70.
Levine, N.D., Corliss, J.O., Cox, F.E.G., Deroux, G., Grain, J., Honigberg, B.M., Leedale, G.F., Loeblich, A.R., Lom, J., Lynn, D., Merinfeld, E.G., Page, F.C., Poljansky, G., Sprague, V., Vavra, J., & Wallace, F.G. 1980. A newly revised classification of the protozoa. *J. Protozool.* 27: 37-58.
Linnenbach, M., Hausmann, K. & Patterson, D.J. 1983. Ultrastructural studies on the food vacuole cycle of a heliozoon (Feeding by *Actinophrys*. III). *Protoplasma* 115: 43-51.
Mignot, J-P. 1979. Etude ultrastructurale de la pédogamie chez *Actinophrys sol* (Heliozoaire). 1. La division progamique. *Protistologica* 15: 387-406.
- 1980a. Etude ultrastructurale de la pédogamie chez *Actinophrys sol* (Heliozoaire) II. Les divisions de maturation. *Protistologica.* 16: 205-25.
- 1980b. Etude ultrastructurale de la pédogamie chez *Actinophrys sol* (Heliozoaire) III. Gamétogénèse, fécondation, enkystement. *Protistologica* 16: 533-47.
Moestrup, Ø. 1982. Flagellar structure in algae: a review, with new observations particularly on the Chrysophyceae, Phaeophyceae (Fucophyceae), Euglenophyceae, and *Reckertia*. *Phycologia* 21: 427-528.
Old, K.M. & Darbyshire, J.F. 1980. *Arachnula impatiens* Cienk., a mycophagous giant amoeba from soil. *Protistologica* 16: 277-87.
Ostroff, C.R. & van Valkenburg, S.D. 1978. The fine structure of *Pseudopedinella pyriforme* Carter (Chrysophyceae). *Br. Phycol. J.* 13: 35-49.
Pascher, A. 1917. Flagellaten und Rhizopoden in ihren gegenseitigen Beziehungen. *Arch. Protistenk.* 38: 1-79.
Patterson, D.J. 1979. On the organization and classification of the protozoon, *Actinophrys sol* Ehrenberg, 1830. *Microbios* 26: 165-208.
- 1980. Contractile vacuoles and associated structures: their organization and function. *Biol. Rev.* 55: 1-46.
- 1981. Contractile vacuole complex behaviour as a diagnostic character for free-living amoebae. *Protistologica.* 17: 243-8.
- 1982. Photomicrography using a dedicated electronic flash. *Microscopy* 34: 437-42.
- 1983. On the organization of the naked filose amoeba, *Nuclearia moebiusi* Frenzel, 1897 (Sarcodina, Filosea) and its implications. *J. Protozool.* 30: 301-7.
- 1984. The genus *Nuclearia* (Sarcodina, Filosea): species composition and characteristics of the taxa. *Arch. Protistenk.* 128: 127-139.

Patterson, D.J. & Fenchel, T. 1985. Insights into the evolution of heliozoa (Protozoa, Sarcodina) as provided by ultrastructural studies on a new species of flagellate from the genus *Pteridomonas. Biol. J. Linn. Soc.* 34: 381–403.
Patterson, D.J. & Hausmann, K. 1981. Feeding by *Actinophrys sol* (Protista, Heliozoa): 1 Light microscopy. *Microbios* 31: 39–55.
– 1982. Morulate bodies in Actinophryid heliozoa: A fixation artefact derived from microtubules? *Cell Struct. Funct.* 7: 341–8.
Patterson, D.J. & Thompson, D.W. 1981. Structure and elemental composition of the cyst wall of *Echinosphaerium nucleofilum* Barrett (Heliozoea, Actinophryida). *J. Protozool.* 28: 188–92.
Penard, E. 1890. Ueber einige neue oder wenig bekannte Protozoen. *Jahrb. Nassauischer Vereins f. Naturkunde* 43: 73–91.
Penard, E. 1904. *Les Héliozoaires d'Eau douce.* Kundig, Geneva.
Penard, E. 1921. Studies on some flagellata. *Proc. Acad. Nat. Sci. Philadelphia* 73: 105–68.
Pienaar, R.N. 1980. Chrysophytes. *In:* Cox, E.R. [Ed.] *Phytoflagellates.* Elsevier North-Holland, New York, pp. 213–42.
Ride W.D.L. 1982. Nomenclature of organisms treated both as plants and animals. *Biol. Intern.* 6: 15–16.
Rieder, C.L. 1979. High-voltage electron microscopy reveals new components in the axonemes and centroplast of the centrohelidian *Raphidiophrys ambigua. J. Cell Sci.* 40: 215–34.
Scherffel, A. 1901. Kleiner Beitrag zur Phylogenie einiger Gruppen niederer Organismen. *Bot. Zeit.* 8: 143–59.
Schouteden, H. 1907. Notes sur quelques flagelles. *Arch. Protistenk.* 9: 108–36.
Silva, P.C. 1980. Remarks on algal nomenclature VI. *Taxon* 29: 121–47.
Skuja, H. 1948. Taxonomie des Phytoplanktons einiger Seen in Uppland, Schweden. *Symbolae Botanicae Upsalienses* 9: 1–399.
Sleigh, M.A. 1964. Flagellar movement of the sessile flagellates *Actinomonas, Codonosiga, Monas,* and *Poteriodendron. Quart. J. Micro. Sci.* 105: 405–14.
Smith, R. McK. & Patterson, D.J. 1986. Analyses of helizoan interrelationships: an example of the potentials and limitations of ultrastructural approaches to the study of protistan phylogeny. *Proc. Roy. Soc. Lond., B:* 227: 325–66.
Suzaki, T., Shigenaka, Y., Watanabe, S. & Toyahara, A. 1980. Food capture and ingestion in the large heliozoan, *Echinosphaerium nucleofilum. J. Cell Sci.* 42: 61–79.
Swale, E.M.F. 1969a. A study of the nannoplankton flagellate *Pedinella hexacostata* Vysotskii by light and electron microscopy. *Br. Phycol. J.* 4: 65–86.
– 1969b. The fine structure of a species of the amoebo-flagellate *Pseudospora* Cienk. *Arch. Mikrobiol.* 67: 71–90.
Taylor, F.J.R. 1976. Flagellate phylogeny, a study in conflicts. *J. Protozool.* 23: 28–40.
Throndsen, J. 1971. *Apedinella* gen. nov. and the fine structure of *A. spinifera* (Throndsen) comb. nov. *Norw. J. Bot.* 18: 47–64.
Tilney, L.G. & Porter, K.R. 1965. Studies on microtubules in Heliozoa I. *Protoplasma* 60: 317–44.
Trégouboff, G. 1953. Classe des Héliozoaires. *In:* Grassé, P.P. [Ed.] *Traité de Zoologie,* Tome I, Fascicule II. Masson, Paris. pp. 437–89.
Valkanov, A. 1928. Protistenstudien. II. Notizen uber die Flagellaten Bulgariens. *Arch. Protistenk.* 63: 419–50.

PART II

Taxonomy

5

Classification of four genera of Chrysophyceae bearing silica scales in a family separate from *Mallomonas* and *Synura*

HANS R. PREISIG

Institute of Systematic Botany, University of Zurich
Zollikerstr. 107, CH-8008 Zurich, Switzerland

AND

DAVID J. HIBBERD[1]

Institute of Terrestrial Ecology, The Culture Centre of Algae and Protozoa
36 Storey's Way, Cambridge CB3 0DT, England

Chrysophyceae bearing silica scales have previously always been classified within the single family Mallomonadaceae.[2,3] Following the recent survey of Bourrelly (1981) this family is comprised of nine genera: *Mallomonas, Mallomonopsis, Synura, Paraphysomonas, Conradiella, Microglena, Catenochrysis, Chrysosphaerella,* and *Spiniferomonas* (=*Chromophysomonas*) (see Nicholls 1985). The classification of these genera within a single family was based mainly on light microscopic observations because thorough electron microscopic data on internal cell structure were known only for a few genera, i.e., *Mallomonas, Mallomonopsis,* and *Synura.* For all other genera, such data were either completely lacking or, as in *Paraphysomonas* and *Chrysosphaerella,* the published micrographs lacked detail or dealt mainly with scale formation (for references see Preisig and Hibberd 1983). We have examined cell morphology of *Paraphysomonas* and *Chrysosphaerella* more thoroughly, including in our studies the genus

[1] Present address: Axle Tree Cottage, Starvecrow Lane, Peasmarsh, Rye, East Sussex TN31 6XL, England.

[2] The name Mallomonadaceae takes precedence over the name Synuraceae, generally used in earlier papers (see Silva 1980).

[3] Some species producing silica scales have also been included in the family Aurosphaeraceae, but this family has an uncertain position within the Chrysophyceae (Parke and Dixon 1976).

The work for this paper was done during tenure of a postdoctoral fellowship by H.R.P. at the Culture Centre of Algae and Protozoa (CCAP) in Cambridge. This author would like to express his thanks to the Director and staff of CCAP and to the Institute of Terrestrial Ecology for their hospitality and for providing facilities for this work. Financial support for the fellowship was provided by the Swiss National Science Foundation.

Spiniferomonas as well as the new genus *Polylepidomonas* (see also Preisig and Hibberd 1982a,b, 1983).

The internal cell structure of all four genera was found to be very similar, apart from the fact that *Paraphysomonas* is colorless and possesses a leukoplast in place of a chloroplast, whereas *Spiniferomonas, Chrysosphaerella,* and *Polylepidomonas* are usually pigmented with one to two chloroplasts. The characteristics of these genera clearly contrast with those of *Mallomonas, Mallomonopsis,* and *Synura* (Mallomonadaceae *sensu stricto*). This is illustrated in Figure 5.1 in which A shows the general features of the Mallomonadaceae *sensu stricto* and B shows those of the four genera studied here. Figure 5.1C shows a member of the Ochromonadaceae, representing the type of organization considered as basic for the Chrysophyceae (cf. Hibberd 1976).

It should be noted that the three illustrated cell types represent generalized models showing the characteristic features of the three groups of Chrysophyceae discussed. Although inter- and intrageneric differences occur within each group, their taxonomic significance is certainly not beyond generic level, e.g., different lengths of the short flagellum in *Mallomonas, Mallomonopsis,* and *Synura* (group A), presence of a leukoplast in place of a chloroplast in *Paraphysomonas* (group B) and *Spumella* (group C), and presence or absence of an eyespot (stigma) in species of *Spiniferomonas* (group B) (cf. Preisig and Hibberd 1983).

Comparing the three cell types shown in Figure 5.1, the following characteristics are evident:

B corresponds with C but differs from A regarding (1) orientation and insertion of the flagella, (2) length of the short flagellum relative to that of the long flagellum, (3) absence of flagellar scales, and (4) nature of the swelling on the short flagellum and its position relative to eyespot or chloroplast. In addition to the features illustrated in Figure 5.1, it also appears that the genera of both B and C (but not A) possess a microtubular root which passes around the flagellar depression (see Preisig and Hibberd 1983).

B corresponds with A in the presence of silica scales that are lacking in C. The development of the scales inside the cells, however, is clearly different in A and B. In A the scales are formed near the chloroplast in association with the chloroplast endoplasmic reticulum, whereas in B formation usually occurs near the Golgi apparatus in association with a vesicle of rough endoplasmic reticulum.

B differs both from A and C in lacking a chloroplast girdle lamella.

In all features discussed here, *Paraphysomonas, Spiniferomonas, Chrysosphaerella,* and *Polylepidomonas* (group B) thus correspond with the Mallomonadaceae *sensu stricto* (group A) only in the presence of silica scales, but the formation of these scales differs between the groups. Furthermore,

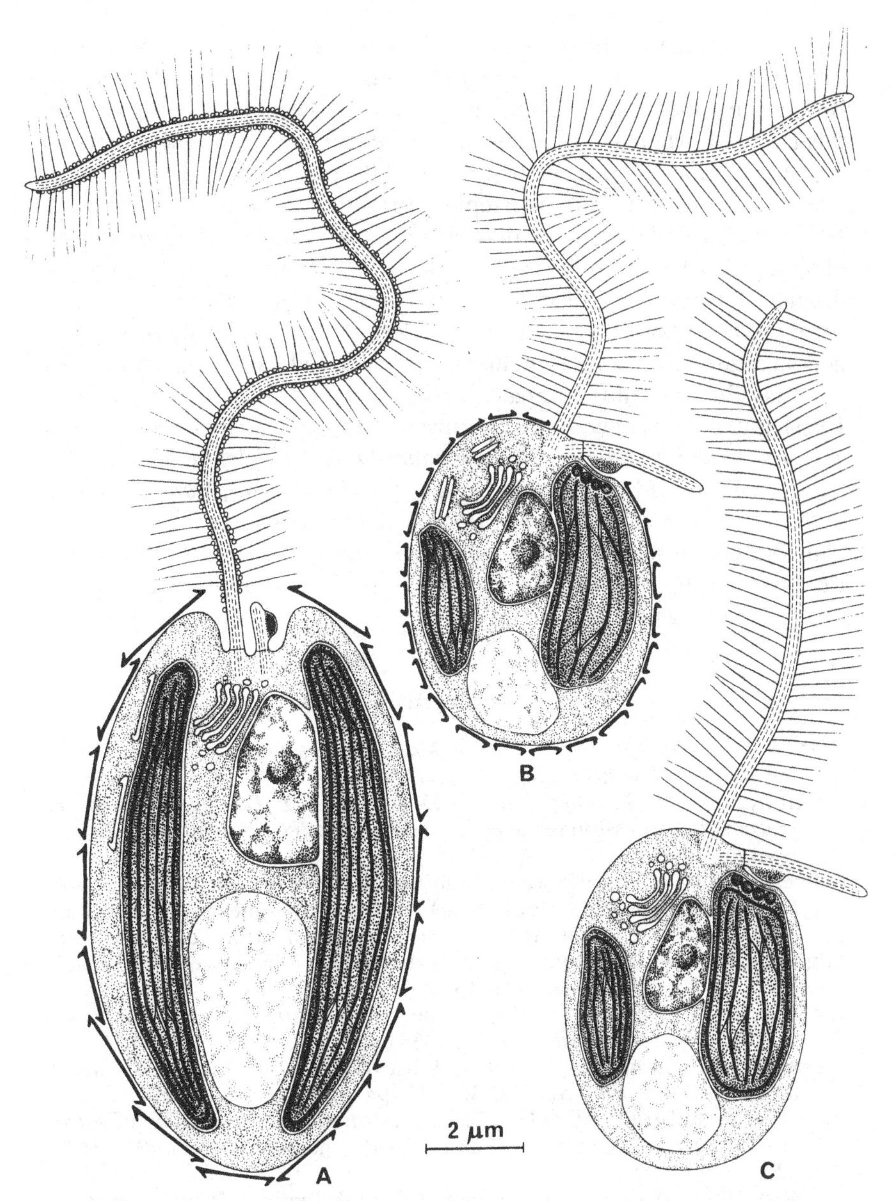

Fig. 5.1. Schematic drawings showing the characteristic features of the chrysophycean families Mallomonadaceae (A), Paraphysomonadaceae (B) and Ochromonadaceae (C). [A modified after Belcher (1969) and Mignot and Brugerolle (1982).]

it is also evident that the scales of the Mallomonadaceae *sensu stricto* may vary in shape according to their position on the cell surface (Mignot and Brugerolle 1982) and are regularly imbricated (Leadbeater, Chapter 9 of this treatise), whereas the scales of *Paraphysomonas* and related genera are more homogeneous in shape and while they may overlap to some extent, they do not appear to be regularly imbricated (Preisig and Hibberd 1983). Because of these clear differences, classification of the four genera of group B within the Mallomonadaceae is considered untenable. Although they have many features in common with the Ochromonadaceae (group C), on the other hand, production of silica scales by these four genera argues against their inclusion within this already large and ill-defined family. The third alternative, adopted by us (Preisig and Hibberd 1983), is to create a new family Paraphysomonadaceae to include *Paraphysomonas, Spiniferomonas, Chrysosphaerella,* and *Polylepidomonas.* The family Mallomonadaceae now includes only *Mallomonas, Mallomonopsis, Synura, Conradiella, Microglena,* and *Catenochrysis,* but examination with the electron microscope will be necessary to determine whether *Conradiella, Microglena,* and *Catenochrysis* have a natural place in this family (Preisig and Hibberd 1983).

References

Belcher, J.H. 1969. Some remarks upon *Mallomonas papillosa* Harris and Bradley and *M. calceolus* Bradley. *Nova Hedwigia* 18: 257–70.

Bourrelly, P. 1981. *Les Algues d'Eau Douce,* tome 2, *Les Algues Jaunes et Brunes.* Réimpression revue et augmentée. Société Nouvelle des Éditions Boubée, Paris.

Hibberd, D.J. 1976. The ultrastructure and taxonomy of the Chrysophyceae and Prymnesiophyceae (Haptophyceae): a survey with some new observations on the ultrastructure of the Chrysophyceae. *Bot. J. Linn. Soc.* 72: 55–80.

Mignot, J.-P. and Brugerolle, G. 1982. Scale formation in chrysomonad flagellates. *J. Ultrastructure Res.* 81: 13–26.

Nicholls, K.H. 1985. The validity of the genus *Spiniferomonas* Takahashi (Chrysophyceae). *Nord. J. Bot.* 5: 403–6.

Parke, M. and Dixon, P.S. 1976. Check-list of British marine algae – third revision. *J. Mar. Biol. Ass. U.K.* 56: 527–94.

Preisig, H.R. and Hibberd, D.J. 1982a. Ultrastructure and taxonomy of *Paraphysomonas* (Chrysophyceae) and related genera 1. *Nord. J. Bot.* 2: 397–420.

– 1982b. Ultrastructure and taxonomy of *Paraphysomonas* (Chrysophyceae) and related genera 2. *Nord. J. Bot.* 2: 601–38.

– 1983. Ultrastructure and taxonomy of *Paraphysomonas* (Chrysophyceae) and related genera 3. *Nord. J. Bot.* 3: 695–723.

Silva, P.C. 1980. Names of classes and families of living algae with special reference to their use in the Index Nominum Genericorum (Plantarum). *Regnum Veg.* 103: 1–156.

6

A proposed phylogenetic sequence for the loricate rhizopodial Chrysophyceae

RICHARD L. MEYER
Department of Botany & Microbiology
University of Arkansas
Fayetteville, Arkansas 72701, U.S.A.

Introduction

The phylogeny of the Chrysophyceae with a dominant rhizopodial cell contained within a lorica has received limited attention. These infrequently reported chrysophytes have been given various levels of treatment in the literature. The assignments of the several taxa by various authors has resulted in a confusing array of interrelationships.

This chapter reviews the earlier taxonomic placement of these little known genera. Specific reference is given to the better known and indicator genera *Chrysopyxis, Derepyxis, Kybotion,* and *Lagynion.* It also summarizes pertinent features of known life histories. Finally, based upon our present state of knowledge, a phylogenetic scheme is presented.

Discussion

The loricate Chrysophyceae that characteristically have rhizopodia extending from apertures have been assigned to one, two, or three ordinal ranks. The early researchers were influenced by Scherffel's (1901) emphasis on the presence of a flagellum in the vegetative cell or zoospore. However, 10 years later, Lauterborn (1911) grouped the ameboid forms into an independent family, the Chrysopodaceae. This family included both naked, e.g., *Chrysamoeba* and *Palatinella,* and loricate genera, e.g., *Chrysopyxis.* The influence of these and other contemporary protistologists was reflected in Pascher's (1913) scheme. Pascher's summary of the known Chrysophyceae placed *Chrysopyxis* in the Chromulinales, *Derepyxis* in the Isochrysidales, and *Lagynion* and *Stylococcus* in the Rhizochrysidinae

The author is most grateful to the Department of Botany of the University of Minnesota and to Dr. David Parmalee, Director, Itasca Forestry and Biology Station, for the time and facilities made available for this study.

with rhizopodial taxa (Table 6.1). The last taxon, however, did not contain order or family designations.

Several more recent researchers have followed the general format proposed by Pascher (1913): Huber-Pestalozzi (1941), Matvienko (1954), Bourrelly (1957, 1968, 1981), and Starmach (1980) have divided these taxa into three orders. However, distinct differences are noted in the number of genera present in each order and the family designations. The principal taxa that most frequently are transferred between orders and/or families include *Chrysopyxis, Derepyxis, Kybotion,* and *Lagynion*. Certain associated genera are reassigned with the principal taxon, i.e., *Platytheca* and *Salpingorhiza* are transferred with the reassignment of *Kybotion*.

Huber-Pestalozzi (1941) placed *Chrysopyxis* in association with all uniflagellated, naked, and loricate cells in the family Euchromulinaceae (Table 6.1). This assignment is based upon Iwanoff's (1899) description of a zoospore of *C. iwanoffii* Lauterb. showing a single flagellum. Similarly, he placed *Derepyxis* in the family Syncryptaceae of the Isochrysidales. Huber-Pestalozzi is the only author that divided the loricate Rhizochrysidales into two families. The families are separated from one another on the basis of the mechanism of attachment. The Stylococcaceae contains the stalked genera *Rhizaster* and *Stylococcus,* while the Lagyniaceae includes those directly appressed to the substrate, e.g., *Chrysocrinus, Heterolagynion,* and *Lagynion*.

The admixture of naked and loricate taxa of Huber-Pestalozzi was recognized by Matvienko (1954) in the placement of both *Chrysopyxis* and *Derepyxis*. The only difference is in the spelling of the ordinal title – Chromulinadales. She, however, employed the earlier family name Isochrysidaceae of Pascher (1913). The Rhizochrysidales contains but one family (Rhizochrysidaceae), which includes naked and stalked or appressed loricate genera (Table 6.1).

Bourrelly (1957, 1968, 1981) and Starmach (1980) each used similar basic schemes but differ in detail (Table 6.1). Each recognized a new family, the Kybotionaceae, in the Chromulinales. They differed, however, in the placement of *Chrysopyxis*. Bourrelly (1957) associated *Chrysopyxis* with those chromulinoids having a dominant, free-swimming loricate phase. He presented no argument for maintaining the separation of the rhizopodial forms, *Chrysopyxis* and *Kybotion,* into two independent families. Bourrelly (1981) also suggested that *Lagynion* be transferred to the Chrysococcaceae based upon the report by Juriš (1977) of a uniflagellate zoospore in *L. fulvum* (Scherff.) Bourr. Starmach's (1980) assignment is the inverse of Bourrelly (1981). That is, he transferred *Chrysopyxis* to the family Stylococcaceae of the Stylococcales (Rhizochrysidales), while retaining the position of *Lagynion*. He did, however, recognize Bourrelly's family Derepyxidaceae for *Derepyxis* and associated taxa.

Smith (1933, 1950) and Fott (1959) restricted the distribution of taxa to

Table 6.1. *Distribution of the loricate rhizopodial Chrysophyceae into three ordinal categories*

Pascher (1913)	Chromulinales	Isochrysidales	Rhizochrysidinae[a]
	Kytochromulineae[a]	Isochrysidaceae	
	Chrysopyxis	*Derepyxis*	*Lagynion*
Huber-Pestalozzi (1941)	Chromulinales	Isochrysidales	Rhizochrysidales
	Euchromulinaceae	Syncryptaceae	Stylococcaceae
	Chrysopyxis	*Derepyxis*	Lagyniaceae
			Lagynion
Matvienko (1954)	Chromulinadales	Isochrysidales	Rhizochrysidales
	Euchromulinaceae	Isochrysidaceae	Rhizochrysidaceae
	Chrysopyxis	*Derepyxis*	*Lagynion*
			Tylochrysis
			(Kybotion)[b]
Bourrelly (1957, 1968, 1981)	Chromulinales	Isochrysidales	Rhizochrysidales
	Chrysococcaceae	Derepyxidaceae	Stylococcaceae
	Chrysopyxis	*Derepyxis*	*Lagynion*[c]
	Kybotionaceae		
	Kybotion		
Starmach (1980)	Chromulinales	Isochrysidales	Stylococcales
	Kybotionaceae	Derepyxidaceae	Stylococcaceae
	Kybotion	*Derepyxis*	*Chrysopyxis*
			Lagynion

[a] Independent of the present codes of nomenclature.
[b] Present epithet.
[c] In 1981 Bourrelly (p. 431) recommended transfer to the Chrysococcaceae.

Table 6.2. *Distribution of the loricate rhizopodial Chrysophyceae into two ordinal categories*

Smith (1950)	Chrysomonadales Syncryptaceae *Derepyxis*	Rhizochrysidales Rhizochrysidaceae *Chrysopyxis* *Kybotion* *Lagynion*
Fott (1959)	Chrysomonadales Dinobryaceae *Chrysopyxis* Isochrysidaceae *Derepyxis*	Rhizochrysidales Lagyniaceae *Lagynion*

two orders (Table 6.2). Each retains the names Chrysomonadales and Rhizochrysidales. They differed in the listing of *Chrysopyxis* and the family name for *Derepyxis.* Smith (1933) placed the former genus in the Rhizochrysidaceae with both naked and loricate taxa and designated a new family, the Syncryptaceae, under the Chrysomonadales for *Derepyxis.* Fott retained *Derepyxis,* in conjunction with other genera with assumed homodynamic flagella in the Isochrysidaceae, but transferred these organisms to the Chrysomonadales. Similarly, he included *Chrysopyxis* with all other heterodynamic or uniflagellated cells contained within a lorica in the Dinobryaceae.

Classification systems that contain a single order have been published by Fritsch (1935), Matvienko (1965), and Meyer and Brook (1969) (Table 6.3). Two divergent system are evident. Fritsch placed all of the taxa under the single order Chrysomonadales, while utilizing two existing families and designating a new one. First, *Chrysopyxis* was included within the Chromulinaceae with other naked and/or loricated taxa with a single flagellum or heterodynamic flagella. Second, all of the assumed isodynamic cells were placed in the Isochrysidaceae, including *Derepyxis.* Third, those taxa with a dominant rhizopodial phase were associated together in the Lagynionaceae. The systematic scheme of Matvienko (1965) included the rhizopodial genera within a single order, the Chrysamoebales, and distributed the loricate forms into three families based upon mechanism of attachment. Meyer and Brook (1969) also included all of the genera with a predominant rhizopodial phase into the Rhizochrysidales. They recognized two families: (1) the Rhizochrysidaceae, containing the naked genera and (2) the Lagyniaceae for those enclosed within loricas.

In 1980, Ettl reported several additions to the Danish flora that included several loricate rhizopodial taxa. These included two species of *Chrysopyxis* and single species of *Lagynion, Stephanoporos,* and *Derepyxis.* How-

Table 6.3. *Distribution of the loricate rhizopodial Chrysophyceae into a single ordinal category*

Fritsch (1935)	Chrysomonadales Chromulinaceae *Chrysopyxis* Isochrysidaceae *Derepyxis* Lagyniaceae *Lagynion*	
Matvienko (1965)		Chrysoamoebales Chrysopyxidaceae *Chrysopyxis* Stylochrysalidaceae *Derepyxis* Lagyniaceae *Kybotion* *Lagynion*
Meyer & Brook (1969)		Rhizochrysidales Lagyniaceae *Chrysopyxis* *Derepyxis* *Kybotion* *Lagynion*

ever, the latter taxon, *Derepyxis,* was transferred to the Prymnesiophyceae. This assignment has been retained by Kristiansen (1980).

The preceeding presentation indicates the diversity of opinion concerning the importance of the dominant phase of the organism's life cycle. It also provides insight into the weight the individual authors gave to the infrequent or unconfirmed sightings of zoospores. There has been only a single report of zoospores in *Chrysopyxis* by Iwanoff, in *Kybotion* by Thompson (in Smith, 1950), and in three species of *Lagynion* (cf. Starmach 1980). Recently, Hibberd (1976) has observed two basal bodies present in vegetative cells of *Lagynion.* In 1971, Meyer further supported Matvienko's (1965) inclusion of *Derepyxis* with other ameboid taxa. He described cell division in four species of *Derepyxis* via ameboid daughter cells and uniflagellated zoospores. He also demonstrated that the appendages extending through the lorica aperture were rhizopodia and not flagella. The appendages are capable of capturing bacteria, encapsulating them within food vacuoles, and translocating the vacuoles to the main cell body where digestion occurs. (Selected portions of these events were shown during the oral presentation of this paper. A manuscript reviewing *Derepyxis* is in preparation). It is, therefore, suggested that *Derepyxis* should not be retained in either the Chrysomonadales or as a remnant of the Isochrysidales. [For a discussion of the present position and contents of the Isochrysidales within the class Prymnesiophyceae refer to Hibberd (1976) and Boney & Green (1982).]

This historical review of the systematics of selected members of the Chrysophyceae provides a perspective for the following considered proposal. The proposal includes three components. The first component addresses the transfer of certain taxa to the Rhizochrysidales. The second component deals with the selection of an appropriate family name. The last component describes a phylogenetic scheme for all of the loricate rhizopodial Chrysophyceae.

Based upon our present state of knowledge and the ongoing discoveries it is anticipated that additional phases in the life history of these taxa will be described. Flagellated phases in other genera or additional species within genera with known zoospores will be reported as research continues. Also, the importance of the temporary existence of the zoospore must be weighed. Obviously Smith (1933, 1950), Matvienko (1954, 1965), Meyer and Brook (1969), and Starmach (1980) considered the dominant rhizopodial vegetative cell of greatest importance in assigning *Chrysopyxis* to the Rhizochrysidales. The separation of *Kybotion* from *Chrysopyxis* is now without foundation, since both have sessile loricas, dominant rhizopodial vegetative cells, and transient uniflagellate zoospores. *Derepyxis* can also be included in the Rhizochrysidales based upon the lack of homodynamic flagella in the vegetative cell, the presence of a rhizopodial vegetative cell, and the presence of a uniflagellate zoospore. In summary, *Chrysopyxis, Kybotion,* associated genera, and *Derepyxis* are included in the Rhizochrysidales along with *Lagynion.*

The loricate genera of the Rhizochrysidales have been separated into the Stylococcaceae, Rhizochrysidaceae, and/or Lagyniaceae. The rules of priority require that the name Stylococcaceae given by Lemmermann (1899) has precedence over the names Rhizochrysidinae of Pascher (1913) and Lagyniaceae of Fritsch *ex* Huber-Pestalozzi (1941).

Bourrelly presented the first phylogenetic scheme in his 1957 publication showing the affinities of the members of the family. He cited certain examples of parallelism and emphasized the need for further examples of life histories. The additional information on *Derepyxis* zoospores and the inclusion of the additional genera in the Stylococcaceae requires modification of the phylogenetic scheme developed by Bourrelly.

The scheme proposed here contains three lines of development radiating from the morphologically variable sessile genus *Lagynion* (Figure 6.1). [Note: The colorless forms are included in association with their plastid containing equivalents.] One line differentiates into those genera with bilateral symmetry by the formation of extensions to the lorica. This line may have developed from an organism similar to *L. macrotrachelum* Pasch. into *C. iwanoffii.* The reduction of the encircling setae and concom-

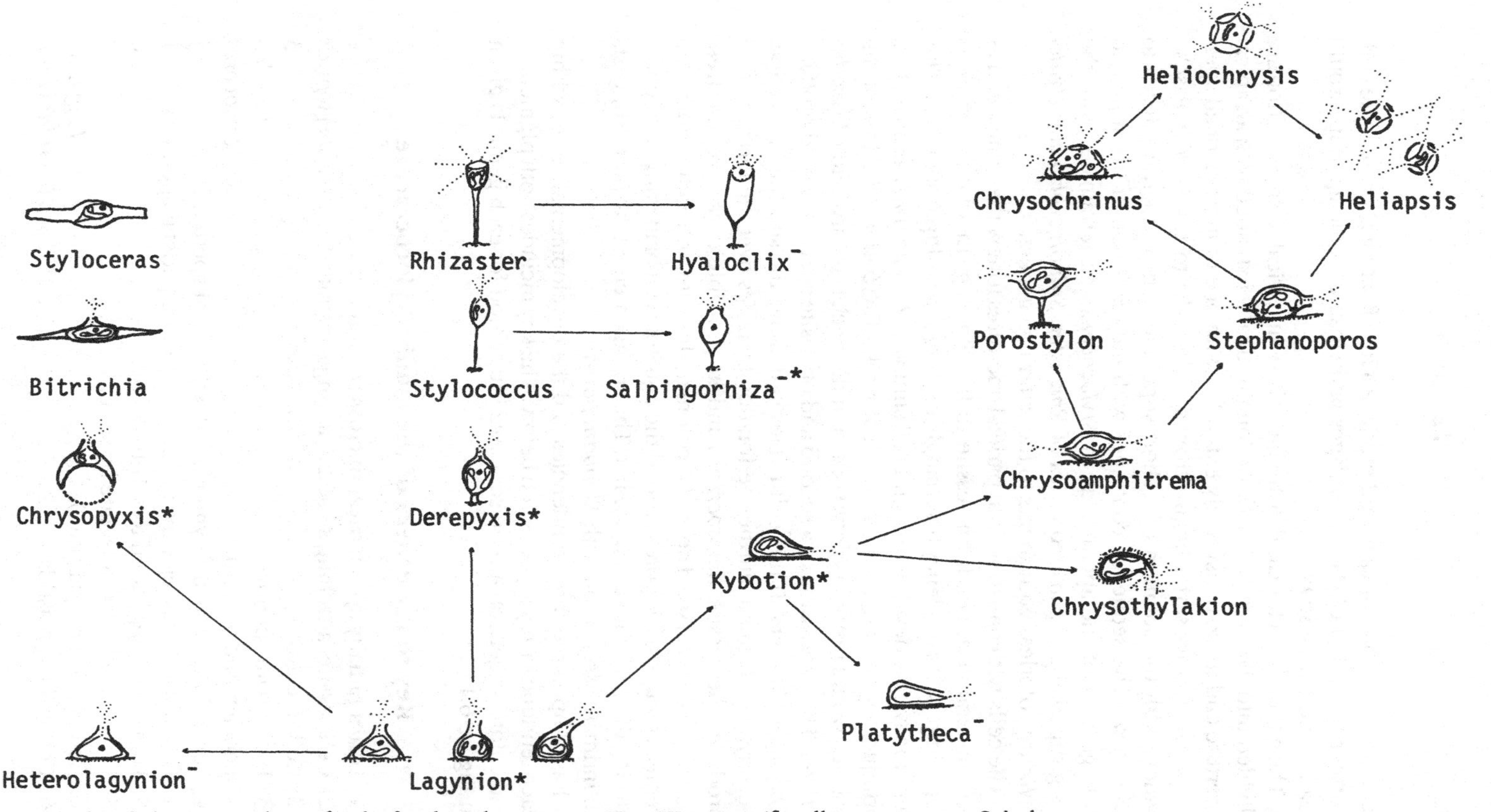

Fig. 6.1. Phylogenetic scheme for the family Stylococcaceae. Key: *Known uniflagellate zoospores. ⁻Colorless.

itant release of the lorica would produce a free-floating *Bitrichia*-like cell. Further reduction of the appendages and closure of the aperture would lead to the genus *Styloceras.*

The genera with loricas that are radially symmetrical in cross section are divided into two sequences: those longitudinally extended with an apical aperture and those transversely extended with one or more parietal apertures. The former series begins from a cell similar to *L. scherffelii* Pasch., which, with the addition of a short stipe and narrow aperture, leads to *Derepyxis.* The sequence continues with the extension of the stipe and enlargement of the aperture as in *Stylococcus* and terminates with the long stiped and cupshaped lorica of *Rhizaster. Salpingorhiza* and *Hyalocylix* are colorless equivalents to the latter two genera.

The cells with one to several parietal apertures follow the same trend as the bilateral taxa, i.e., from sessile to free floating. In addition, there is superimposed a tendency from single to paired to multiple apertures. This series may have developed from a cell similar to *L. cystodinii* Pasch. with its oblique neck to the totally horizontal neck of *Kybotion.* The loss of the plastids would result in *Platytheca,* or if the neck is recurved to *Chrysothylakion.* The expression of a second and opposite aperture would produce *Chrysoamphitrema.* If this cell is elevated above the substrate by a stipe, *Porostylon* is formed. Further perforation by pores over the lorica surface produces the sessile *Chrysocrinus* and its probable derivative, the free-floating *Heliochrysis.* The only colonial, loricate, rhizochrysophycean genus, *Heliapsis,* has pores around the equator with branching rhizopodia uniting the cells into a delicate net. The arrangement of the pores suggests an intimate association with *Stephanoporos.*

Based upon the three tendencies and the developmental series within each tendency a key is presented below. The key includes both pigmented and colorless genera. A representative illustration for each genus is given in Figure 6.1.

Key to the genera of the family Stylococcaceae

1. Lorica radially symmetrical in cross section. 4
2. Cell sessile, attaching setae encircling substrate. . . . . . *Chrysopyxis*
2. Cell free floating. 3
3. Lorica with aperture. . . . . *Bitrichia*
3. Lorica without aperture. . . . . *Styloceras*
4. Lorica longitudinally symmetrical with apical aperture. 5
4. Lorica with transverse axis with one or more parietal apertures. . .11
5. Lorica base appressed to substrate. 6
5. Lorica attached to substrate by stipe. 7
6. Cell containing plastid(s). . . . . *Lagynion*
6. Cell lacking plastids. . . . .*Heterolagynion*

7. Lorica attached by a short gelatinous stipe; with narrow aperture. . . . . *Derepyxis*
7. Lorica attached by a narrow, elongated stipe of lorica wall material. 8
8. Aperture less than maximum diameter of lorica. 9
8. Aperture equal to maximum diameter of lorica. 10
9. Protoplast containing plastid(s). . . . . *Stylococcus*
9. Protoplast lacking plastids. . . . . *Salpingorhiza*
10. Protoplast containing plastids. . . . . *Rhizaster*
10. Protoplast lacking plastids. *Hyalocylix*
11. Single lateral aperture. 12
11. Multiple lateral apertures. 13
12. Neck straight. 14
12. Neck curved, protoplast containing plastids. . . . . . *Chrysothylakion*
13. Protoplast containing plastid(s). . . . . *Kybotion*
13. Protoplast lacking plastids. *Platytheca*
14. Two oppositely opposed apertures. 15
14. More than two apertures. 16
15. Lorica sessile. . . . . *Chrysoamphitrema*
15. Lorica elevated by stipe. . . . . *Porostylon*
16. Sessile. 17
16. Free floating. 18
17. Apertures equatorial. . . . . *Stephanoporos*
17. Apertures dispersed over surface. . . . . *Chrysocrinus*
18. Unicellular. . . . . *Heliochrysis*
18. Colonial. . . . . *Heliapsis*

Summary

The family Stylococcaceae includes all of the pigmented and colorless Chrysophyceae that have a dominant rhizopodial vegetative phase contained within a sessile or free lorica. Three developmental tendencies radiate from the central genus *Lagynion*. These tendencies are exemplified by the genera *Chrysopyxis, Derepyxis,* and *Kybotion*. In each of these four genera one or more species are known to produce zoospores. For the derived genera, zoospores have not been reported. *Chrysopyxis* typifies those genera with bilateral morphologies, while *Derepyxis* and *Kybotion* typify tendencies with radial symmetry. The orientation of the principal axis distinguishes the *Derepyxis* and *Kybotion* sequences. The former has a longitudinal axis and an apical aperture, while the latter has a transverse axis with one or more equatorially or randomly distributed pores. In the bilaterally symmetrical forms and the radially symmetrical forms with transverse axis, the progression is from sessile to free-floating habit. The tendency sequence in each series is used to produce the key to genera.

References

Boney, A.D. & Green, J.C. 1982. Prymnesiophyceae (Haptophyceae): Introduction and Bibliography. *In:* Parker, B. & Rosowski, J. [Eds.] *Selected Papers in Phycology,* Vol. II, University of Nebraska Press, Lincoln, pp. 705–11.

Bourrelly, P, 1957. Recherches sur les Chrysophycées. Morphologie, phylogénie, systématique. *Rev. Algol. Mém. Hors-Sér.* 1: 1–412.

– 1968. *Les Algues d'Eau Douce. Tome II. Algues Jaunes et Brunes.* Boubee, Paris.

– 1981. *Les Algues d'Eau Douce. Tome II. Algues Jaunes et Brunes.* 2nd ed. Boubee, Paris.

Ettl, H. 1980. Beitrag zur Kenntnis der Süsswasseralgen Dänemarks. *Bot. Tidsskr.* 74: 179–223.

Fott, B. 1959. *Algenkunde.* Fischer, Jena.

Fritsch, F.E. 1935. *The Structure and Reproduction of Algae,* Vol. 1. Cambridge University Press, Cambridge.

Hibberd, D.J. 1976. The ultrastructure and taxonomy of the Chrysophyceae and Prymnesiophyceae (Haptophyceae): a survey with some new observations on the ultrastructure of the Chrysophyceae. *Bot. J. Linn. Soc.* 72: 55–80.

Huber-Pestalozzi, G. 1941. Das Phytoplankton des Süsswasser. Systematik und Biologie. *In:* Thienemann, A. [Ed.] *Die Binnengewasser* Bd. 16, T. 2(1). Schweizerbart'sche, Stuttgart.

Juriš, S. 1977. Contributions to the knowledge of the algae of Slovakia. I. *Zb. Slov. Act. Rer. Nat. Mus. Nat. Slop. Bratislava* 23: 91–102.

Kristiansen, J. 1980. *A Checklist of Danish Freshwater Chrysophyceae. Also including Prymnesiophyceae and Bicosoecophyceae.* Institute of Plant Anatomy & Cytology, University of Copenhagen.

Iwanoff, L. 1899. Beitrag zur Kenntnis der Morphologie und Systematik der Chrysomonadinen. *Bull. Acad. Sci. St. Petersbourg* 11: 247–62.

Lauterborn, R. 1911. Protozoen-Studien, IV. Flagellaten aus dem Gebiete des Oberrheins. *Zeitschr. Wiss. Zool.* 65: 369–91.

Lemmermann, E. 1899. Das Phytoplankton sächischer Teiche. *Forschungsber. Biol. Stat. Plön* 7: 96–135.

Matvienko, A.M. 1954. *Zolotistie Vodorosli* (*Chrysophyta*). Opredelitel presnovodnikh, Moscow.

– 1965. *Zolotisti Vodorosti-Chrysophyta.* Viznacnik prisnovodnich, Kiev.

Meyer, R.L. 1971. The life-history and systematic position of *Derepyxis* (Chrysophyceae). *J. Phycol.* (*Suppl.*) 7: 14–15.

Meyer, R.L. & Brook, A.J. 1969. Freshwater algae from the Itasca State Park, Minnesota. II. Chrysophyceae and Xanthophyceae. *Nova Hedwigia* 17: 105–12.

Pascher, A. 1913. Chrysomonadinae. *In:* Pascher, A. & Lemmerman, E. [Eds.] *Die Süsswasser-Flora Deutschlands, Österreichs und der Schweiz. 2: Flagellatae II.* Fischer, Jena, pp. 7–95.

Scherffel, A. 1901. Kleiner Beitrag zur Phylogenie einiger Gruppen niederer Organismen. *Bot. Zeit.* 59: 143–58.

Smith, G.M. 1933. *The Fresh-Water Algae of the United States.* McGraw-Hill, New York.

– 1950. *The Fresh-Water Algae of the United States,* 2nd ed. McGraw-Hill, New York.

Starmach, K. 1980. Chrysophyceae-Złotowiciowce. *In:* Starmach, K. & Siemińska, J. [Eds.] *Flora Słodkowodna Polski* Vol. 5, Państuowe Wydawnictwo Naukowe, Warszawa, pp. 1–775.

7

New species of the genus *Mallomonas* (Mallomonadaceae, Chrysophyceae) from New Zealand

MONIKA DÜRRSCHMIDT
Institut für Pflanzenökologie, Justus-Liebig Universität
Heinrich-Buff-Ring 38, D-6300 Giessen, Federal Republic of Germany

Introduction

Examinations of the freshwater algal flora of New Zealand are rather numerous (Cassie 1980, 1981). However, there have been comparatively few records of Chrysophyceae, in particular those referred to as the family Mallomonadaceae, and members of this family have as a rule only been noted in passing in phytoplanktonic studies (e.g., Johnston 1972, Green 1974, Pearl et al. 1979). All previous records of scale-bearing chrysophytes from New Zealand, with the exception of two reports (Thomasson 1974, Flint 1975), are based on light microscopy only, which is for many species of limited value (Wee 1982). In contrast to the now reasonably well-studied areas in Europe, North America, and Japan, the only electron microscopic (EM) verified observations from the Southern Hemisphere are those by Vigna (1979, 1981) from Argentine, Thomasson (1963) and Dürrschmidt (1980, 1982) from Chile, and Croome and Tyler (1983) from Australia. It was, therefore, of great interest to examine the mallomonads of New Zealand to add to the knowledge of their composition and distribution in the Southern Hemisphere (see Figure 7.0).

During a 4-week visit in July 1982, 76 samples were collected from freshwater habitats on North and South Island. The results of this survey are divided between two papers. This chapter deals with seven hitherto unknown species. Their description is based exclusively on electron microscopic observations. No living cells were studied.

The author wishes to express her sincere thanks to Berit Asmund for her kind help and valuable discussions, U. Richter for technical assistance, and E. Kaus for translating the diagnosis into Latin; the author is also grateful to I.M. Gilchrist for correcting the English grammar.

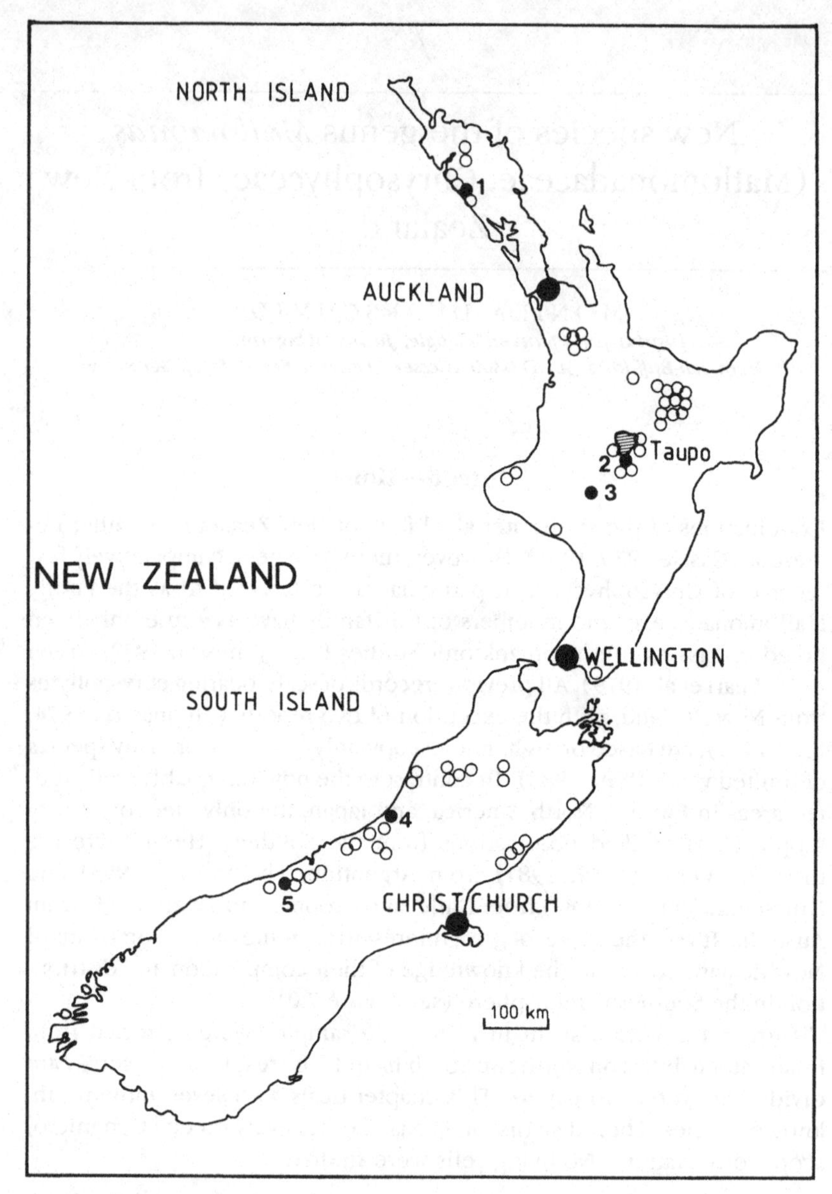

Fig. 7.0. The investigated area. The localities are indicated by circles, while black dots symbolize type habitats.

Material and methods

Phytoplankton was collected with nets (mesh diameter 5–20 μm) and fixed with 3% glutaraldehyde in 0.1 M phosphate buffer, pH 7.2. For transmission electron microscopy (TEM), the samples were rinsed with distilled water, one drop placed on formvar-coated grids, dried, and then studied directly under a Philips EM 300. For scanning electron microscopy (SEM) these grids were mounted on aluminum stubs, sputtered with gold, and examined in a Cambridge S4 Stereoscan. Scanning and transmission electron microscopy was carried out at the "Strahlenzentrum" of the Justus Liebig University in Giessen.

Results and discussion

Mallomonas novae-zelandiae *sp. nov. (Figures 7.1–7.6)*

Cellulae parvae, ovoideae vel globosae. Squamae tripartitae. Squamae corporis cupulis asymmetricis instructae. Ramus costae submarginalis ad cupulam porrigitur structuram aliformem faciens. Extremum anticum in spinam desinens. Scutum aliquot lacunas habet. Area cucullo contecta 1–4 poris ornata est. Rami costae V-formis circulariter obtusae brevibus costis internis regulariter dispositis virgati sunt. Margines antici angusti tigilla valida exhibent. Margo posticus laevis. Labrum posticum striatum. Squamae apicales cupulas alata ornatas habent. Setae parum flexae tantum ex latere convexo serratae in apicem acutum diminuuntur. Cystae ignotae.

Dimensio cellulae: 10–15 × 8–10 μm; squamaea apicales: 3.0–3.6 × 2.4–2.8 μm; squamae corporis: 4–4.8 × 2–2.9 μm; setae: 5–12 μm.

Holotypus die 16 m. Iulii a. 1982 in palude sita apud litus australe lacus, qui Lake Taupo (2) appellatur, in insula septentrionali Novae Zelandiae lectus est. Figuris 7.1–7.6 monstratur, qui in Herbario Universitatis Christchurchensis (CHR) Novae Zelandiae depositus est.

The epithet was chosen because this species was found for the first time in New Zealand.

The cells are small, ovoid to spherical, covered entirely by siliceous scales and bristles (Figure 7.1). Two types of scales occur: five to six forward pointing apical scales surround the proximal part of the flagellum and imbricate, spirally arranged larger scales cover the rest of the cell. All scales are tripartite.

The body scales (Figure 7.4) are oval, provided with prominent, slightly asymmetrically shaped domes. One branch of the anterior submarginal rib extends to the dome that then forms a small vertically raised winglike structure. The anterior edge often extends beyond the dome, ending in a sharp thorn (Figure 7.6). The surface of the dome is ornamented anteriorly by regularly spaced longitudinal ribs with short cross-ribs in between, while a roughly triangular patch – perforated by minute pores – is present posteriorly. The shield reveals irregularly spaced, annular pits – irregular in frequency – which are set into an unequally dense secondary layer (Figure 7.4). The pits, situated in the center of the shield, are surrounded by thinner areas (Figure 7.6). Posteriorly the secondary layer is more weakly developed or missing. One to four pores are present in the area

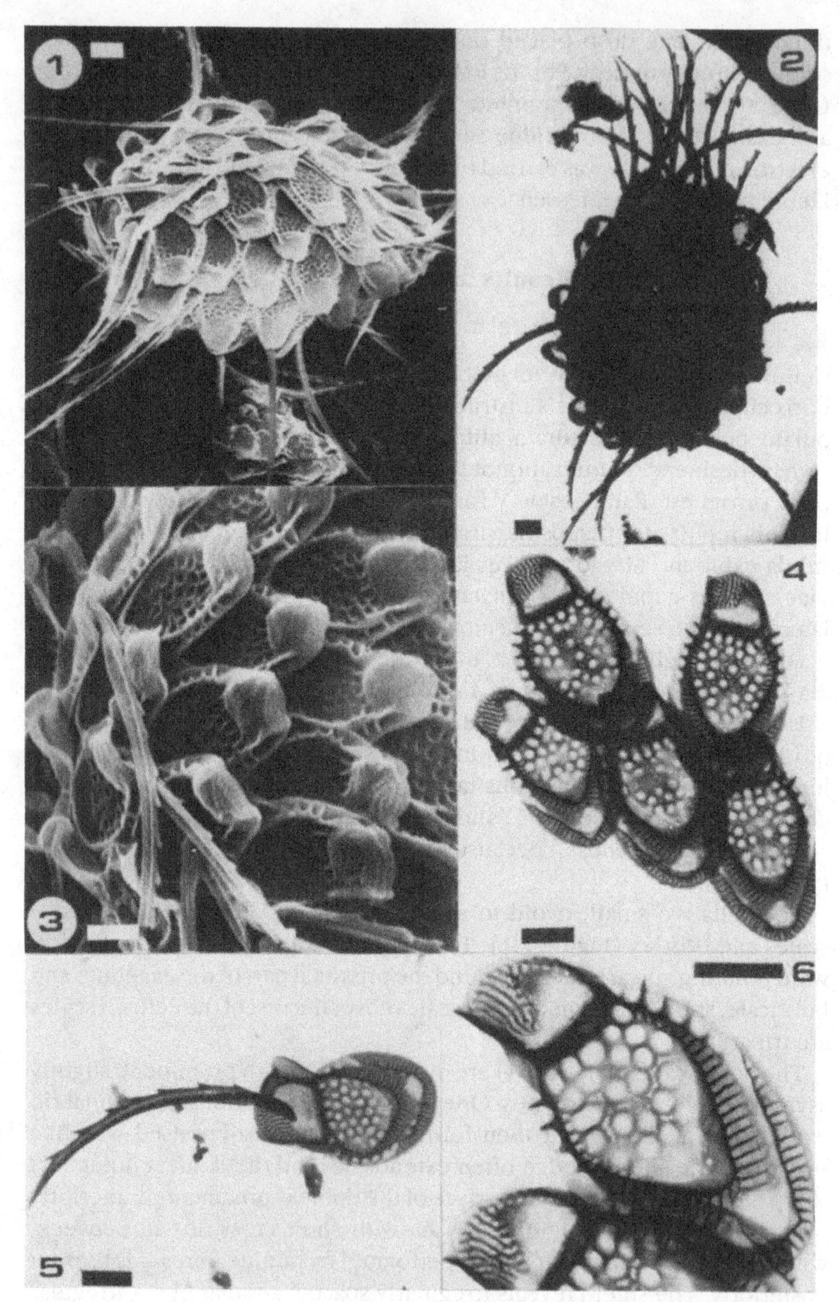

Figs. 7.1–7.6. *Mallomonas novae-zelandiae* sp. nov. ***Fig. 7.1.*** Entire cell, SEM. ***Fig. 7.2.*** Entire cell, TEM. ***Fig. 7.3.*** Part of the cell armor, SEM. ***Fig. 7.4.*** Groups of body scales, TEM. ***Fig. 7.5.*** Apical scale with bristle, TEM. ***Fig. 7.6.*** Body scale, TEM. Scale = 1 μm.

obscured by the hood. The V-rib is sharply rounded at the base end. The branches are striped with short, regularly spaced internal cross-ribs. This conspicuous feature diminishes gradually along the anterior submarginal ribs and terminates shortly before the dome. The anterior flanges are narrow with widely spaced struts radiating from the anterior submarginal rib. The posterior flange is smooth, the proximal area being less electron dense than the rest of the flange. The proximal border is striated.

The apical scales (Figure 7.5) are smaller than the body scales and are asymmetric. Their domes are broad with a lateral wing.

A front tuft consists of six to eight forward pointing bristles, shorter than the body bristles (Figure 7.2). Both front and body bristles are slightly bent and unilaterally serrated on their convex side, tapering into a pointed apex. Cysts were not observed.

Mallomonas novae-zelandiae belongs certainly to the sect. *Mallomonas* (Momeu & Péterfi 1979) in that the scales are tripartite, consisting of a hooded V-rib, shield, and flange. However, it is not possible to place this species in any of the five series described by Momeu and Péterfi (1979). From the ser. *Mallomonas* and *Paludosa,* it is distinguished by having smooth posterior flanges, no finely perforated shield, and no reticulate thickenings forming a network of large annular meshes. Species of the ser. *Intermediae* deviate in possessing smooth, perforated shields and no lateral extensions. Those of ser. *Papillosae* have nonprominent domes and a basically papillate shield pattern. Thus, for the moment, *M. novae-zelandiae* may be considered to be an isolated species within the sect. *Mallomonas.*

The type specimen was collected from a swamp on the southern shore of Lake Taupo. At the date of sampling, water temperature was 11.8°C, pH 4.7, and conductivity 70 μS_{20}. The water was rich in humic acids. Besides *M. novae-zelandiae,* many other Mallomonadacean species were present: *M. plumosa* Croome & Tyler, *M. maculata* Bradley, *M. eoa* Takahashi, *M. alata* Asmund et al., *Synura sphagnicola* Korsh., *S. curtispina* (Petersen & Hansen) Asmund, *S. mammillosa* Takahashi, and others. *Mallomonas novae-zelandiae* occurred frequently in several other localities on North and South Island.

Mallomonas tongarirensis *sp. nov. (Figures 7.7–7.12)*

Cellulae ovoideae vel cylindricae. Squamae in 4 typos. Omnes squamae costam V-formen cucullatam praebent. Scutum papillis conicis lacunisque annularibus late dispositis, quarum plurimae secundum costas V-formen ac submarginalem anteriorem sitae, ornatum est. Unaquaeque lacuna papillam mediam aut aream incrassatam poro solitario instructam exhibet. Margo anticus angustus unica papillarum serie ornatus est. Margo posticus laevis. Squamae apicales structura aliformi praebent. Squamae posticae

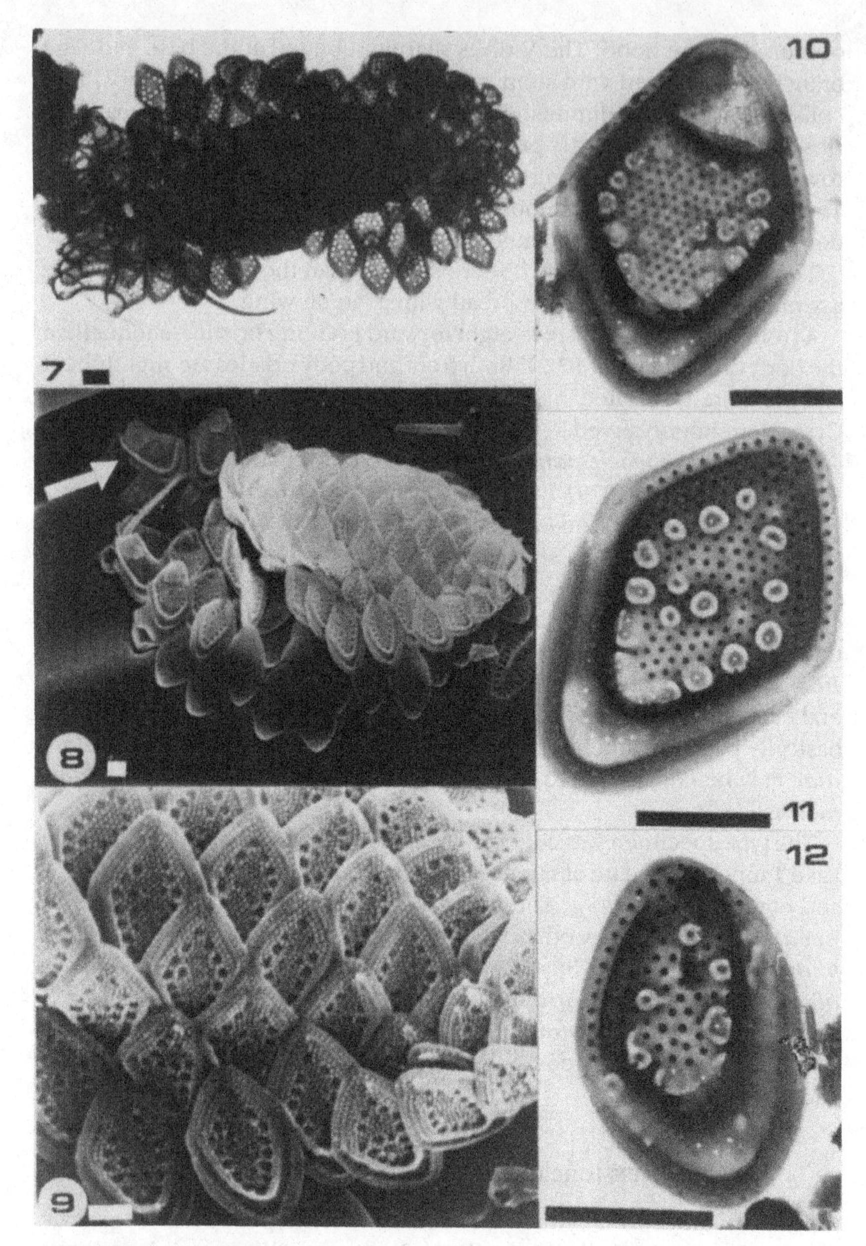

Figs. 7.7–***7.12. Mallomonas tongarirensis*** sp. nov.***Fig.*** *7.7.* Complete cell armor showing four types of scales and the needlelike bristles, TEM. ***Fig.*** *7.8.* Entire cell, apical scale (arrow), SEM. ***Fig.*** *7.9.* Rhomboidal body scales, SEM. ***Fig.*** *7.10.* Tripartite body scale, TEM. ***Fig.*** *7.11.* Rhomboidal body scale. TEM. ***Fig.*** *7.12.* Rear scale, TEM. Scale = 1 μm.

parvis spinis habent. Setae laeves, leviter curvatae ac diminutae in apice dentem subapicalem exhibent. Cysta ignota.

Dimensio cellulae: 16–24 × 5–9 μm; squamae apicales 4 × 2 μm; squamae tripartitae: 3.5 × 2.1 μm; squamae corporis rhomboides: 3–3.5 × 1.5–1.8 μm; squamae posteriores: 1.5–2 × 1–1.2 μm; setae 7–10 μm.

Holotypus die 30 m. Iulii a. 1982 in stagno vadoso prope viam a Wanganui (3), vico septentrionalis insulae Novae Zelandiae, ad lacum Taupo ferentem paulo post pontem Makatotensem in angulo austro-occidentali regionis Tongarirensis ad rerum naturam conservandam destinatae lectus est. Figuris 7.7–7.12 monstratur, qui in Herbario Universitatis Christchurchenis (CHR) Novae Zelandiae depositus est.

The specific epithet refers to the type locality, which is situated on the southwest border of the National Park Tongariro.

The cells are ovoid or cylindrically elongated (Figures 7.7 and 7.8). There are four types of scales: (i) approximately six forward pointing apical scales, (ii) a few tripartite body scales, (iii) numerous domeless body scales, and (iv) small asymmetrically shaped scales posteriorly.

All scales are provided with a hooded V-rib, which has a rounded base. The shield is marked by regularly arranged, conical papillae in addition to widely spaced, annular pits. The latter are chiefly situated along the V-rib and the anterior submarginal ribs; however, some are irregularly scattered all over the shield (Figure 7.9 and 7.11). At the bottom of each pit the base plate is visible. Each pit has a central papilla or a thickened area with a solitary pore. The anterior flange is narrow, and ornamented with a single row of papillae. The posterior flange is smooth; however, one row of pores runs at a certain distance from the V-rib. Thinner parts of scale material are present at the proximal end of the flange.

The apical scales are asymmetrically shaped with smooth, pronounced domes (Figures 7.7 and 7.8, arrow). One of the anterior submarginal ribs is extended forward, past the sides of the dome and forms a small wing that projects with a sharp point beyond the end of the scale. The marking of the shield is less elaborate than that of the body scales.

Tripartite body sales (Figure 7.10) have small domes, ornamented with papillae and a triangular patch of minute perforations. The anterior submarginal ribs terminate within the dome area.

In the rhomboidal body scales, the anterior submarginal ribs do not join at the distal edge of the scale (Figure 7.11).

Rear scales have small spines distally, formed by the ends of the submarginal ribs (Figure 7.12).

The bristles are smooth, slightly curved, and tapering; their tip is provided with one more or less developed subapical tooth. Cysts were not observed.

Mallomonas tongarirensis belongs to the ser. *Papillosae* Momeu & Péterfi (1979) of the sect. *Mallomonas* (Momeu & Péterfi 1979). According to the four scale types, the appearance of the bristles, and the papillate shield pattern, it seems to be a twin brother of *M. annulata* (Harris & Bradley) Harris. Distinguishing features are the lack of pores on the base plate and the absence of a mesh pattern formed by groups of papillae. The arrangement of scales and bristles differ in that tripartite scales are principally found at both ends in *M. annulata* but are absent on the posterior end in *M. tongarirensis*. The most striking character for *M. tongarirensis* is the presence of almost annular pits.

Mallomonas tongarirensis was found only in a single locality, a shallow pool on the road from Wanganui to Lake Taupo, shortly after the Makatote Viaduct on the southwest border of the Tongariro National Park, North Island. The water was slightly yellowish, water temperature 6°C, pH 5.8, and conductivity 190 μS_{20}. Other members of the Mallomonadaceae were *M. papillosa* Harris & Bradley and *M. maculata*.

Mallomonas perpusilla *sp. nov. (Figures 7.13–7.18)*

Cellulae perparvae globosae vel ovoideae. Squamae antisimae cupulas latas atque oblongas exhibent, quae in antica parte una pluribusve seriebus papillarum transversis ornatae sunt, in postica autem laeves. Scutum 3–4 costis transversis concurrentibusque irregulariter laqueatum est. Margines posticus atque anticus angusti laevesque. Labrum posticum in angulum proximalem translatum. Squamae corporis partibus anticis parvis et demissis in cupulam praeditae sunt. Margines antici tigilla exhibent, quorum bina solitarium porum includunt. Costa V-formis cucullata. Setae breves, tenues, acutae. Cysta ignota.

Dimensio cellulae: 4.5–12 × 2.8–5 μm; squamae apicales: 2 × 1.3 μm; squamae corporis: 2.0–2.5 × 1.2–1.7 μm; setae: 1.5–4 μm.

Holotypus die 21 m. Iulii a. 1982 in stagno parvo (4) tescorum *Nothofagi* prope viam Hokitika ad lacum, qui Lake Kaniere dicitur, ferentem in ora occidentali insulae australis Novae Zelandiae lectus est. Figuris 7.13–7.18 monstratur, qui in Herbario Universitatis Christchurchensis (CHR) Novae Zelandiae depositus est.

The species was named *"perpusilla"* referring to its very small cell size.

The cells are very small, spherical or ovoid, observed by two different types of scales and needlelike bristles (Figure 7.14). Scanning electron micrographs (Figure 7.13) show the transverse arrangement of the imbricate scales, except toward the anterior region, where these are longitudinally oriented in surrounding the base of the hairy flagellum. Bristles are closely set at the anterior end of the cell, but occur sporadically on the midsection.

Scales of the extreme anterior part are nearly rectangular with broad oblong domes (Figure 7.17). The front part of the domes reveals one to several transverse rows of papillae, while the posterior portion is smooth,

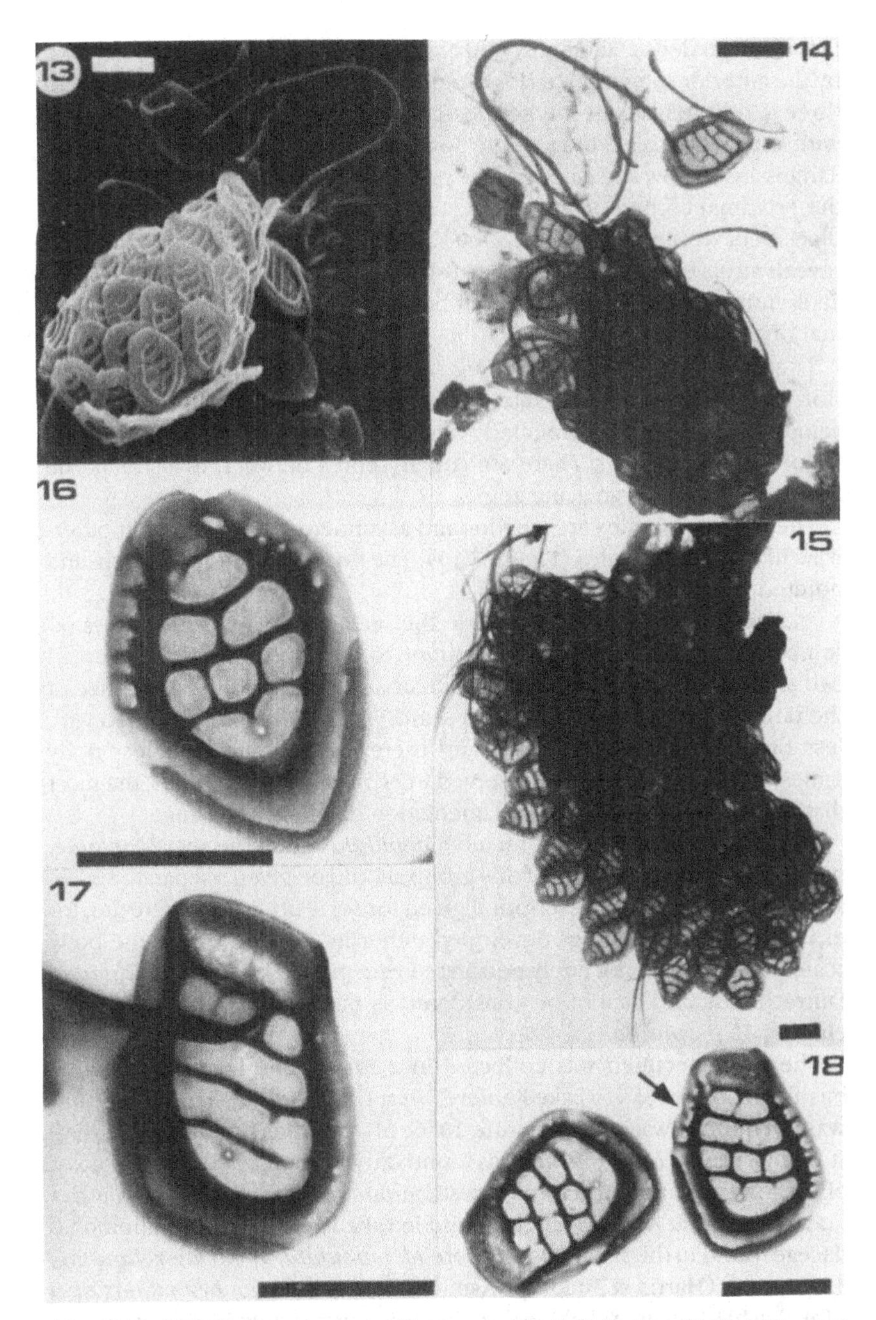

Figs. 7.13 - 7.18. Mallomonas perpusilla sp. nov. ***Fig. 7.13.*** Entire cell, SEM. ***Fig. 7.14.*** Entire cell, TEM. ***Fig. 7.15.*** Complete cell armor showing two types of scales and the needlelike bristles, TEM. ***Fig. 7.16.*** Body scale, TEM. ***Fig. 7.17.*** Apical scale, TEM. ***Fig. 7.18.*** Transitional scales showing struts (arrow) and a somewhat displaced proximal border of the scale, TEM. Scale = 1 μm.

less electron dense, and finely perforated. The lateral edges are thickened by the anterior submarginal ribs. The shield is irregularly panelled, due to three to four transversal ribs connected by a varied number of cross-ribs or with struts. Each scale has a single pore posteriorly. Anterior and posterior flanges are narrow and smooth. The proximal border is displaced to one of the proximal corners. Transitional scales showing different stages in displacement were also observed. One scale, pictured in Figure 7.18 (arrow), reveals struts on the anterior flange (characteristic for body scales) and has an asymmetrically placed proximal border as well as an asymmetric submarginal rib.

The body scales (Figure 7.16) are rhomboidal with small, shallow anterior parts not clearly elaborated as a dome (Figure 7.13). The anterior flanges have somewhat irregularly developed struts radiating from the anterior submarginal rib. There are solitary pores between two struts. The hooded V-rib forms an acute angle.

The posterior scales are smaller and asymmetrically shaped but otherwise like the body scales (Figure 7.15). The bristles are short, delicate and pointed. Cysts were not observed.

Mallomonas perpusilla belongs to the sect. *Mallomonas* and shows resemblance to *M. pillula* Harris. Common to both species is the very small cell size, the possession of needlelike bristles, and the shape and size of the domed body scales. The apical scales in *M. pillula* are provided with less broad, less oblong domes, and there are genuine domeless body scales, which have not been observed in *M. perpusilla*. However, the most distinguishing feature is the ornamentation of shield and flange.

Momeu and Péterfi (1979) include *M. pillula* within the ser. *Papillosae,* but, in my opinion, species of this group should be given a separate series, as they do not fit in the description given for ser. *Papillosae*. There are, for example, no lateral wings developed either in the apical or in the body scales, the shield is not finely perforated except in *M. pillula* f. *valdiviana* Dürrschmidt, and cannot be considered as basically papillose, when including *M. perpusilla*.

The type specimen was collected in a small pond in the *Nothofagus* bushland on the road to Lake Kaniere, West Coast, South Island. The water was polyhumic, water temperature 10°C, pH 4.9, and conductivity 41 μS_{20}. It is notable that *M. perpusilla* was found only in habitats on the west coast of the South Island during this investigation: a swamp near Greymouth, a gravel ditch near Kumara Junction, and in Lake Paringa. Other Mallomonadaceae found in the same sample were *M. maculata, M. novae-zelandiae, M. pumilio* (Harris & Bradley) Asmund et al., *Chromophysomonas abei* (Takahashi) Preisig & Hibberd, *Chrysodidymus synuroides* Prowse. The phytoplankton was rich in desmids.

Mallomonas grossa *sp. nov. (Figure 7.19–7.27)*

Cellulae ovoideae vel ellipsoideae. Squamae in 3 typos. Omnes squamae costa submarginali manifeste definita carent. Contecta sunt squamae strato secundario denso laevique maculam medii scuti tenuiore circumdante. 2-6 cristae transversae structuram scalariformem significant. Linea tenuiore a postica parte illius maculae usque ad extremum proximale se extendit. Squamae apicales cupulas camino similes exhibent. Extensa aliformia cupulae se coniungunt. Squamae corporis virgam praebent, quae brevi spatio a margine exteriore postica est. Squamae posticae limbo levi cupullaque camino similes exhibent. Setae breves, rectae, acutae. Cysta ignota.

Dimensio cellulae: 12–20 × 6–8 μm; squamae apicales: 2.3 × 1.4 μm; squamae corporis: 2.5–3.2 × 1.8–2.0 μm; squamae posticae 2.2 × 1.8 μm; setae: 2.5–7 μm.

Holotypus die 11 m. Iulii a. 1982 in rivulo (1), qui paludi herbae Monocotylae aquam deducit, prope Mamaranui in insula septentrionali Novae Zelandiae sito lectus est. Figuris 7.19–7.27 monstratur, qui in Herbario Universitatis Christchurchensis (CHR) Novae Zelandiae depositus est.

The epithet *grossa* refers to the dense secondary layer of the shield.

The cells are ovoid to ellipsoidal (Figures 7.19 and 7.21), covered by three different types of scales: (i) "dome"-bearing apical scales with bristles, (ii) spirally arranged body scales, and (iii) "domed" rear scales. All types of scales lack clearly defined submarginal ribs, which would divide the scale into shield and flange. Instead, the scales are covered by a smooth, dense secondary layer, surrounding an almost oval patch of thinner scale material in the middle of the scale. There are two to six transversal ridges, which give a steplike impression. A line of thinner scale material runs directly from the posterior part of this patch toward the proximal end of the scale.

Apical scales (Figure 7.22) are slightly asymmetric with chimneylike domes. These are short and cylindrical, with an apparently open basal part backward, although this has to be demonstrated by SEM. There is a hood, partly covering the steplike pattern. Anteriorly, large, vertically raised winglike extensions, which are formed by the dense secondary layer, join the dome. Transitional scales with rudimentary domes were also present (Figure 7.27).

The body scales are rhomboidal (Figure 7.24). Along the anterior margin of the scale they have a stripe of less dense material, with a thin, smooth rib running a short distance from the outer margin. As seen in SEM, the foremost part of this rib is somewhat raised (Figure 7.20). Posteriorly the scales have a smooth rim.

The rear scales are ornamented like the body scales, but have distally

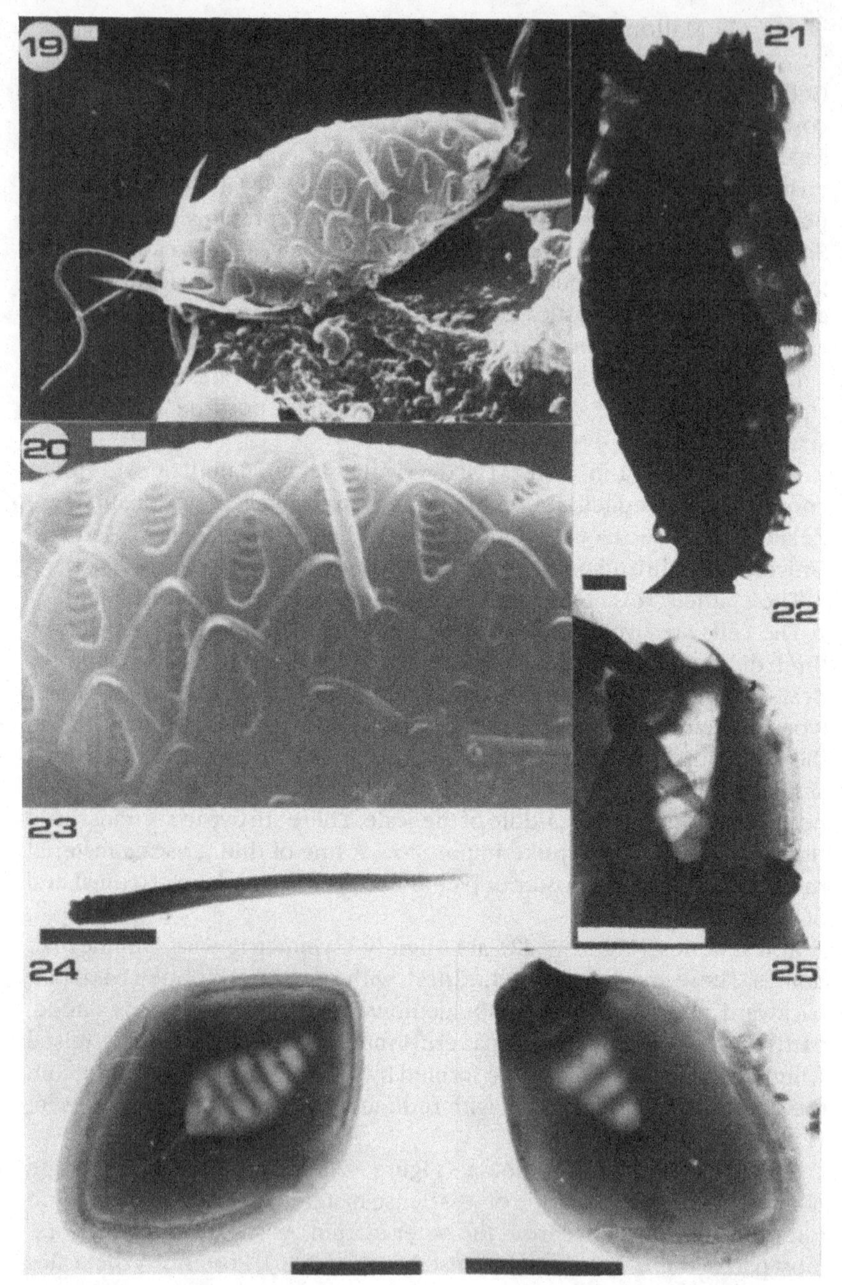

Figs. 7.19–7.25. ***Mallomonas grossa*** sp. nov. ***Fig. 7.19.*** Entire cell, SEM. ***Fig. 7.20.*** Part of the cell armor, SEM. ***Fig. 7.21.*** Nearly complete cell armor, indicating three different types of scales, TEM. ***Fig. 7.22.*** Apical scale, TEM. ***Fig. 7.23.*** Bristle, TEM. ***Fig. 7.24.*** Rhomboidal body scale, TEM. ***Fig. 7.25.*** Rear "dome"-bearing scale, TEM. Scale = 1 μm.

chimneylike domes (Figures 7.25 and 7.26). Contrary to the apical scales, they lack hood and anterior lateral extensions. An anterior rib like that of the body scales is absent.

The bristles are short, straight, and pointed (Figure 7.23). Cysts were not observed.

Mallomonas grossa is the first example of a *Mallomonas* species to possess a new type of dome, which appears as a short chimneylike structure. Because of the formation of a hood and winglike extensions on the apical scales, the dense part of the scale could be interpreted as a submarginal rib, although a considerably transformed one. Consequently, the ribbed patch would correspond to the shield.

The only species to which *M. grossa* bears some resemblance seems to be *M. annulata*. Common to both species is the possession of bristle-bearing scales on both ends and rhomboidal body scales on the midsection of the cell. In other respects, e.g., the scale ornamentation, the lack of papillae, the smooth tapering bristles, and the shape of the dome, they are very different. By reason of its deviating scale structure, *M. grossa* cannot be grouped in any of the so far described sections. It is placed therefore under the "isolated species."

Mallomonas grossa occurred frequently in samples from a small stream that drains a swamp near Mamaranui, North Island. Water temperature was 7°C, pH 6.05, and conductivity 210 μS_{20}. Other chrysomonads were *M. eoa*, *M. elliptica* Matvienko, and *Synura petersenii* Korsh. Besides the type locality, it was frequent in several habitats on both islands.

Mallomonas conspersa *sp. nov. (Figures 7.28–7.32)*

Cellulae ovoideae. Squamae tripartitae. Cupula parva papillis maculaque ex poris minutis praedita spinam tenuem exhibet. Scutum reticulo praeditum est. Lamina basalis cupula exclusa poros habet. Costa V-formis cucullata acutum angulum format. Costa submarginalis anterior in antica parte leniter evoluta papillae habent. Margo posticus latus perforatusque exstat, labrum proximale laeve. Squamae apicales minores parum asymmetricae. Setae tenues laeves in apica unum dentem subapicalem habent. Cysta ignota.

Dimensio cellulae: 8–15 × 5–7 μm; squamae apicales: 2.4 × 1.9 μm; squamae corporis: 2.8–3.5 × 1.5–2 μm; setae: 5–10 μm.

Holotypus die 22 m. Iulii a. 1982 in stagnis silvae (5) semipalustris *Podocarpi* prope lacum, qui Lake Whapo appelatur, Novae Zelandiae, sitae lectus est. Figuris 7.28–7.32 monstratur, qui in Herbario Universitatis Christchurchensis (CHR) Novae Zelandiae depositus est.

The epithet *conspersa* refers to the somewhat irregular shield pattern.

The cells are ovoid, entirely covered with bristles. All scales are tripartite with few variations in size and fine structure along the cell (Figure 7.28).

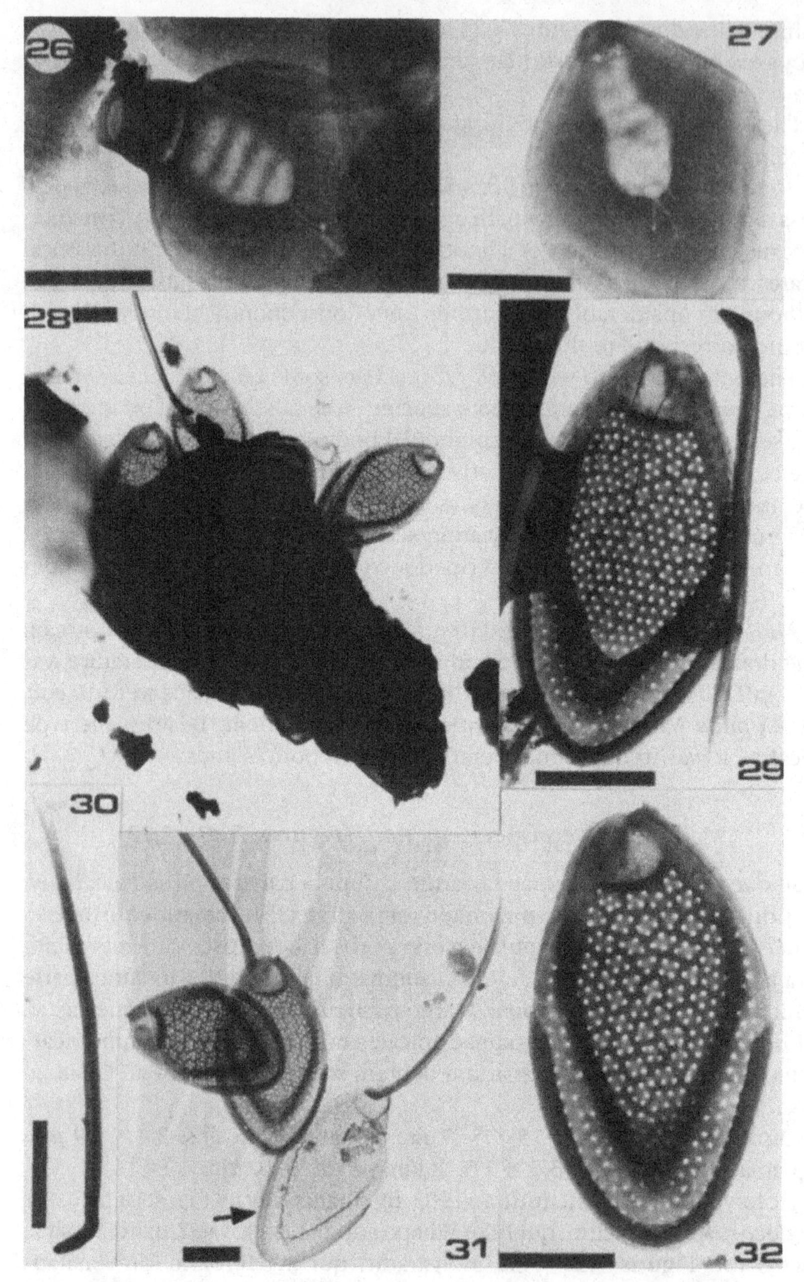

Figs. 7.26 and 7.27. *Mallomonas grossa* sp. nov. ***Fig. 7.26.*** Rear scale with "chimneylike" dome, TEM. ***Fig. 7.27.*** Scale with rudimentary dome, TEM.

Figs. 7.28–7.32. *Mallomonas conspersa* sp. nov. ***Fig. 7.28.*** Incomplete cell armor, TEM. ***Fig. 7.29.*** Body scale with bristle, TEM. ***Fig. 7.30.*** Bristle, TEM. ***Fig. 7.31.*** Body scale lacking a secondary layer (arrow), TEM. ***Fig. 7.32.*** Body scale, TEM. Scale = 1 μm.

The body scales are suboval (Figure 7.32). The dome is small, shallow, marked with more or less elaborated papillae and a patch of closely spaced minute pores. A small spine, formed apparently by one of the anterior submarginal ribs, projects beyond the distal edge of the dome. The shield is ornamented by a reticulum of roughly triangular or rectangular meshes, posteriorly more weakly developed or missing. The base plate has, with the exception of the dome, small closely spaced pores. The V-rib is hooded and forms an acute angle. The anterior submarginal rib is weakly elaborated in the foremost part and not clearly delimited at the anterior flange. More or less developed papillae are irregularly scattered on the anterior flange and the anterior submarginal rib. The posterior flange is broad and perforated. The proximal border is smooth.

The apical scales are smaller than the body scales and slightly asymmetric. Incomplete or irregular scale ornamentation, such as partial or total lack of the secondary layer were observed in several scales (Figure 7.31, arrow).

The slender bristles (Figure 7.30) are smooth, slightly curved and nontapering; their tip is provided with one more or less developed subapical tooth. Cysts were not observed.

Mallomonas conspersa shows close relationship to *M. calceolus* Bradley: for instance, in the lack of domeless scales, the uniformity in appearance of all body scales, the ornamentation of the small nonprominent domes, the appearance of the V-rib, the presence of papillae on the anterior flange and submarginal rib, and the shape of the bristles. In other respects, e.g., the perforated base plate and the reticulate shield pattern, they are different.

Momeu and Péterfi (1979) arranged *M. calceolus* within the sect. *Paxillata* of the subgenus *Mallomonopsis*. Consequently, *M. conspersa* should be grouped in the same section, although a second smooth flagellum, as observed by Belcher (1969) in *M. calceolus*, has not been found in *M. conspersa*, but it might have been lost during fixation.

Micrographs of scales belonging to *M. conspersa* have already been published by Takahashi (1959, Figure 88) from Japan.

Mallomonas conspersa occurred in very small numbers in puddles of a *Podocarpus* semiswamp forest near Lake Whapo, Westland National Park, West Coast, South Island. The water was humic, water temperature 9°C, pH 6.5, and conductivity 90 μS_{20}. It was also present in samples from a small swamp near Greymouth, West Coast, South Island.

Mallomonas villosa *sp. nov. (Figures 7.33 and 7.34)*

Cellulae ovoidea vel ellipsoidea. Scutum squamarum corporis papillis praebet. Costa submarginalis antica papillis habet a margine anteriore non perspicue terminata tigilla aliquantum obscura radiatim effundit. Costa submarginalis postica cucullata margo posticus labrumque proximale laeve. Squamarum collorum cupulae cacumen validum acutiusque habent. Scutum praebet reticulum ex papillis saepius congregatis. Squamae posticae singulis spinis ornatae sunt. Setae laeves, tenues, acutae.

Dimensio cellulae: 9–15 × 2.9–5 μm; squamae colli: 2.8 × 1.3 μm; squamae rhomboides: 2.1 × 1.8 μm; squamae posteriores 1.5 × 1.2 μm; setae 4–6 μm.

Holotypus die 11 m. Iulii a. 1982 in palude prope Mamaranui (1) in insula septentrionali Novae Zelandiae sita lectus est. Figuris 7.33, 7.34 monstratur, qui in Herbario Universitatis Christchurchensis (CHR) Novae Zelandiae depositus est.

The specific epithet refers to the downy appearance of the shield surface.

Mallomonas villosa belongs to the sect. *Torquatae* Momeu & Péterfi (1979). For general description of this section see also Asmund & Cronberg (1979).

The cells are small, ovoid or elongate ellipsoidal (Figure 7.33). The shield of the rhomboidal body scales is marked by closely spaced papillae that are connected by offshoots so that systems of somewhat parallel curved or straight ribs are formed. The secondary layer is absent posteriorly on the shield, and the base plate with a solitary pore is conspicuous. The distal part of the anterior submarginal rib is not clearly defined against the anterior flange; this has depressions of thinner scale material (Figure 7.34, arrow) separated by somewhat blurred struts, radiating from the anterior submarginal rib. More or less developed papillae are present along the anterior submarginal rib and on the struts. One row runs along the outside edge of the anterior flange. The proximal part of the posterior submarginal rib is hooded. The posterior flange and the proximal border are smooth.

The collar scales (Figure 7.33) are elongate and subtriangular. Their domes, each with a stout, sharply pointed peak, are provided with a patch of minute pores and papillae. The shield reveals an irregular mesh pattern formed by groups of papillae.

The rear scales are small and asymmetric, each with one to several spines projecting vertically from the scale surface (Figure 7.33).

Bristles are smooth, delicate, and pointed (Figure 7.33). Cysts were not observed.

Judged by the presence of papillae, this species might be place within the *M. mangofera* group. It has also a certain likeness to *M. grata* Takahashi (1963). As in that species, the shield is marked by papillae, but depressions such as those in *M. grata* are absent in *M. villosa.*

Mallomonas villosa occurred in very small numbers together with *M. roscida* in a swamp near Mamaranui. The water was polyhumic, water temperature 5°C, pH 5.0, and conductivity 290 μS_{20}. Cohabitants were *M. papillosa, M. eoa, M. calceolus. M. pumilio, M. mangofera* Harris & Brad-

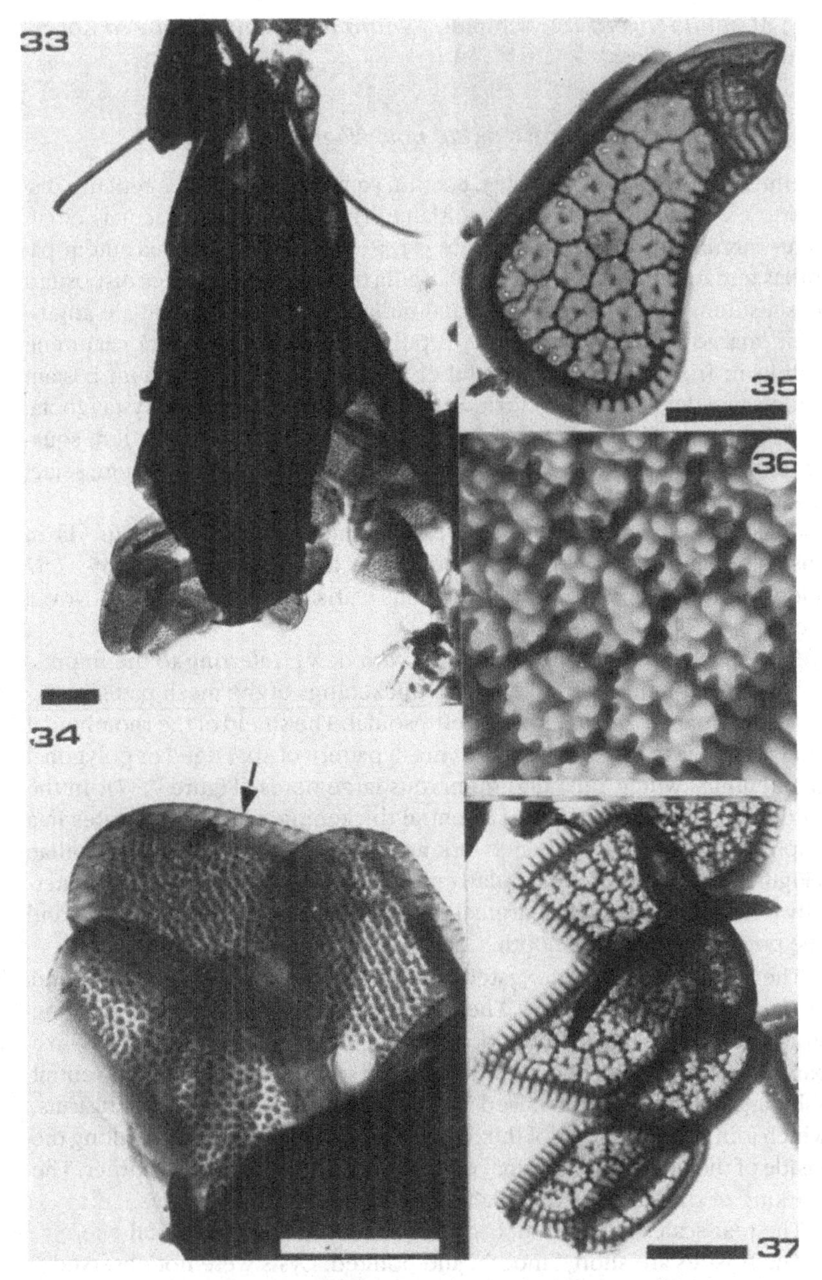

Figs. 7.33 and 7.34. Mallomonas villosa sp. nov. ***Fig. 7.33.*** Entire cell, TEM. ***Fig. 7.34.*** Rhomboidal scales with depressions on the anterior flange (arrow), TEM.

Figs. 7.35 – 7.37. Mallomonas roscida sp. nov. ***Fig. 7.35.*** Collar scale, TEM. ***Fig. 7.36.*** Part of the reticulate shield pattern, TEM. ***Fig. 7.37.*** Group of rhomboidal body scales, TEM. Scale = 1 μm.

ley, *M. multisetigera* Dürrschmidt, *Synura petersenii, S. spinosa* Korsh., and *S. mammillosa.*

Mallomonas roscida *sp. nov. (Figures 7.35–7.37)*

Cellula ovoidea vel ellipsoidea. Scutum squamarum corporis reticulo distincto ex spissatis sexangulis multiangulisve, quae plures lacunas circulares circumdat, instructo ornatum est. Reticulum in extremis autem papillas guttae similes praebet. Multa tigilla transversa regulariter disposita a costa submarginali anterior utroque radiantur. Labrum proximale angustum, margo posticus laevis est. Cupula squamarum collarum cacumine valido in frontem spectante exhibet. Series pororum secundum costam submarginalem dorsalem currit. Setae breves, laeves, acutae. Cysta ignota.

Dimensio cellulae: 8–16 × 4–7 μm; squamae colli: 3.3 × 2 μm; squamae rhomboides: 2.8 × 1.7 μm; squamae posteriores: 2.1 × 1.1 μm; setae: 5–10 μm.

Holotypus die 11 m. Iulii a. 1982 in palude prope Mamaranui (1) in insula septentrionali Novae Zelandiae sita lectus est. Figuris 7.35–7.37 monstratur, qui in Harbario Universitatis Christchurchensis (CHR) Novae Zelandiae depositus est.

The species was named *roscida* (=with dew), referring to the impression of the droplike papillae on the thickenings of the mesh pattern.

The cells are ovoid to elongate ellipsoidal. The shield of the rhomboidal body scales is marked by a distinct mesh pattern of six-edged or polygonal thickenings, which surround numerous large pores (Figure 7.37). In the middle of each mesh there is a central thickening, which culminates in a papilla. The edges of this network are furnished with droplike papillae (Figure 7.36). Numerous regularly spaced transverse struts radiate bilaterally from the anterior submarginal rib. The proximal border is narrow, and the posterior flange is smooth.

The collar scales are elongated and curved, thus having a "dorsal" and "ventral" edge (Figure 7.35). Their dome is small, with a forward pointing stout peak. The surface of the dome is decorated with about five ribs in a parallel arrangement, and a few irregularly scattered papillae. The ventral submarginal rib is steeply raised in the anterior part. There are short struts, which join the rib pattern of the dome. A row of small pores runs along the inside of the dorsal submarginal rib including the lower dorsal corner. The marking of the shield is less elaborated than in the body scales.

The rear scales are asymmetric with a small spine at the distal end.

The bristles are short, smooth, and pointed. Cysts were not observed.

Mallomonas roscida also belongs to the sect. *Torquatae* Momeu & Péterfi (1979) and closely resembles members of the *M. pumilio* group. As in *M. pumilio* var. *munda* Asmund, Cronberg, & Dürrschmidt (1982), the anterior flanges of the body scales have regularly spaced struts; the shield has a distinct mesh pattern, and there are short spines on the rear scales. In other respects, *M. roscida* is distinguishable, e.g., in the appearance of the shield: the meshes are significantly greater and they surround a greater number of pores. Moreover, the presence of droplike papillae on the crest of the thickenings and the possession of the central papilla are unique for *M. roscida*.

Mallomonas roscida was rare in samples from the Mamaranui swamp.

References

Asmund, B. & Cronberg, G. 1979. Two new taxa of *Mallomonas* (Chrysophyceae). *Bot. Not.* 132, 409-18.

Asmund, B., Cronberg, G. & Dürrschmidt, M. 1982. Revision of the *Mallomonas pumilio* group (Chrysophyceae) including description of three new taxa. *Nord. J. Bot.* 2, 383-95.

Belcher, J.H. 1969. Some remarks upon *Mallomonas papillosa* Harris and Bradley and *M. calceolus* Bradley. *Nova Hedwigia* 18, 257-70.

Cassie, V. 1980. Bibliography of the freshwater algae of New Zealand. 1849-1980. *N. Z. J. Bot.* 18, 433-47.

- 1981. Additions to the "Bibliography of the freshwater algae of New Zealand 1849-1980." *N. Z. J. Bot.* 19, 389-91.

Croome, R.L. & Tyler, P.A. 1983. *Mallomonas plumosa* (Chrysophyceae), a new species from Australia. *Br. Phycol. J.* 18, 151-8.

Dùrrschmidt, M. 1980. Studies on the Chrysophyceae from Río Cruces, Prov. Valdivia, South Chile by scanning and transmission microscopy. *Nova Hedwigia* 33, 353-88.

- 1982. Studies on the Chrysophyceae from South Chilean inland waters by means of scanning and transmission electron microscopy, II. *Arch. Hydrobiol.* Suppl. 63/2 (*Algol. Stud.* 31), 121-63.

Flint, E. 1975. Phytoplankton in some New Zealand lakes. *In:* Jolly, V.H. & Brown, J.M.A. [Eds.] *New Zealand Lakes,* Chapter 12. Oxford University Press, pp. 163-92.

Green, J.D. 1974. The limnology of a New Zealand reservoir with particular reference to the life histories of the copepods *Boeckella propinaqua* Sars and *Mesocyclops leuckarti* Claus. *Int. Revue ges. Hydrobiol.* 59, 441-87.

Johnston, I.M. 1972. Limnology of Western Springs, Auckland, New Zealand. *N. Z. J. Mar. Freshw. Res.* 6, 298-328.

Momeu, L. & Péterfi, L.S. 1979. Taxonomy of *Mallomonas* based on the fine structure of scales and bristles. *Contrib. Bot.* Cluj-Napoca 13-20.

Pearl, H.W., Payne, G.W., McKensie, A.L., Kellar, P.E., & Downes, M.T. 1979. Limnology of nine Westland beech forest lakes. *N. Z. J. Mar. Freshw. Res.* 13, 38-47.

Takahashi, E. 1959. Studies on genera *Mallomonas, Synura,* and other plankton in fresh-water by electron microscope (1). *Bull. Yamagata Univ. Agr. Sci.* 3, 117-51.

- 1963. Studies on genera *Mallomonas, Synura* and other plankton in freshwater with the electron microscope. IV. *Bull. Yamagata Univ. Agr. Sci.* 4, 169-87.

Thomasson, K. 1963. Araucanian Lakes. *Acta Phytogeogr. Suec.* 47, 1–139.
– 1974. Rotorua phytoplankton reconsidered (North Island of New Zealand). *Int. Revue Ges. Hydrobiol.* 59, 703–27.
Vigna, M.S. 1979. Ultrastructura de las escamas de *Synura petersenii* Korsh. f. *bonaerensis* nov. forma (Ochromonadales). *Bol. Soc. Argent. Bót* 18, 47–51.
– 1981. Estudio con microscopio electrónico de barrido de nuevos taxones de la familia Synuraceae (Chrysophyceae). *Physis (Buenos Aires), Sec. B.* 39, 15–20.
Wee, J.L. 1982. Studies on the Synuraceae (Chrysophyceae) of Iowa. *Bibliotheca Phycologica* 62, 1–183.

8

Some new observations on *Saccochrysis piriformis* Korsh. emend. Andersen (Chrysophyceae)

ROBERT A. ANDERSEN
Department of Biological Sciences
De Paul University
1036 Belden Avenue
Chicago, Illinois 60614, U.S.A.

Introduction

Saccochrysis is an unusual and perhaps unique colony of flagellate cells. Unlike most flagellates that swim through the water, *Saccochrysis* forms a gelatinous envelope and swims within this envelope. Korshikov (1941) described the genus and two species in 1941. The type species, *S. piriformis* Korsh., has since been observed by two other workers (Bourrelly 1957, Skuja 1964). The second species, *S. breviciliata* Korsh., has not been reported again, but as Bourrelly (1957) points out, the invalid *S. socialis* Korsh. et Matv. (Matvienko 1952) resembles it.

I have collected *S. piriformis* from Arkansas, Michigan, Minnesota, and Illinois, and established cultures from these collections. This chapter describes the observations made from culture and field material and adds new aspects to the life history of the species. These new findings are summarized in emended descriptions of the genus and the species.

Materials and methods

Saccochrysis was collected and cultured from the following sites in the United States: Rush Creek, Arkansas; a small spring-fed pool near Winslow, Arkansas; a peatlands near Red Lake, Minnesota; Tobacco River, Michigan; two different roadside pools near Gay, Michigan; a small pond near East Lansing, Michigan; and Volo Bog, Illinois. *Saccochrysis* was observed soon

I would like to thank Drs. Jan Janssens and John Kingston for collecting the Minnesota sample, Stephen Kump for collecting two northern Michigan samples and Dr. James Wee for taking me to the East Lansing, Michigan site. I would also like to thank Dr. G. B. Bouck and Jon Charlesworth for use of the electron microscope facilities at the University of Illinois-Chicago. Supported by NSF Grant #DEB 81-119527.

after collection in only the Tobacco River, Volo Bog, and East Lansing samples. The remainder of the samples were maintained in a culture chamber for approximately 6 months before *Saccochrysis* cells were found in the samples. Unialgal cultures have been established from the Rush Creek, Minnesota bog, and Tobacco River samples; the remainder were observed in mixed algal cultures. Unialgal cultures were maintained in a modified Chu #10 medium (Andersen 1982), soil – water medium with or without the addition of the vitamin solution described in the modified Chu #10 medium or DY-III medium (Lehman 1976). Cultures were maintained at various temperature and light regimes, including room conditions without obvious differences in growth.

For electron microscopy, cells were either fixed with 2% gluteraldehyde in a 0.05 M sodium cacodylate buffer (pH 7) for 15 minutes and post-fixed with 1% osmium tetroxide for 2 hours (Figures 8.25, 8.27, 8.28, and 8.29), or fixed with a combination of 1% gluteraldehyde and 1% osmium tetroxide in the same buffer for 30 minutes (Figure 8.26). Cells were then prestained with 0.5% uranyl acetate, dehydrated through an ethanol series, and embedded in Spurr's resin (Spurr 1969).

Results

The vegetative cells are typically 7 – 10 μm long, 6 – 10 μm wide, and pyriform in shape (Figure 8.1). A broad band-shaped chloroplast fills most of the anterior end. The posterior end is usually filled with a large, slightly bluish-silver chrysolaminarin droplet. There are usually two or three lipid droplets located in the middle of the cell, and these stain with Sudan III. The narrow, band-shaped chloroplast illustrated by Korshikov (1941) was observed in field collections from Michigan and Illinois. Once in culture, almost all cells produce a broader chloroplast band. Furthermore, the chloroplast band in cultured cells is not as strongly helixed from anterior to posterior as illustrated by Korshikov (1941). In old cultures, some or most cells may lack the typical chrysolaminarin droplet, and a granular inclusion appears in the cell posterior. This inclusion is birefringent and has the appearance of a minature sand grain.

Colonies of radiating cells in a spherical gel (Figure 8.2) were observed twice in field collections (Tobacco River and Volo Bog). These were radially arranged as illustrated by Korshikov (1941, see Figure 9a), Bourrelly (1957, Plate VIII, see Figure 10) and Skuja (1964, see Figure 10). When Tobacco River colonies were isolated and placed in culture tubes, large, amorphous colonies formed (Figure 8.3). The cells swam freely within the amorphous gel, but each cell maintained a small "home range" rather than swimming throughout the gel. The cells were not oriented in any particular direction, and each cell continually changed its orientation.

Small spherical colonies formed in new cultures when DY-III medium

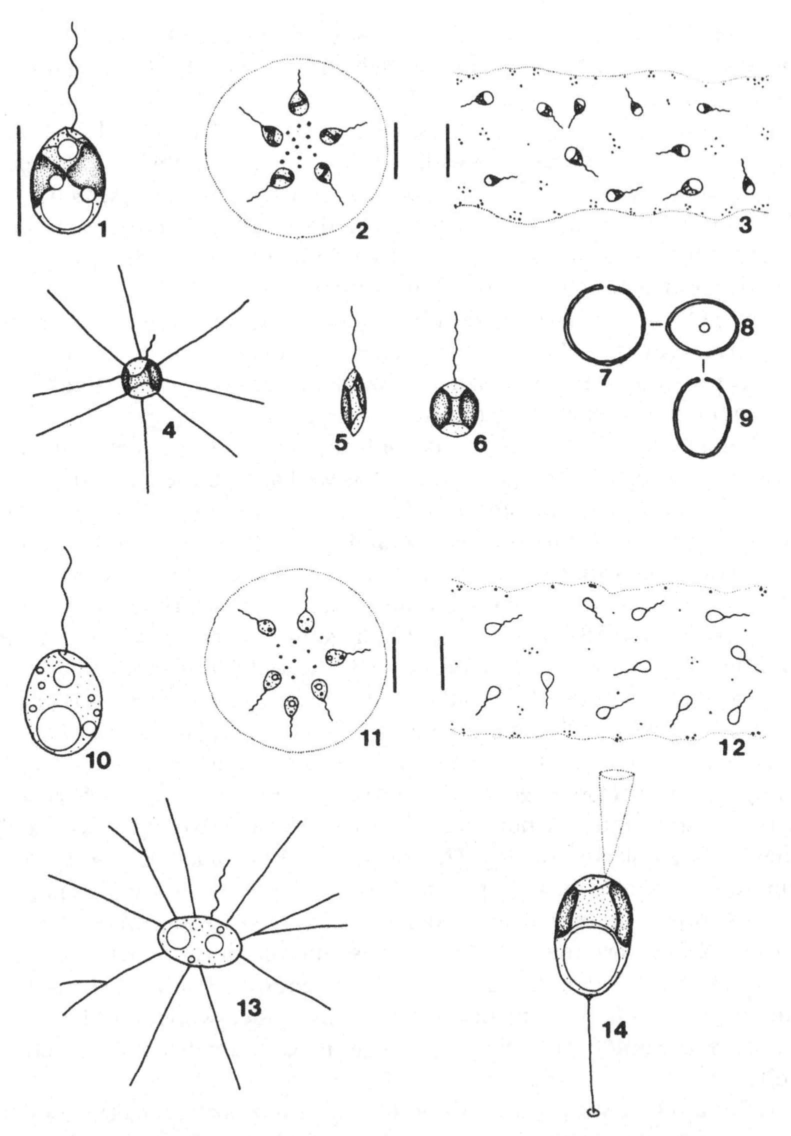

Figs. 8.1–8.14. Stages in the life history of *Saccochrysis piriformis.* The following are based upon light microscopy and therefore the second flagellum is not illustrated. Figure 8.1 = 10 μm, Figures 8.2 and 8.10 scale = 20 μm, Figures 8.3 and 8.12 Scale = 30 μm; all other figures use Figure 8.1 scale = 10 μm. ***Fig. 8.1.*** Vegetative cell. ***Fig. 8.2.*** Classic colony form with radially arranged cells. ***Fig. 8.3.*** Typical amorphous colony with randomly oriented cells. ***Fig. 8.4.*** Ameboid cell. ***Fig. 8.5.*** Typical spindle-shaped zoospore. ***Fig. 8.6.*** Spherical zoospore. ***Fig. 8.7–8.9.*** Statospore in front, top, and side views, respectively. ***Fig. 8.10.*** Colorless vegetative cell. ***Fig. 8.11.*** Small colony of radially arranged colorless cells. ***Fig. 8.12.*** Amorphous colony with randomly arranged colorless cells. ***Fig. 8.13.*** Colorless ameboid cell. ***Fig. 8.14.*** Stalked vegetative cell with long flagellum beating so as to form a vortex.

was used. The cells were not radially arranged as described above, but were oriented in a random manner with each cell continually changing direction.

The composition of the gel itself has not been determined. Indeed, unless the gel is demonstrated with India ink or stain, it is invisible. The gel readily stains with basic stains, such as basic fuchsin and Janus Green B. It does not stain with acid stains. When stained, the gel appears to have a thin, tough, outer skin that is able to hold a thinner, less viscous gel in the interior, just as Korshikov (1941) first described.

The gel was partially digested by a 0.005% aqueous solution of crude pectinase. Additional attempts to repeat the experiment using a more highly purified pectinase preparation were unsuccessful, suggesting there was an unknown enzyme in the cruder preparation.

Within the gel are numerous coccoid bacteria, 0.5–1.0 μm in diameter. These are scattered throughout the gel as well as near the gel periphery.

In old cultures or in defined media deficient in nitrogen, some vegetative cells will transform from free-swimming cells to ameboid cells (Figure 8.4). These cells are smaller, usually about 6–7 μm in diameter, and lack the distinctive chrysolaminarin droplet. Fine, unbranched filapodia extend away from the cell for up to 20 μm. The flagellum is present, but it is much shorter, about 2–4 μm, and it beats in a slow, undulating motion.

Zoospores are formed when the gel is severely torn or when it is enzymatically digested. The more common form is spindle shaped (Figure 8.5), but spherical zoospores have also been observed (Figure 8.6). The morphogenesis from flagellated vegetative cell to zoospore is dynamic. The cell apparently anchors itself to a substrate (microscope slide) and then begins to shake violently. The cell gradually becomes smaller, and the chrysolaminarin droplet disappears. As the cell reduces in size, the chloroplast becomes more helically arranged, and the shape becomes spindle shaped. Often, near the end of this process, the anterior and posterior ends form small nipplelike protrusions. Once the morphogenesis is complete, the zoospore, now 5–7 μm long and 2–2.5 μm wide, swims away. Because the zoospore is not within the gel, it swims much faster than the vegetative cell.

Culture tubes, with zoospores produced by enzymatic breakage of the gel, were wrapped with aluminum foil. A small hole was punctured in the foil and light was shined onto the hole. Within a few minutes, zoospores congregated in this region. The same experiment was repeated with typical, flagellated vegetative cells, but produced no congregation.

Statospores were rarely observed, and they could not be induced by changing the temperature, hydrogen ion concentration, light regime, or concentrations of nitrogen and phosphorus. The statospores were not spherical, but noticeably compressed laterally and slightly compressed anteroposteriorly (Figures 8.8–8.9). The diameter in the broadest plane was 7.5–8 μm. The pore was 0.7 μm in diameter and plain, without a neck or collar.

Colorless cells were observed in field collections from Volo Bog, Illinois, and in cultures (Figure 8.10). In these cells, no trace of the typical golden-green color was observed. The cytoplasm of these cells had the same appearance as colorless flagellates. There were usually two or more larger vesicles and several variously sized smaller vesicles. The morphology of the cell was identical to its colored analogue.

A colony of six colorless cells was observed in the Volo Bog field collection. At first the cells were radially arranged (Figure 8.11), but soon they swam without regard to orientation and continually changed direction. The colorless cells found in cultures are usually isolated cells among typical pigmented cells, but rarely one will see a small group of colorless cells with a small amorphous gel (Figure 8.12).

One of the six colorless cells of the Volo Bog collection broke free from the gel envelope and swam rapidly across the slide in a nearly straight line. Suddenly, the cell appeared to burst and the protoplasm seemed to flow out. This process was similar to the sight one sees when any naked cell disrupts. Yet, within 10 seconds, it became apparent that the cell had not disrupted, but had undergone an instantaneous change in cell form. Pseudopodia formed and the rapidly swimming cell had transformed into a colorless ameba (Figure 8.13) in less than 1 minute. The flagellum was still present, but it was beating in a slow, undulating fashion just as observed with the pigmented, ameboid forms.

After approximately 5 minutes, the ameba began to change form again into a typical flagellated cell, but it was now moving back and forth, continually changing directions. This characteristic movement suggested the cell was within a gel envelope again, but the presence of the gel was not confirmed with India ink or stains. This entire process may suggest a rapid, but yet unknown, means of forming a new gel envelope.

Cell division occurs 5–8 hours into the light period in cultures on a 14–10 hour light–dark schedule. Dividing cells are first recognized by a change in their movement and morphology. Dividing cells show little swimming activity, and there are numerous lipid droplets in the trough of the chloroplast. Two flagella are soon apparent, and the anterior of the cell has a papillalike appearance (Figure 8.15). The flagella soon separate, slipping down opposite sides of the papilla. Once this occurs the papilla disappears, and the flagella continue to separate, each sliding along the surface of the now spherical cell (Figure 8.16). As the flagella become situated on opposite sides of the cell, the cell begins to elongate (Figure 8.17). It constricts at or near the center, and two daughter cells are formed. These cells, now with a typical morphology, are held together by a very fine cytoplasmic thread for 5–10 minutes (Figure 8.18). The cells separate, and the division is complete. The entire process takes 15–30 minutes.

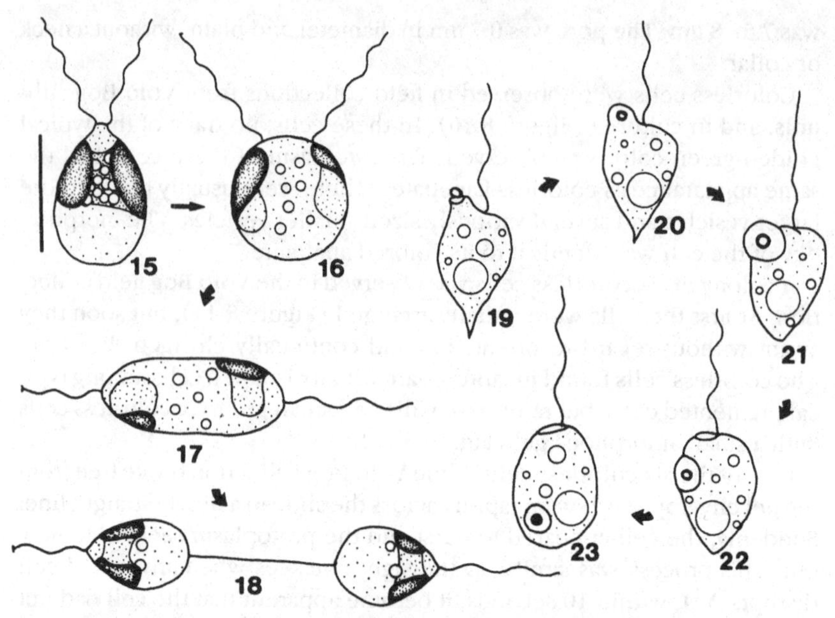

Figs. 8.15–8.23. Stages in the life history of *Saccochrysis piriformis* continued. Scale bar = 10 μm. ***Figs. 8.15–8.18.*** Cell division stages. ***Fig. 8.15.*** Early division stage with two long flagella and numerous lipid droplets in the chloroplast trough. ***Fig. 8.16.*** Spherical cell with two chloroplasts and the flagella separating. ***Fig. 8.17.*** Oblong cell with one long flagellum at each end. ***Fig. 8.18.*** Two daughter cells connected by a fine protoplasmic strand. ***Figs. 8.19–8.23.*** Phagocytosis in colorless cells. ***Fig. 8.19.*** Early stage of engulfment; a proboscislike process surrounds the bacterium and the posterior of the cell concomitantly becomes pointed. ***Fig. 8.20.*** Vacuole forms around the bacterium. ***Figs. 8.21 and 8.22.*** Vacuole moves down the side of the cell, becoming smaller as it moves posteriorly; the posterior end becomes more rounded. ***Fig. 8.23.*** Typical cell with dark food vacuole near the base.

The fate of the nucleus is not known. It is difficult to observe the nucleus in living cells because it lies in the trough of the chloroplast. Nuclear staining was not attempted.

The pigmented cells can be generally regarded as autotrophic based upon their growth in a defined mineral medium, but under some conditions they may also be heterotrophic. Cells grown in phosphorus-deficient media, or occasionally cells found in old cultures, attach their posterior end to the substrate, either directly or by means of a slender cytoplasmic stalk (Figure 8.14). The flagellum beats so as to form a vortex, and bacteria and particles are drawn into the whirlpool. Yet, in these pigmented cells, phagocytosis was not clearly observed.

Phagocytosis has been observed in colorless cells (Figures 8.19–8.23). As a bacterium approaches the cell surface, a proboscislike process is quickly formed and pushed out to engulf the particle. A large vesicle containing the particle forms at the anterior end of the cell, and then quickly moves down the surface of the cell. By the time it reaches the posterior end of the cell, a period of 30–60 seconds, the vesicle has reduced in size and the contents have become more dense. Accompanying the entire process is a remarkable change in the shape of the cell. As the proboscislike process forms, the posterior of the cell becomes very pointed. The pointed posterior gradually rounds as the vesicle moves down the cell.

Colorless cells from the Volo Bog collection were observed feeding on bacteria at a rate of 2–3/minute. The cells swam within the gel envelope in the characteristic pattern of everchanging directions, until they were in contact with a bacterium. At this time, the cells would stop, extend the proboscislike process, engulf the particle, and then begin swimming again.

Ultrastructural observations have been limited because a good method of fixation has not been found. Apparently the gel retards fixatives, and cells almost always burst before they can be fixed. Yet, two significant observations have been made. A second flagellum is present, but it is only about 750 nm long (Figures 8.24 and 8.25). Another noticeable feature is the lack of a girdle lamella in the chloroplast (Figure 8.26). No ultrastructural evidence of a pyrenoid was observed. Finally, the prokaryotes within the gel lack thylakoids and are bacteria and not symbiotic blue-green algae (Figures 8.27 and 8.28).

Discussion

The observations here demonstrate that *Saccochrysis piriformis* has a variety of life history stages. Not only is the typical flagellate, vegetative cell capable of transforming into an ameba, an attached stalked cell, a zoospore, or a resting statospore, but it can also lose its pigmentation, form colorless swimming and ameboid cells, and become phagocytic.

It should also be pointed out that the presence of distinct zoospores formed by a specific morphogenesis is clear proof that the colonial habit of the vegetative cells is not a temporary stage of an otherwise *Chromulina*-like organism. This type of life form is apparently not represented by any other living organism. It cannot be called a colonial flagellate in the sense of *Synura, Eudorina,* or *Volvox.* These cells are joined together, and the flagella beat in the water of their environment; *Saccochrysis* cells are independent of one another and swim in their own gelatinous matrix. Likewise, *Saccochrysis* does not represent a palmelloid state, such as *Chrysocapsa* or *Gloeocystis. Saccochrysis* represents a colonial life form somewhere between the flagellated and palmelloid life forms. I term this a pseudopalmelloid life form.

Certainly, a characteristic feature of this organism is the swimming pat-

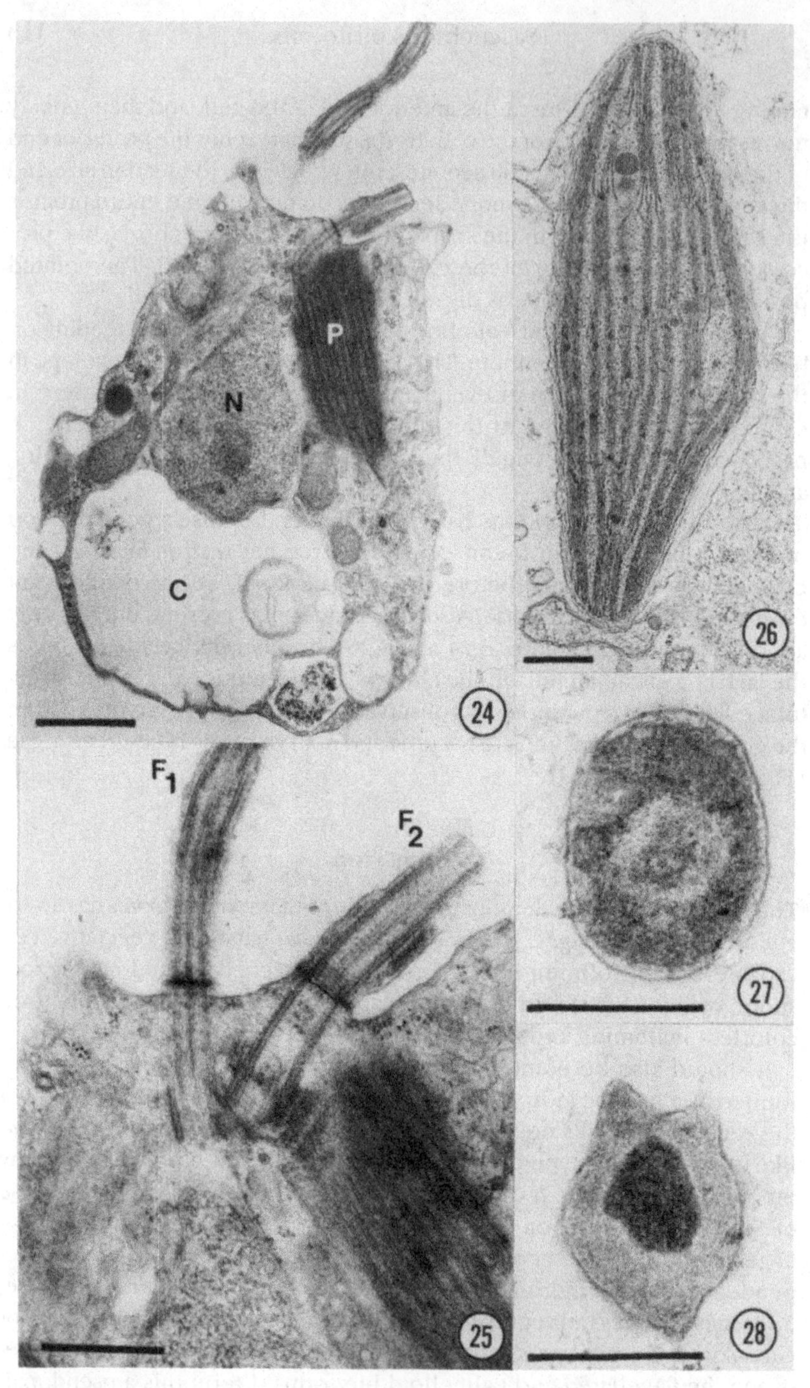

Figs. 8.24–8.28. Ultrastructure of *Saccochrysis piriformis* and bacteria in the colonial gel. Figure 8.24, Scale bar = 1.0 μm, Figures 8.25–8.28, Scale bar = 0.5 μm. ***Fig. 8.24.*** Vegetative cell illustrating the plastid (P), nucleus (N), and chrysolaminarin vacuole (C). ***Fig. 8.25.*** Same cell as Figure 8.24 (two serial sections later) illustrating the two flagella, long flagellum (F_1), and short flagellum (F_2). ***Fig. 8.26.*** Chloroplast showing the lack of a girdle lamella. ***Figs. 8.27 and 8.28.*** Sections through bacteria that occur in the colonial gel.

tern within the gel. Korshikov (1941) described the swimming pattern as "crawling and nuzzling in every direction." I would describe this as a neverending obsession of lazily swimming about their immediate area, seemingly intent about doing something, yet apparently doing nothing. One must ask the question of why they swim at all. Seemingly secure within the gel, why do they not stop swimming and form immobile colonies such as *Chrysocapsa* or *Gloeocystis?* Perhaps the gel is not entirely secure, and swimming is necessary to continually check and repair the gel envelope. Some support for this hypothesis is that the activity of the cells increases in intensity when the gel is enzymatically broken apart.

The small spherical colonies that appear in field collections but rarely in cultures are probably due to rugged conditions in nature or sample handling that prevent larger, irregular gel masses from forming or being maintained. The radiating cell arrangement that has been illustrated (Korshikov 1941, Bourrelly 1957, Skuja 1964) is somewhat misleading, because as Korshikov and Bourrelly point out, this occurs only in small colonies with few cells. It is much more common, in field and culture studies, to find the cells swimming without regard to direction or arrangement.

The occurrence of a short, second flagellum is not unexpected because all Chrysophyceae with the exception of *Rhizochromulina marina* Hibberd et Chretiennot-Dinet (Hibberd & Chretiennot-Dinet 1979) have been found to possess two flagella. The lack of a girdle lamella is an unusual feature for a chrysophyte, but has been reported missing in two ameboid forms, *R. marina* (Hibberd & Chretiennot-Dinet 1979) and *Chrysamoeba radians* Klebs (Hibberd 1971) as well as two silica-scaled genera, *Chrysosphaerella* and *Chromophysomonas* (Preisig & Hibberd 1983).

The one characteristic in the original description (Korshikov 1941) and two later studies (Bourrelly 1957, Skuja 1964) that does not agree with my observations is the length of the long flagellum. These studies describe the length as 1.5 to 2(3) times the cell length. I always found the flagellum to be approximately equal to the cell length. This discrepancy cannot be explained, but it alone seems insufficient to warrant any separate taxonomic treatment. The flagellum length of *S. piriformis* that I observed is similar to that described for *S. breviciliata* (Korshikov 1941), but it has a distinct pyrenoid. No pyrenoid was observed in *S. piriformis* from field collections or cultures, as viewed with light and electron microscopes.

A third possible species of *Saccochrysis* is the invalid *S. socialis* Korsh. et Matv. reported in Matvienko (1952). Matvienko acknowledges Korschikov's discovery of the genus *Saccochrysis,* but she believed that Korshikov had died before he had a chance to publish the description and that all his illustrations were lost. Consequently, she attempted to publish a generic description of the already described *Saccochrysis* Korsh. 1941. The species description, although invalid, is certainly not synonymous with *S. piriformis,* and it may represent a species yet to be validly described. Bourrelly

(1957) has commented on the resemblance of *S. breviciliata* and *S. socialis,* and regarding the flagellated vegetative state of *S. socialis* this seems true. However, the description in Matvienko (1952) includes synchronized cell division (2, 4, 8, 16, 32, . . .) of immobile cells within a common mucus prior to the occurrence of typical flagellated vegetative cells. Synchronized division was not observed in *Saccochrysis* by our laboratory and was not described by Korshikov (1941), Bourrelly (1957) or Skuja (1964). In certain ways *S. socialis* is similar to *Chrysosphaera magna* Belcher (=*Mucosphaera* Dop et Van Beem) (Dop 1980). Thus, any attempt to validly describe the organism Matvienko (1952) called *S. socialis* will have to delineate clearly its life history and address the distinguishing features that separate *Saccochrysis* and *"Mucosphaera."*

It seems that an organism with such distinct and varied life forms should have attracted the attention of more workers than it has. Some of this may be due to the failure to recognize the gelatinous envelope, and therefore the monads are identified as a *Chromulina* sp. One species of *Chromulina, C. nebulosa* Cienk., has a chloroplast arrangement that is very reminiscent of *Saccochrysis piriformis* (Cienkowsky 1870). Indeed, Cienkowsky may have actually observed the organism we now recognize as *Saccochrysis piriformis,* as is indicated by the following quote from his description of *C. nebulosa:*

> Das zweite monadenartige Wesen, bei welchem ich dieselbe Cystenbildung entdeckt habe, will ich *Chromulina nebulosa* Cnk. nennen. Ihre Zoosporen leben in grossen Gesellschaften zarte, braune Haute bildend; in Form nebelartiger Massen hullen sie untergetauchte Gegenstande ein in Torfmoos-Pfutzen des nordlichen Russland (Yarosllaff). Sie sind von keiner sichtbaren Substanz zusammengehalten und gehen bei leisestem Drucke leicht auseinander.

The morphology of the cell and chloroplast illustrated by Cienkowsky is very similar to drawings of *S. piriformis.* The major emphasis of Cienkowsky's paper was statospore formation, and it is here where a marked difference seems to exist between the two descriptions. Cienkowsky illustrates cysts with very distinctive ridges. These are not illustrated by Korshikov (1941), Bourrelly (1957), Skuja (1964) or this chapter. It is possible that the cysts in the *Saccochrysis* papers were not fully formed. Several papers (Hibberd 1977, Andersen 1982, Sandgren, 1983) have shown that cysts do not always develop the external ornamentation. If *S. piriformis* is shown to have mature statospores like those of *C. nebulosa,* then the two should be considered synonymous. If this happens, a new genus must be formed for all free-swimming motile cells with a single flagellum because *C. nebulosa* is the type species for the genus *Chromulina.*

Emended descriptions

Saccochrysis Korsh. *Arch. Protistenk.* 95:35 (1941) emend Andersen.

Flagellated, vegetative cells naked; with two flagella, one longer and visible with light microscope, the other very short and visible only with electron microscope; chloroplast single, band-shaped; girdle lamella absent in chloroplast; large, posteriorly placed chysolaminarin droplet usually present; contractile vacuole(s) anterior; cell division vegetative, not coenobial; zoospores and siliceous cysts produced. Flagellated vegetative cells swimming constantly within a gelatinous envelope; cell arrangement within the gel random, rarely radial.

Saccochrysis piriformis Korsh. *Arch. Protistenk.* 95:35–36 (1941) emend Andersen.

Flagellated vegetative cells pyriform, 7–14 μm long, 6–10 μm wide; long flagellum 1–2(3) times cell length; chloroplast narrow or broad band, slightly or greatly spiraled in cell; stigma lacking; zoospores spindle-shaped, rarely spherical, 5–7 μm long, 2–3 μm wide; stalked cells, ameboid cells, colorless flagellated cells, colorless ameboid cells present in life history; cysts laterally compressed, 7–10 μm in diameter.

References

Andersen, R.A. 1982. A light and electron microscopical investigation of *Ochromonas sphaerocystis* Matvienko (Chrysophyceae): the statospore, vegetative cell and its peripheral vesicles. *Phycologia* 21: 390–8.

Bourrelly, P. 1957. Recherches sur les Chrysophyceés. Morphologie, Phylogénie, Systématique. *Rev. Algol., Mémoire Hors-Série.* 1: 1–412.

Cienkowsky, L. 1870. Ueber Palmellaceen und einige Flagellaten. *Arch. Mikroskop. Anat.* 6: 421–38.

Dop, A.J. 1980. Benthic Chrysophyceae from the Netherlands. Thesis. University of Amsterdam.

Hibberd, D.J. 1971. Observations on the cytology and ultrastructure of *Chrysamoeba radians* Klebs (Chrysophyceae). *Br. Phycol. J.* 6: 207–23.

– 1977. Ultrastructure of cyst formation in *Ochromonas tuberculata* (Chrysophyceae). *J. Phycol.* 13: 309–20.

Hibberd, D.J. and Chretiennot-Dinet, M.-J. 1979. The ultrastructure and taxonomy of *Rhizochromulina marina* gen. et sp. nov., an amoeboid marine chrysophyte. *J. Mar. Biol. Ass. U. K.* 59: 173–93.

Korshikov, A.A. 1941. On some new or little known flagellates. *Arch. Protistenk.* 95: 22–44.

Lehman, J.T. 1976. Ecological and nutritional studies on *Dinobryon* Ehrenb.: Seasonal periodicity and phosphate toxicity problem. *Limnol. Oceanogr.* 21: 646–58.

Matvienko, A.M. 1952. Chrysomonadineae e viciniis urbis Charkov. *Not. Syst. Cryptog. Inst. Bot. Nom. V. L. Komarovii, Acad. Sci. URSS.* 8: 16–33.

Preisig, H.R. and Hibberd, D.J. 1983. Ultrastructure and taxonomy of *Paraphysomonas* (Chrysophyceae) and three related genera 3. *Nord. J. Bot.* 3: 695–723.

Sandgren, C.D. 1983. Morphological variability in populations of chrysophycean resting cysts. I. Genetic (interclonal) and encystment temperature effects on morphology. *J. Phycol.* 19: 64–70.
Skuja, H. 1964. Grundzüge der Algenflora und Algenvegetation der Feldgegenden um Abisko in Schwedisch-Lappland. *Nova Acta R. Soc. Scient. Upsal., Ser. IV.* 18(3): 1–465.
Spurr, A.R. 1969. A low viscosity epoxy resin embedding medium for electron microscopy. *J. Ultrastruct. Res.* 26: 31–43.

PART III

Cell biology

9

Scale case construction in *Synura petersenii* Korsch. (Chrysophyceae)

BARRY S.C. LEADBEATER
Department of Plant Biology, University of Birmingham
Edgbaston, Birmingham B15 2TT, England

Introduction

Members of the family Mallomonadaceae Diesing 1866 (=Synuraceae Lemmermann 1899), to which *Synura petersenii* belongs, are characterized by the possession of a "case" composed of scales, usually silicified. As construed by Takahashi (1978) and Kristiansen (1979), this family contains such pigmented genera as *Synura* Ehr., *Mallomonas* Perty, *Mallomonopsis* Matvienko, *Chrysophaerella* Lauterborn, and *Chromophysomonas* Preisig & Hibberd, together with the colorless genus *Paraphysomonas* De Saedeleer. However, species of *Synura, Mallomonas,* and *Mallomonopsis* can be grouped together and distinguished from the other genera mentioned above on the basis of ultrastructure (Preisig & Hibberd 1983) as well as scale morphology and the details of scale case construction. Scale and scale case characters are described, and their significance is discussed in this chapter.

Early attempts to identify species of *Synura* and *Mallomonas* according to cell size and shape were unreliable and were soon superseded by the use of scale morphology as a taxonomic character, once the existence of an enveloping scale case had been established for these taxa by the exemplary work of Petersen (1918) followed by Korshikov (1929) and others. The importance of scale characters to the taxonomy of *Synura* and *Mallomonas* at the species level has been further reinforced by the addition of characters obtained from transmission and scanning electron microscopy. There are now many illustrations of scales and scale cases obtained by these methods to be found in the literature (e.g., Manton 1955, Harris & Bradley 1956, Petersen & Hansen 1956, Takahashi 1978). However, in spite of this, the precise details of how scales are held together to form a continuous

I am grateful to Emeritus Professor I. Manton F.R.S. and Dr. Sheila Crosbie for reading this manuscript during preparation and for many helpful comments. I also wish to thank Mr. Stephen Price for technical assistance. Acknowledgment is also made to the Royal Society, London for permission to reproduce Figures 9.7, 9.8, 9.10, and 9.11.

case have not been established. The results presented here include a detailed description of scale case construction in one particular species of *Synura* in culture, together with a preliminary explanation of how the scale case is assembled and modified in shape and size throughout the cell cycle.

Material and methods

Synura petersenii was obtained from the Culture Collection of Algae and Protozoa, Cambridge and was grown in biphasic (soil–water) medium.

For electron microscopy (EM), cells were fixed with 2% osmium tetroxide in 0.1 M cacodylate buffer at pH 7.0. Subsequent preparation of cells for embedding and sectioning was by standard techniques described elsewhere (Leadbeater 1979). Whole mounts were also prepared by standard techniques and were shadowcast with gold–palladium. Transmission electron micrographs were taken on a Philips EM 300 microscope. For scanning electron microscopy, a suspension of cells was fixed with 1% osmium tetroxide and then partially concentrated by means of a Millipore filter. The residual suspension on the filter membrane was rinsed in an acetone series, care being taken that the cells at all times remained moist. They were then critically point-dried. The Millipore membranes were scanned in a ICS 200 scanning electron microscope.

Empty, intact, scale cases were obtained by treatment of living cells with detergent as explained in the next section. The cases were subsequently washed with distilled water. For enzyme treatment of the empty cases, the requisite enzyme was dissolved in 0.1 M phosphate buffer at the appropriate pH (see Table 9.1). Enzyme treatment lasted for 15 minutes at 37°C and, after washing with distilled water, the material was processed into whole mounts for electron microscopic examination. For all enzyme experiments, an appropriate control consisting of buffer alone at the correct pH was run concurrently.

Observations

Light microscopy

Synura petersenii is a motile, colonial species, each colony containing up to 50 golden brown cells. Each monad is pear shaped and is attached to other members of a colony by a cytoplasmic stalk. Two unequal, heterokont, flagella emerge from the broad, outwardly directed end of each cell. A translucent layer, the scale case, covers each protoplast.

Empty, intact scale cases can be isolated by treatment of a suspension of living cells with a 1% aqueous solution of Triton X-100 detergent for 5 minutes at room temperature (Figure 9.5). From such specimens the arrangement of scales can be discerned with the light microscope thereby confirming Petersen (1918) (see Figure 9.1).

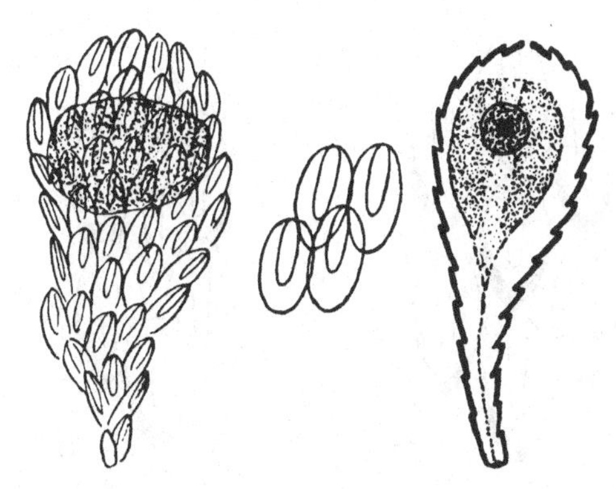

Fig. 9.1. Reproduction of Petersen's (1918) original drawings of *Synura* (*uvella*) *petersenii* showing the morphology, arrangement, and imbrication of scales forming a scale case.

Micromorphology of scales

Scales of *S. petersenii* are oval to elliptical in shape, those on the stalk being narrower than those on the remainder of the cell. All scales in *S. petersenii* have a similar morphology and patterning (Figures 9.3 and 9.6) although, in old cultures, there may be minor variations. Each scale consists of a flat, finely perforate base plate that is more or less smooth on the proximal surface. The outwardly facing surface carries emergent structures of several different sorts, including a central hollow chamber, arranged longitudinally and prolonged into a point anteriorly (Figure 9.6). A series of transverse ridges extend from the central chamber to the scale periphery. An upturned, in-flexed rim is more or less confined to the posterior end of the scale, usually extending forward for a greater distance along one side than along the other. The base plate contains many small perforations, but larger holes are present in the wall of the central chamber.

In median transverse section, the base plate of a scale is slightly concave (Figures 9.4 and 9.7), and, apart from the perforations, the component silica is of more or less uniform density. The substructure of the wall of the hollow chamber is similar in thickness and electron opacity to that of the base plate. However, the rim is trilaminate in substructure, with two thin layers of silicified material separated by a layer of minute spherical chambers (Figures 9.7 and 9.8). On the outwardly facing surface of the rim, there is a superficial deposit of diffuse material arranged in one or more layers (Figure 9.7).

Construction of the scale case

Scanning electron microscopy provides the best obtainable views of scale case construction (Figure 9.3). The scales are arranged in diagonal rows with a precise and consistent imbrication. When a scale case is viewed from

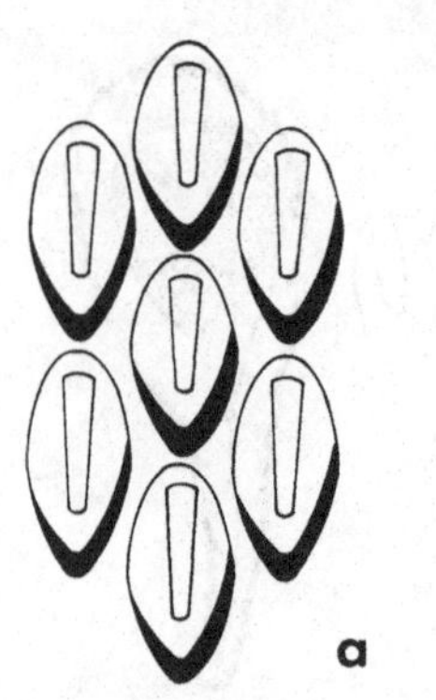

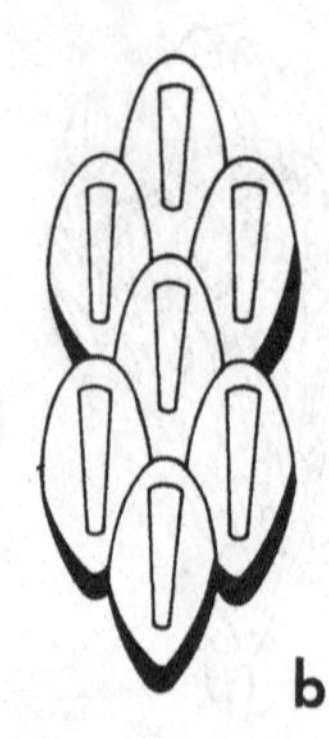

Fig. 9.2. Two illustrations [(a) and (b)] demonstrate the manner in which scales combine to form a case. In both illustrations one scale is characteristically surrounded by six others and the scale rims are shown in black. In (a) the scales have been separated so that they do not touch. In (b) the scales have been brought together so that the anterior edge of the central scale overlaps part of the rim of each scale in front.

the outer surface with the anterior (flagellar) pole forward, the anterior edge of each row of scales can be seen to overlap the posterior edge of the row in front. In most specimens, the rows of scales slope from lower left to upper right, thereby describing a counterclockwise spiral (Figure 9.3). The direction of spiraling is most obvious around the flagellar pole.

The inflexed scale rim plays an important part in holding scales together. Sections of overlaps between adjacent scales (Figure 9.8, arrow) show that mutual contact is confined to the upper surface of the rim of one scale and a peripheral band on the lower surface at the anterior end of the conjoint adjacent scales. The diffuse material on the rim surface is thus sandwiched between overlaps of adjacent scales, and it probably represents the adhesive substance that holds them together.

The way in which scales are arranged to form a case is perhaps best understood by considering an individual scale surrounded by six others to which it is, at first, not attached (Figure 9.2a). The position of the rim is clearly indicated on each scale in this figure. If these scales are then brought together as in Figure 9.2b, the anterior edge of the central scale overlaps part of the rim of each of three adjacent scales in front. This produces the imbrication characteristic of a scale case when seen in scanning EM (Figure 9.3) and in whole mounts (Figure 9.6). The diagonal deflection of the rows of scales is probably brought about by the unequal forward extension of the upturned rim on the two sides of every scale.

The adhesive material between scales

From the foregoing account it would appear probable that an adhesive material of some kind must hold adjacent scales together to form a continuous case. Isolated scale cases, obtained as described above, can be agitated

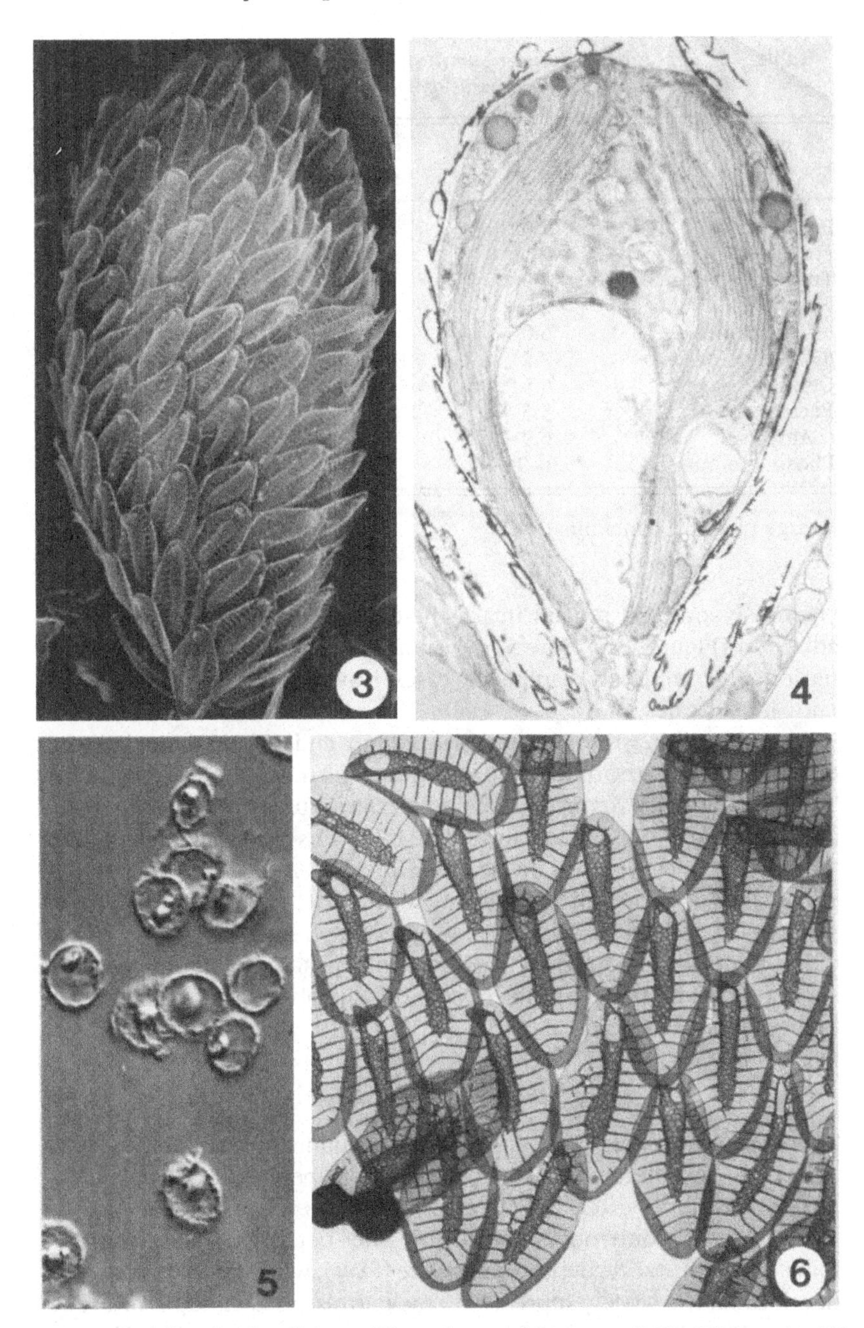

Figs. 9.3–9.6. A selection of views of the scale case of *S. petersenii.* ***Fig. 9.3.*** Scanning EM of scale case. × 2300. ***Fig. 9.4.*** Longitudinal section of a cell showing the internal arrangement of chloroplasts, parts of developing scales on the right side of the protoplast and the continuous scale case which covers the protoplast surface. × 5600. ***Fig. 9.5.*** Empty, intact scale cases isolated after treatment of living cells with detergent. × 900. ***Fig. 9.6.*** A spread of scales seen as a whole mount. × 5600.

Table 9.1. *Effects of various enzymnes (0.5 mg/ml) at 37 °C on isolated intact scale cases of* Synura petersenii

Enzyme	pH	Scale cases remain intact	Scales released from scale-case
Pronase	7.5	−	+
Protease	7.5	−	+
Trypsin	7.9	−	+
Pepsin	2.0	−	+
β-Glucuronidase	5.0	+	−
β-Galactosidase	7.2	+	−
Cellulase	5.5	+	−
Pectinase	5.5	+	−
α-Amylase	6.9	−	+[a]
Phosphate buffer	5.0–7.9	+	−

[a] Large protease contamination.

vigorously and still remain intact, thereby showing the strength of the adhesion (Figure 9.5). However, treatment with concentrated sulfuric or nitric acids completely disrupts the cases, leaving only separate but apparently intact scales.

A number of enzymes, in particular protein- and carbohydrate-digesting enzymes, were tested on isolated scale cases and the results are recorded in Table 9.1. For all enzyme treatments, a control using buffer alone at the appropriate pH was run concurrently. The results show that only the proteolytic enzymes released the scales from cases. After such disruption of scale cases by means of proteolytic enzymes, it was found by means of sections that the diffuse layer, formerly present on each scale rim, had been either removed or considerably altered in appearance (Figure 9.9). This supports the idea that the diffuse layer normally plays some significant part in mediating adhesion.

Biogenesis of scales

There are now several well-documented accounts of scale biogenesis in *Synura* in the literature (Schnepf & Deichgräber 1969, Hibberd 1978, McGrory & Leadbeater, 1981, Mignot & Brugerolle 1982, Leadbeater 1984), and it is only necessary to give brief details here. Each scale is initiated within a scale deposition vesicle (SDV), which, at an early stage of development, is closely appressed to the periplastidial endoplasmic reticulum (ER) that covers the chloroplast envelope (Figure 9.10). Microtubules and microfilaments are intimately associated with the developing SDV and are probably in some way responsible for shaping it, thereby controlling the morphology of the future scale. In addition, a diverticulum

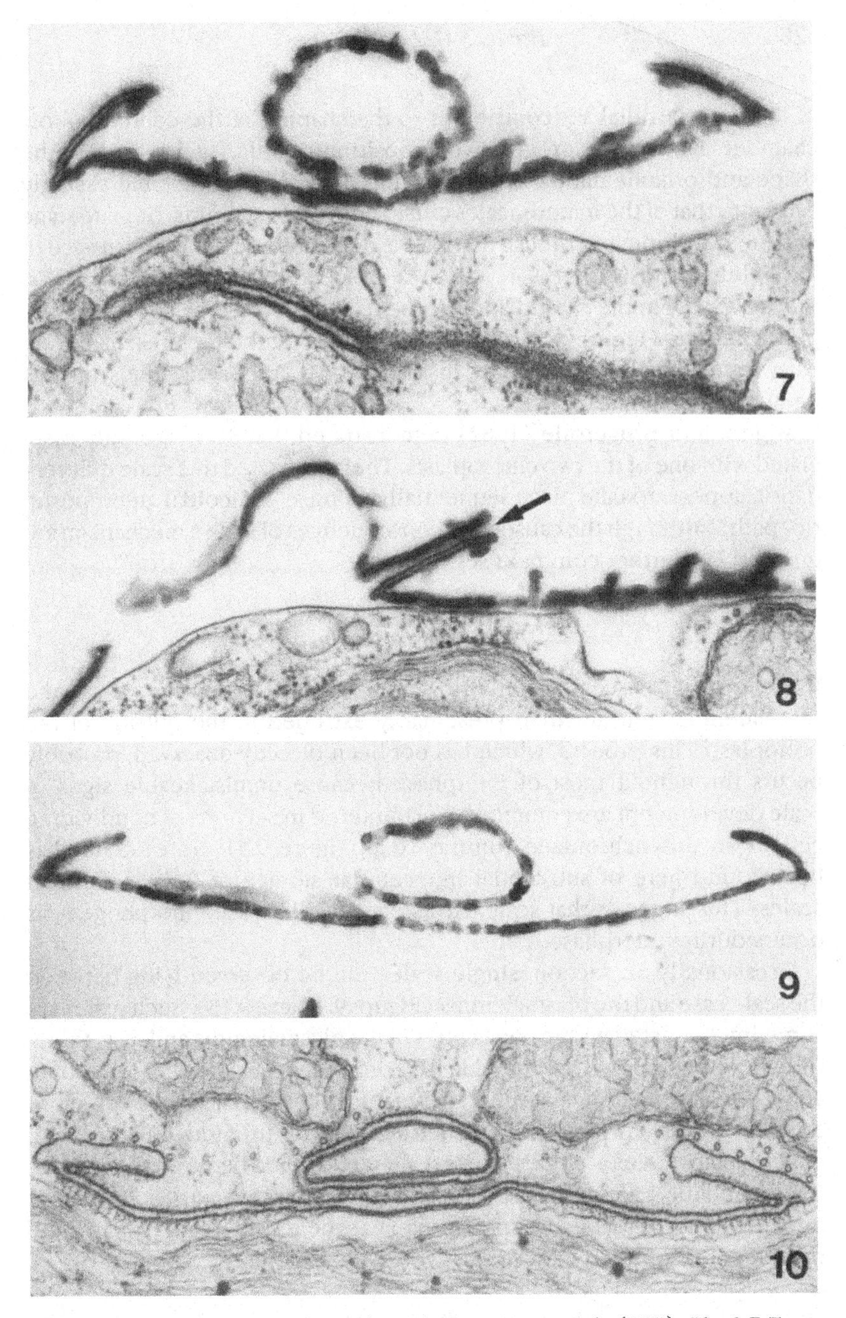

Figs. 9.7–9.10. Sections of scales and a scale deposition vesicle (SDV). ***Fig. 9.7.*** Transverse section of a scale outside the plasmalemma and a SDV within protoplast. Note diffuse layer of material on outer surface of the scale rim. × 46,500. ***Fig. 9.8.*** Section through overlap (arrow) between two adjacent scales showing diffuse material sandwiched between scales. × 37,200. ***Fig. 9.9.*** Transverse section of a scale after treatment with pronase. Note absence of diffuse material on outer surface of scale rim. × 37,200. ***Fig. 9.10.*** Transverse section of a SDV located on chloroplast surface. Note overlying microtubules and an underlayer of compound filaments. × 37,200.

of the periplastidial ER contributes to the shaping of the central hollow chamber. During the progressive deposition of silica within a SDV, the shape and ornamentation of the developing scale becomes increasingly similar to that of the mature scales outside the plasmalemma. At a late stage in scale development, diffuse material is deposited on the outer surface of the inflexed rim (Figure 9.11). Mature nascent scales, still within SDVs, move away from the plastid and are then usually located just beneath the plasmalemma (Figures 9.4 and 9.13). The microtubules, present during early stages of silica deposition, disappear once a recognizable scale has been formed (Figure 9.11).

Mignot and Brugerolle (1982) demonstrated that SDVs are only associated with one of the two chloroplasts. They also stated that scale differentiation appears to take place sequentially along a "helicoidal anteroposterior path," although the causes and consequences of such a mechanism are unclear. For further comment see below.

Exocytosis of scales

When mature, nascent scales, still within their SDVs, move toward the plasmalemma and are then presumably extruded to the outside of the protoplast. This process, which has not been directly observed, probably occurs throughout most of interphase because unmistakeable signs of scale development are commonly encountered in sections of nondividing cells from unsynchronized cultures (e.g., Figure 9.4). No evidence has been found here of substantial intracellular storage of fully completed scales. This suggests that scale extrusion as well as scale morphogenesis occurs during interphase.

Occasionally, in section, single scales can be observed lying between the scale case and the plasmalemma (Figure 9.12 and 9.13). Such scales are sometimes situated within concavities of the plasmalemma and may, therefore, have been newly extruded by the protoplast (Figure 9.13). In some cells, developing scales at different stages of maturity can be seen close together within a protoplast. Just such a group is illustrated in Figure 9.13, which also contains a single mature scale sandwiched between the plasmalemma and the scale case in a position suggesting recent extrusion. Such an arrangement could be interpreted as a sequence of intracellular scale movements ending in exocytosis.

Discussion

The basic morphology of scales, each consisting of a flat plate with asymmetrical inflexed rim, and the construction of the scale case in *Synura petersenii* are characters typical of all species of *Synura, Mallomonas,* and *Mallomonopsis.* Since these characters are distinctive and are not

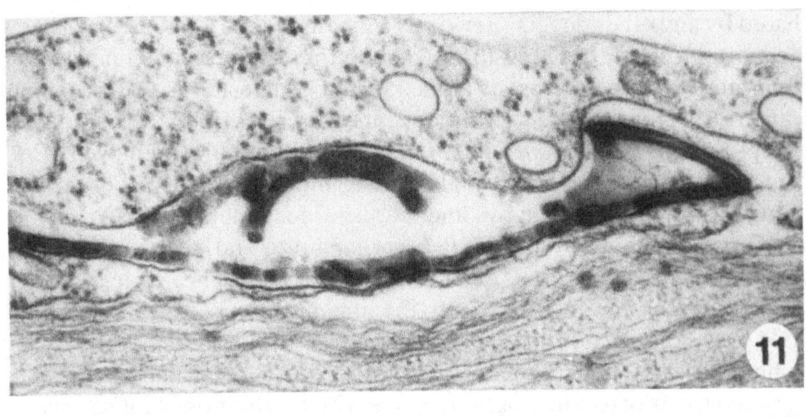

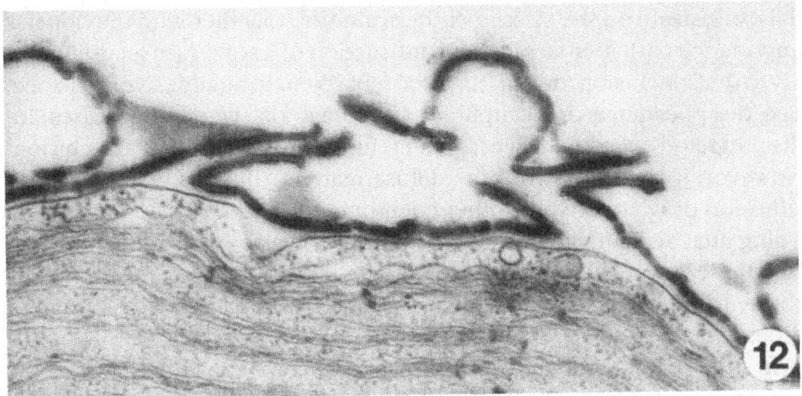

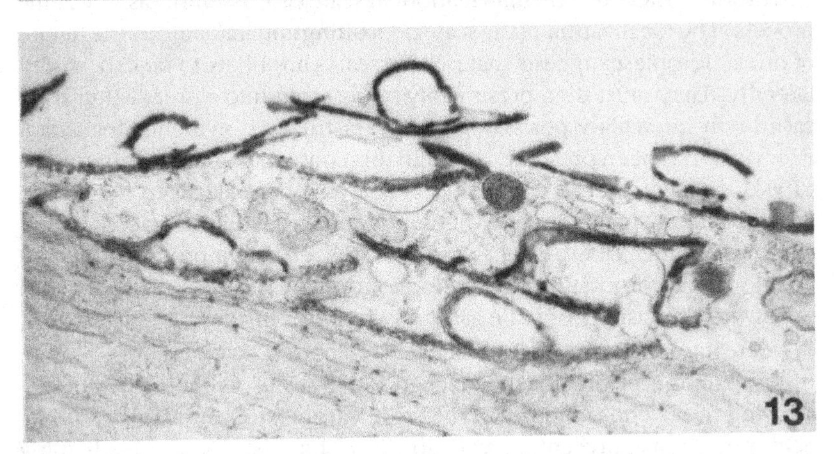

Figs. 9.11 - 9.13. Sections of scales within and outside the protoplast. ***Fig. 9.11.*** Late stage in scale development. Note diffuse material on outer surface of scale rim. × 46,500. ***Fig. 9.12.*** Single scale sandwiched between scale case and plasmalemma. × 37,200. ***Fig. 9.13.*** Group of developing scales within protoplast and a single mature scale in a concavity of the plasmalemma sandwiched between the plasmalemma and the scale case. × 18,600.

shared by any other genera previously included in the Mallomonadaceae, there are justifiable grounds for separating these genera into a newly defined family for these reasons alone. Such a conclusion is in agreement with Preisig and Hibberd's (1983) recent recommendation that all genera except *Synura, Mallomonas* and *Mallomonopsis* should be removed from the Mallomonadaceae. Accordingly, Preisig and Hibberd (1983) have now created a new family (Paraphysomonadaceae) for genera such as *Paraphysomonas, Chromophysomonas, Polylepidomonas,* and *Chrysosphaerella,* which have an *Ochromonas* type of cell but produce silica scales.

As has been demonstrated here, the apparently successful analysis of scale case construction has depended on recognition of the mode of attachment of one scale to its neighbors. In this context, the role of the in-flexed rim is of paramount importance. The localization of diffuse material exclusively on the surface of the scale rim, together with the obvious importance of this surface in the construction of a scale case leads inevitably to the conclusion that this material must contain an adhesive substance. The disappearance or disruption of this material when scale cases are disintegrated by chemical treatment further supports this conclusion. However, it is not known if the diffuse material alone is responsible for adhesion or whether some other component located on the immediately facing attached surface of the adjacent scale is also involved. Little is, in fact, understood about the nature of the adhesive material except that its ability to hold scales together is destroyed by treatment with a range of proteolytic enzymes.

In spite of improved understanding of scale case construction as a static end product, there is little information on scale case assembly as a dynamic process. The localization of the scale depositing apparatus along the length of one chloroplast suggests that mature scales are likely to be exocytosed laterally. They must then presumably be inserted into a preexisting scale case in an intercalary position. Some ultrastructural evidence for lateral exocytosis has been presented here. In this context, Mignot & Brugerolle's (1982) concept mentioned above is relevant because if a "helicoidal, anteroposterior path" of scale differentiation were to be aligned with the external symmetry of the scale case, this in itself might predetermine the positions of incorporation of new scales into an existing case in the manner observed. This aspect of scale assembly requires further investigation.

Intercalary insertion of new scales, if it occurred, would require temporary separation of previously attached older scales, without thereby destroying the overall integrity of the scale case. Exactly how this could be achieved is not at present known. However, if it is accepted that adjoining scales in a scale case can separate and then rejoin in a new configuration after insertion of an additional scale, this could explain several of the known properties of scale cases. These include enlargement to accommodate to protoplast growth, a capacity to adjust to cytokinesis without scale case disruption, and the ability of two scale cases to coalesce into a single case during syngamy. All these attributes have been documented from light microscopy of living cells (Conrad 1926, Wawrik 1972). Thus, it would appear that the scale case, far from being a rigid constraining armor as scanning electron micrographs might suggest, is capable of constant dynamic modifications throughout the cell cycle. Experimental work is now being undertaken to investigate further some of the outstanding problems associated with these phenomena.

References

Conrad, W. 1926. Recherches sur les flagellates de nos eaux saumâtres, 2e Partie: Chrysomonadines. *Arch. Protistenk.* 56: 167-234.

Diesing, K.M. 1866. Revision der Prothelminthen. Abtheilung: Mastigophoren. *Sitzungsber. Kaiserl. Akad. Wiss. Math.-Naturwiss. Cl.*, Abt 1, 52: 287-401.

Harris, K. & Bradley, D.E. 1956. Electron microscopy of *Synura* scales. *Discovery* 17: 329-32.

Hibberd, D.J. 1978. The fine structure of *Synura sphagnicola* Korsch. (Chrysophyceae). *Br. Phycol. J.* 13: 403-12.

Korschikov, A.A. 1929. Studies on the Chrysomonads I. *Arch. Protistenk.* 67: 253-90.

Kristiansen, J. 1979. Problems in classification and identification of Synuraceae (Chrysophyceae). *Schweiz. Z. Hydrol.* 40: 310-19.

Leadbeater, B.S.C. 1979. Developmental studies on the loricate choanoflagellate *Stephanoeca diplocostata* Ellis. I. Ultrastructure of the non-dividing cell and costal strip production. *Protoplasma* 98: 241-62.

- 1984. Silicification of cell walls of certain protistan flagellates. Mineral phases in biology. *Phil. Trans. Roy. Soc. B,* 304: 529-36.

Lemmermann, E. 1899. Das Phytoplankton sächsischer Teiche. *Forschungsber. Biol. Stat. Plön* 7: 96-135.

Manton, I. 1955. Observations with the electron microscope on *Synura caroliniana* Whitford. *Proc. Leeds Phil. Lit. Soc.* 6: 306-16.

McGrory, C.B. & Leadbeater, B.S.C. 1981. Ultrastructure and deposition of silica in the Chrysophyceae. *In:* Simpson, T.L. & Volcani, B.E. [Eds.] *Silicon and siliceous structures in biological systems,* Springer-Verlag, New York, pp. 201-30.

Mignot, J.P. & Brugerolle, G. 1982. Scale formation in chrysomonad flagellates. *J. Ultrastruct. Res.* 81: 13-26.

Petersen, J.B. 1918. Om *Synura uvella* Stein og nogle andre Chrysomonadiner. *Vidensk. Medd. Dansk. Naturh. Foren. Kbh.* 69: 345-57.

Petersen, J.B. & Hansen, J.B. 1956. On the scales of some *Synura* species. (1). *Biol. Medd. Kongel. Dansk. Vidensk. Selsk.* 23 (2): 1-27.

Preisig, H.R. & Hibberd, D.J. 1983. Ultrastructure and taxonomy of *Paraphysomonas* (Chrysophyceae) and related genera 3. *Nord. J. Bot.* 3: 695-723.

Schnepf, E. & Deichgräber, G. 1969. Uber die Feinstruktur von *Synura petersenii* unter besonderer Berücksichtigung der Morphogenese ihrer Kieselschuppen. *Protoplasma* 68: 85-106.

Takahashi, E. 1978. *Electron Microscopical Studies of the Synuraceae (Chrysophyceae) in Japan.* Tokai University Press, Tokyo.

Wawrik, F. 1972. Isogamie Hologamie in der Gattung *Mallomonas* Perty. *Nova Hedwigia* 23: 353-62.

10

Autofluorescent vesicles in *Ochromonas danica* Prings. (Chrysophyceae)

ANNETTE W. COLEMAN
Division of Biology and Medicine, Brown University
Providence, Rhode Island 02912, U.S.A.

Introduction

Fluorescence microscopy sometimes reveals striking new features in otherwise well-studied organisms. The genus *Ochromonas* has received much attention since its nutrition was first analyzed by Pringsheim (1951). Here we describe unusual vesicles in *Ochromonas* that are filled with an autofluorescent material. The vesicles move to the cell surface and release the autofluorescence into the medium when irradiated with wavelengths of light absorbed by the autofluorescent compound.

Materials and methods

Ochromonas danica (UTEX 1298) was grown in a complex medium (*Ochromonas* medium; Starr, 1978) that contains glucose, tryptone, yeast extract, and liver extract, at 23°C in constant light (75 μeinsteins/m^2/sec), unless otherwise noted. Soil water tubes with or without $CaCO_3$ supplementation (Starr, 1978) were also used. Cells were examined by phase contrast and Nomarski interference contrast light microscopy and by epifluorescent microscopy using a Zeiss Photoscope and Zeiss excitation barrier filter combinations 487702 (excitation peak = 365 nm), 487704 (excitation peak = 405 nm), 487709 (excitation peak = 470 nm), and 487715 (excitation peak = 550 nm) normally used, respectively, for observation of Hoechst, catecholamine, fluorescein, and rhodamine fluorescence.

Fluorescence excitation and emission spectra of whole cell suspensions and fluorescent solutions were measured and recorded using a Spex model 1990 Fluorolog 2 spectrofluorometer.

The advice and assistance of Dr. Samuel I. Beale in obtaining spectra is most gratefully acknowledged. Supported by National Science Foundation Grants PCM 78-15783, PCM 81-08122, and PCM 82-00383.

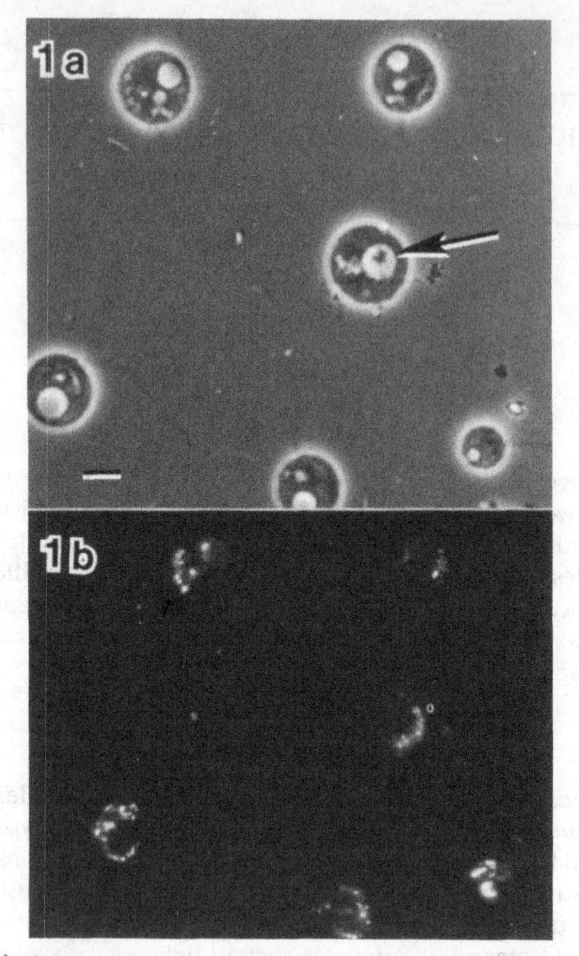

Fig. 10.1. (a) Phase and (b) fluorescence micrographs of living *Ochromonas danica* cells in medium containing 1×10^{-3} M Na_3VO_4. Arrow indicates leucosin granule. Scale = 10 μm.

Results

In *Ochromonas* medium, which contains glucose and complex organic supplements, *Ochromonas danica* grows readily to densities of greater than 10^6 cells/ml. Such living cells contain numerous cytoplasmic vesicles as observed by light microscopy (Figures 10.1 – 10.3). When observed with either the Hoechst or catecholamine filter combination, many of these vesicles are observed to emit an intense blue (green with the catecholamine filter set because of the blocking filter) fluorescence (Figure 10.1b).

Irradiation for a very short time does not appear to affect the behavior of

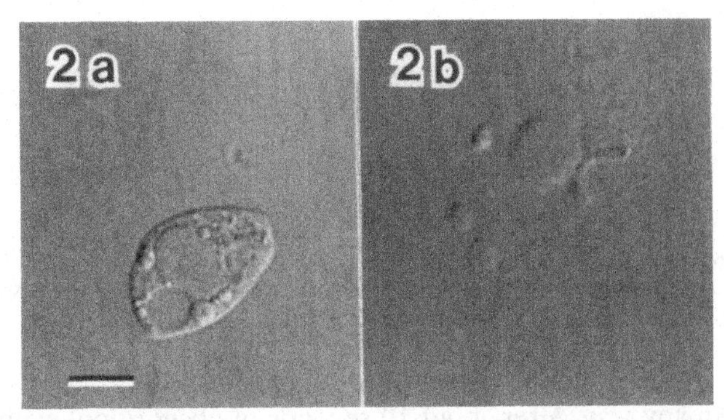

Fig. 10.2. Nomarski micrographs of (a) intact cell and (b) cell lysed by light irradiation using 02-filter combination. Several of the vesicles in (a) were fluorescent. Scale = 10 μm.

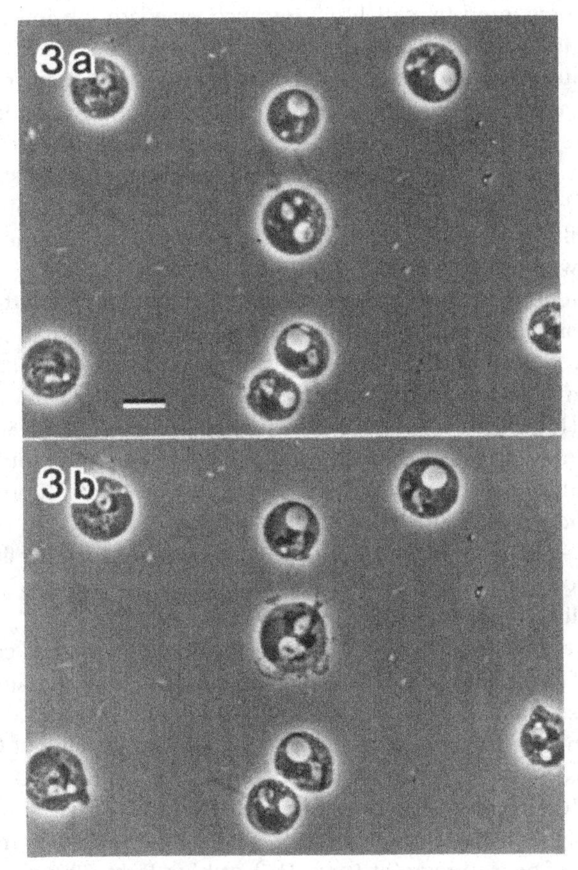

Fig. 10.3. Phase micrographs of (a) intact cells and (b) cells after several fluorescent vesicles have been released. Medium contained 1×10^{-3} M Na_3VO_4. Scale = 10 μm.

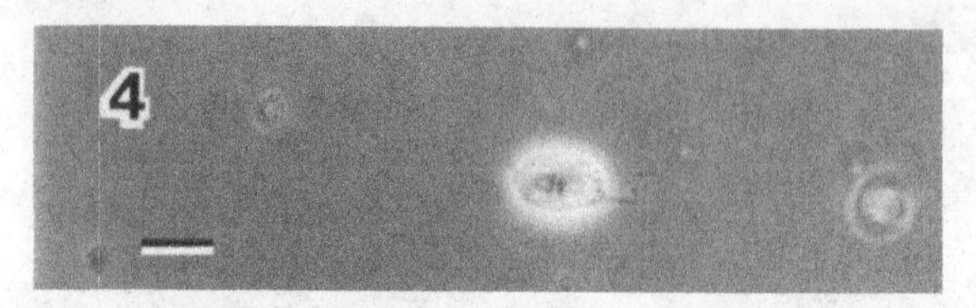

Fig. 10.4. Phase micrograph of cell growing in soil–water medium. Although this cell has numerous small vesicles in its periphery, none were fluorescent and there was no response to blue light irradiation. Scale = 10 μm.

the cells in any way. After about 10 seconds of observations with the 02 filter set and the 40× objective, the vesicles suddenly disappear, and a return to phase contrast microscopy reveals that the cell has lysed (Figure 10.2). The vesicles release their contents into the medium, where the material rapidly disperses. The phenomenon of vesicle firing was not always accompanied by cell lysis. Because vanadate ion interferes with exocytosis in various cell types, cells were exposed to 1×10^{-3} M Na_3VO_4 for 30 minutes prior to observation. Such cells respond more slowly to irradiation and release vesicles singly, often without obvious detrimental effect to the cell (Figure 10.3). One can observe single vesicles escaping intact from a cell; they then slowly swell and lyse while floating in the medium. Occasional cells in *Ochromonas* medium contain only one or a few small fluorescent vesicles, and these cells can release their vesicles and then swim away.

The content of fluorescent vesicles depends upon the growth medium, but not upon the stage of growth or whether cells are grown in the light or in the dark. Cells with many, intensely fluorescent vesicles occur in complex media such as *Ochromonas* medium, even when 5 mM acetate is substituted for the glucose. Cells growing in *Volvox* medium (Starr, 1978) have smaller numbers of fluorescent vesicles. In *Volvox* medium supplemented with acetate, vesicles are larger but less intensely fluorescent, yet still respond to irradiation. Less than 10% of cells growing in soil–water (Figure 10.4) or in soil–water plus $CaCO_3$ have any observable fluorescent vesicles. However, those that do have such vesicles respond to irradiation. These media cover a pH range of 5.1 to 7.0.

Vesicle release can be caused by a variety of chemicals and conditions. Addition of an equal volume of saturated picric acid solution causes immediate release and cell lysis. So also does addition of 2% glutaraldehyde or 3 : 1 (ethanol : glacial acetic acid) fixative. Methylene blue at 0.005% or higher causes vesicle release within 30 min, even in the dark. Warming cells to 52°C or higher also causes complete vesicle release and cell lysis.

The initial observations suggested that vesicle release was induced by absorption of near-ultraviolet (near-UV) or blue light. The phenomenon occurred only when the Hoechst (02) and catecholamine (04) filter com-

Table 10.1. *Effect of sunlight on* Ochromonas danica[a]

Time (hr)	Exposed cells Fluorescent (%)	Exposed cells Intact (%)	Shielded cells Fluorescent (%)	Shielded cells Intact (%)
zero	98	100		
1	2	12	92	100
1½	2	13	96	98
2	0	7	94	98

[a] *Ochromonas* (5×10^5 cells/ml) growing in *Ochromonas* medium exposed to full sunlight (midday, July) in a 32 °C water bath.

binations were used, and heat barriers did not interfere. Since the epifluorescence excitation light is very intense, various visible light sources were tested in the laboratory to see if they could also induce the response. None have yet come close to matching the 10-sec response obtained with epifluorescent near-UV light (02 filter) on a microscope slide. A further test utilized two matched glass test tubes of an *Ochromonas* culture, one tube shielded with aluminum foil, which were floated in a water bath illuminated by the midday sun (Table 10.1). Clearly, cells kept at the water surface on a sunny summer day release their vesicles and lyse.

The absorption and fluorescence spectra of whole cells grown in *Ochromonas* medium and rinsed in dilute salt solution (a dilute medium for growth of freshwater autotrophs, containing a total of 2 mM salts) are shown in Figure 10.5. Since cells heated to 52°C release all their vesicles, yet retain much of their cell structure, such a preparation was made, the cell bodies were removed by centrifugation, washed once, and then resuspended in the same dilute salt solution. Microscopic observation confirmed that the cells had all lysed, and spectrofluorometer measurements showed that essentially all the fluorescent compound was recovered in the lysed cell medium. The peaks of absorption and of fluorescence emission were only very slightly shifted. This lysed cell medium was dialyzed overnight at 2°C. The fluorescent material passed freely through a 12,000–14,000 MW porous Spectrapor dialysis membrane over this time period. The fluorescent material was unchanged by heating, at least up to 80°C.

A number of other algae were examined for autofluorescent blue vesicles. The vesicles were observed in both *Ochromonas* species examined, *O. danica* and *O. minuta* (UTEX 1300), and in the closely related *Poterioochromonas malhamensis* (UTEX 1297). In all cases, the color appeared to be the same and the response to epifluorescent microscope light was identical: explosive vesicle release. Neither *Synura* sp. nor *Dinobryon* sp. had such fluorescent vesicles, nor did any Prymnesiophyceae, Xanthophyceae, or other algae (Coleman 1985).

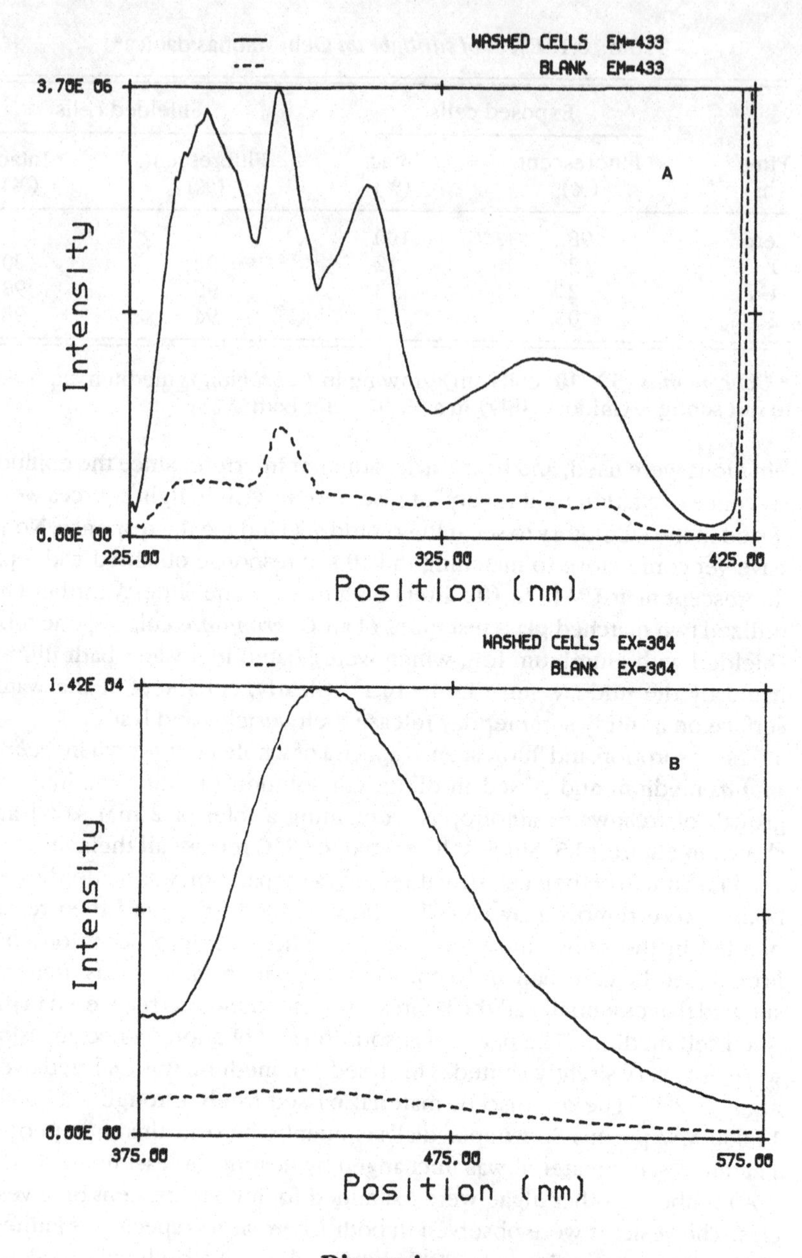

Discussion

Ochromonas cells are well endowed with cytoplasmic vesicles (Aaronson et al. 1971). Some contain materials that stain with fat-soluble dyes (Hovasse and Mignot 1975, Andersen 1982), and others are capable of ejecting

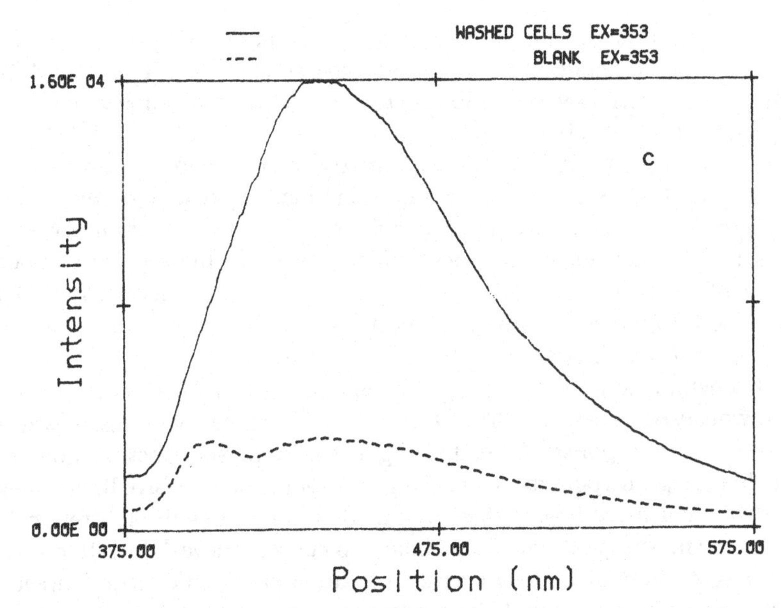

Fig. 10.5. (A) Spectrum of wavelengths that induce fluorescence at 433 nm from *Ochromonas danica* cells. The full emission spectra of the two right-hand peaks are shown in (B) and (C). The two left-hand peaks were not examined further. Fluorescence emission spectrum: excitation at 304 nm. Fluorescence emission spectrum: excitation at 353 nm.

an organized microstructure called a discobolocyst (Hibberd 1970). Many undoubtedly contain the mucouslike material that sometimes surrounds cells in older cultures. The fluorescent vesicles reported here are a discrete subset of the many vesicles seen in *Ochromonas danica.* Vesicles in a cell are either highly fluorescent or completely nonfluorescent; all the fluorescent vesicles are capable of discharge in response to blue light, while it appears that none of the nonfluorescent vesicles are. The fluorescent vesicles range in size from about 2 μm in diameter downward, and are not always spherical, perhaps because two or more have become appressed. Although they may be what have been called mucocysts (Hovasse and Mignot 1975), we have not yet examined them with the electron microscope and we have not observed any evidence of mucouslike material in their discharge, i.e., the discharged material does not exclude India ink particles. The species we examined clearly do not discharge any organized structure such as a discobolocyst. However, the fluorescent vesicles are capable of violent discharge, a characteristic of discobolocysts but not of mucocysts, as defined by Hovasse and Mignot (1975).

The fluorescent material, which is distributed uniformly within the fluorescent vesicles, has two broad absorption peaks in the near-UV wavelengths, at about 300 nm and at 350 nm. Light energy absorbed at either of these peaks is reemitted as fluorescence in a single broad peak around 433 nm. The material is water soluble, of relatively small molecular weight, and remarkably heat stable.

The fluorescent vesicles are discharged in response to long wavelength UV light. The two of the four filter combinations that pass such wavelengths are the same two that excite fluorescence and that induce vesicle discharge. Their excitation spectrum also overlaps the absorption spectrum of the fluorescent compound, both as it exists in intact cells and when release into the surrounding solution. Hence we conclude that near-UV light induces discharge.

No natural source of discharge for mucocysts or related structures has previously been reported. The *Ochromonas* fluorescent vesicles, however, discharge in response to direct sunlight. This suggests a possible function in detecting, and perhaps responding to, a source of excessive light or heat detrimental to cell or perhaps population survival. Although few cells growing in soil–water medium contained fluorescent vesicles, all those in *Volvox* medium did, and at least some natural environments undoubtably contain inorganic salt and vitamin levels comparable to this relatively dilute medium.

The discoveries made with *Ochromonas* should be interesting to pursue in several directions. Hibberd (1970) has suggested that discobolocysts in *O. tuberculatus* might actually be elaborated mucocysts. If so, perhaps discobolocysts and other types of extruded organelles might respond to light irradiation. Perhaps it is only chance that the light absorbing material in the *Ochromonas* species examined is also fluorescent.

Much more should also be learned about the nutritional state that encourages production of fluorescent vesicles, and the nature of the fluorescent material. Finally, the extrusion of *Ochromonas* vesicles may prove to be another example of the widespread phenomenon of exocytosis. If so, *Ochromonas* would provide an exceptionally rapid and easily quantitated assay system with which to study the physiology of exocytosis.

References

Aaronson, S., Behrens, U., Orner, R., and Haines, T.H. 1971. Ultrastructure of intracellular and extracellular vesicles, membranes and myelin figures produced by *Ochromonas danica. J. Ultrastruct. Res.* 35: 418–30.

Andersen, R.A. 1982. A light and electron microscopical investigation of *Ochromonas sphaerocystis* Matvienko (Chrysophyceae): the statospore, vegetative cell and its peripheral vesicles. *Phycologia* 21: 390–8.

Coleman, A.W. 1985. Diversity of plastid DNA configuration among classes of eukaryote algae. *J. Phycol.* 21: 1–16.

Hibberd, D.J. 1970. Observations on the cytology and ultrastructure of *Ochromonas tuberculatus* sp. nov. (Chrysophyceae) with special reference to the discobolocysts. *Brit. Phycol. J.* 5: 119–43.

Hovasse, R. and J.-P. Mignot. 1975. Trichocysts and analogous organelles among the Protista. *Annee Biol.* 14: 397–422.
Pringsheim, E.G. 1951. On the nutrition of *Ochromonas. Quart. J. Microsc. Sci.* 93: 71–96.
Starr, R.C. 1978. The culture collection of algae at the University of Texas at Austin. *J. Phycol.* 14(Suppl.): 47–100.

11

Chloroplast DNA in the Chrysophyceae: copy number, packaging, and gene expression in *Olisthodiscus luteus*

ROSE ANN CATTOLICO,[1] JANE ALDRICH,[2] STEVEN BRESSLER, DUNCAN ERSLAND-TALBOT,[3] SCOTT NEWMAN, AND MICHAEL REITH[4]

Department of Botany KB-15
University of Washington
Seattle, Washington 98195, U.S.A.

Introduction

Chloroplasts are semiautonomous organelles that were suggested by Schimper (1883) to have evolved from blue-green algae. The discovery of the prokaryotic alga *Prochloron* by Lewin and Withers (1975) fanned an already active flame raising much speculation and experimentation with respect to (a) the mechanism of chloroplast acquisition by the host cell, (b) the type of organism that served as a donor of these organelles, and (c) the mechanism by which coordinate control at the macromolecular level is maintained between host and organelle genome.

If we look at the plant community, it is evident that all plants have not evolved equally with respect to their photosynthetic organelles. In an attempt to distinguish among the many chloroplast types, investigators have accumulated information (Whatley 1983) on pigment composition, thylakoid stacking, the number of membranes that limit the plastid, and storage product localization. However, one facet of chloroplast biology remains prejudiciously unexplored. Although extensive data (see Bohnert, Crouse, & Schmitt 1982 for review) on chloroplast DNA structure in "green" chlorophyll *a*- and *b*-containing plants (chlorophytes), "nongreen" plant sys-

[1]To whom correspondence should be sent.
[2]Present address: The Standard Oil Co., 3092 Broaday Avenue, Cleveland, Ohio 44115.
[3]Present address: Agrigenetics Research Park, 5649 East Buckeye Road, Madison, Wisconsin 53716.
[4]Present address: Department of Botany, University of Toronto, Toronto, Ontario M5S 1A1, Canada.
These studies were supported by Grant PCM 8022653A01 from the National Science Foundation, and a University of Washington Graduate Research Fund grant to Rose Ann Cattolico. NIH Training Grant awards were held by Jane Aldrich, Duncan Ersland-Talbot, Scott Newman, and Michael Reith.

tems, those plants that contain chlorophyll *a* and *c* (chromophytes) or chlorophyll *a* and phycobilin pigments (rhodophtyes) remain virtually unstudied. This chapter will discuss characterization, DNA packaging, and gene expression in the Chrysophyceae with reference to other plant systems whenever possible.

Information gathered in this area of endeavor will provide further clues to the complex puzzle of organelle relatedness.

Experimental organism

Olisthodiscus luteus, an organism that has received much attention with respect to its taxonomic placement both in this symposium and in the literature (Loeblich and Fine 1977, Gibbs, Chu, & Magnussen 1980), is excellently suited to organelle biogenesis studies. This eukaryotic cell (Figure 11.1A) is naturally wall-less (Leadbeater 1969), being limited by a plasma membrane. The small discoidal chloroplasts are numerous (Cattolico, Boothroyd, & Gibbs 1976) and peripherally located. These chloroplasts are bounded by four membranes and display electron-translucent areas that contain DNA.

The organism can be synchronized (Figure 11.1B) by a 12-hour light–12-hour dark cycle. Both protein and nucleic acid biosynthetic events of the cell cycle have been well defined by studies in our laboratory (Cattolico and Gibbs 1975, Cattolico et al. 1976, Cattolico 1978, McIntosh and Cattolico 1978, Aldrich and Cattolico 1981, Ersland and Cattolico 1981, Ersland, Aldrich, & Cattolico 1981, Aldrich, Gelvin, & Cattolico 1982, Newman and Cattolico 1985, Reith and Cattolico 1985a). Chloroplast and cell division are temporally separate during synchronous cell growth (Figure 11.1C). The chloroplast division complement can be altered by changing the physiology of the cell, and organelle number can be easily monitored by light microscopy.

Chloroplast DNA characterization

How does Olisthodiscus *chloroplast DNA compare to that of other plant systems?*

Olisthodiscus chloroplasts contain circular DNA molecules that have a molecular weight of 97×10^6. This size was determined by three methods: (a) contour length measurements were made (Aldrich and Cattolico 1982) of DNA molecules that were spread for electron microscopic analysis; (b) a sum molecular weight of product fragments was determined (Aldrich et al. 1981) after digestion of chloroplast DNA (ctDNA) with restriction enzymes; and (c) the reassociation rate of sheared ctDNA was compared (Ersland et al. 1981) to that of λ, T_4, *Escherichia coli,* and *Bacillus subtilis* DNA standards. Tables 11.1 and 11.2 present a comparison of information on chloroplast DNA size among a variety of nongreen and green plant

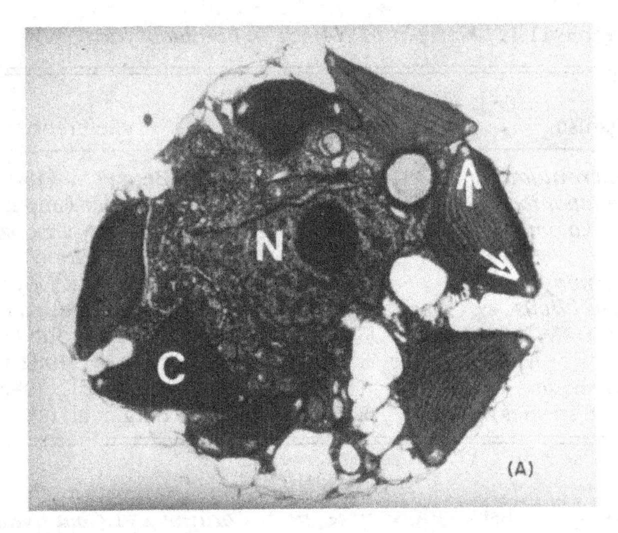

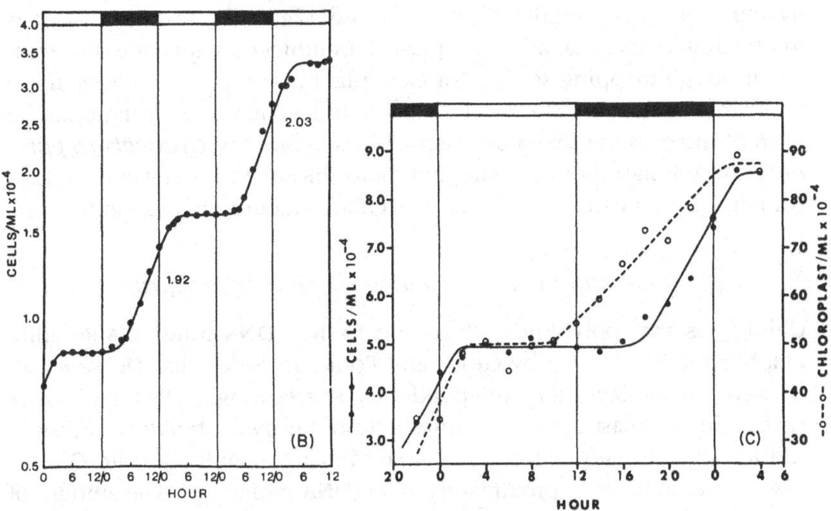

Fig. 11.1. (A) Electron micrograph of *Olisthodiscus luteus.* N, nucleus; C, chloroplast; arrow, nucleoid region of the chloroplast. (B) Synchronous cell division of an *Olisthodiscus luteus* culture when maintained on a 12-hour light–12-hour dark cycle. (C) Temporal separation of chloroplast (o-o-o) and cell division (•-•-•) during the synchronous cell cycle of *Olisthodiscus luteus.*

species. All known chromophytic representatives are included in Table 11.1. These plants have a similar DNA size, falling predominantly in the 70–100 $\times 10^6$ molecular weight range. Chlorophytic plants also have a similar size distribution. Table 11.2 presents a representative, although by no means exhaustive, tabulation of the algal and land plant species examined to date (see Bohnert et al. 1982 for review). Table 11.2 does demonstrate that ctDNA from green algal, liverwort, fern, and angiosperm representatives have an average molecular weight of 90×10^6. Algae provide the

Table 11.1. *Chloroplast DNA size in chromophytic plants*

Organism	Molecular weight of ctDNA size ($\times 10^6$)	Reference
Botrydium granulatum	80	Von Berg et al. (1982)
Chatonella japonica	106	R.A. Cattolico (unpublished)
Dictyota dichotoma	83	K. Kowallik (personal communication)
Odontella sinensis	73	Von Berg et al. (1982)
Olisthodiscus luteus	97	Aldrich & Cattolico (1981)
		Ersland & Cattolico (1981)
		Aldrich et al. (1982)
Tribonema viride	86	Von Berg et al. (1982)
Vaucheria (8 species)	72–77	Von Berg et al. (1982)

end points of the distribution curve, with *Codium* and *Chlamydomonas* having molecular weights of 56×10^6 and 124×10^6, respectively. It is interesting to note that all chloroplast DNA molecules appear to be circular, although mapping studies for each plant type would be necessary to verify this conclusion. Table 11.2 also includes the value for the cyanelle DNA of the presumptive blue-green algal symbiont in *Cyanophora paradoxa.* In summary, both chromophytic and chlorophytic plant chloroplasts contain a genome that is most likely circular in nature and of a limited size.

How many DNA molecules are in a chloroplast?

Chloroplasts are polyploid with respect to their DNA content. Autoradiographic studies in 1973 by Gibbs and Pool concluded that *Ochromonas danica* had at least ten independently segregating ctDNA molecules within this uniplastidic cell. Studies within our own laboratory (Ersland, Aldrich, & Cattolico 1981) demonstrated that the multiplastidic *Olisthodiscus* cell contains approximately 650 ctDNA molecules. The amount of ctDNA in an *Olisthodiscus* cell is constant under a variety of physiological conditions. If *Olisthodiscus* DNA is separated into nuclear and chloroplast components on a CsCl gradient by Model E ultracentrifugation, the proportion of nuclear and chloroplast components is 95 and 5% respectively. Knowing quite precisely the number of chloroplasts per cell (Cattolico 1978) that are present under each growth condition and the size of the chloroplast DNA molecule (Aldrich and Cattolico 1981), we calculate that there is an inverse relationship between the number of chloroplast DNA molecules present and the chloroplast complement. A similar plasticity in chloroplast DNA amount per chloroplast has been reported for pea (Lamppa, Elliot, & Bendich 1980) and spinach (Scott and Possingham

Table 11.2. *Chloroplast DNA size in chlorophytic plants*[a]

Organism	Molecular weight of ctDNA ($\times 10^6$)	Reference
Chlamydomonas reinhardtii	124	Behn & Herrmann (1977)
Codium fragile	56	Hedberg et al. (1981)
Euglena gracilis	89	Manning & Richards (1972)
Sphaerocarpos donellii	77	Herrmann et al. (1980)
Asplenium nidus	89	Herrmann et al. (1980)
Pteris vittata	89	Herrmann et al. (1980)
Narcissus pseudonarcissus	88	Thompson et al. (1981)
Zea mays	76	Kolodner & Tewari (1975)
Antirrhinum majus	92	Herrmann et al. (1975)
Cucumis sativus	102	Palmer (1982)
Petunia hybrida	92	Bovenberg et al. (1981)
Pisum sativum	78	Kolodner & Tewari (1975)
Spinacia oleracea	92	Crouse et al. (1978)
Cyanophora paradoxa (cyanelle DNA)	115	Mucke et al. (1980)

[a] *Acetabularia* appears to be an exception to the observations made on ctDNA from other plant species. The ctDNA of this alga occurs as a linear molecule over 200 μm in length (Green 1978). Unlike the highly polyploid plastid usually found in plants, *Acetabularia* chloroplasts may contain no ctDNA (Woodcock and Bogorad 1970, Coleman 1979) when present in a vegetative cell.

1980). Although the precise number of ctDNA molecules is known for only a few plant species (Bohnert et al. 1982), virtually all plastids are known to be polyploid.

How different are the many ctDNA molecules within the chloroplast?

By restriction analysis (Aldrich et al. 1982) and reassociation kinetic analysis (Ersland et al. 1981), we have shown that only a very small amount of heterogeneity exists within the ctDNA population. Thus every 97×10^6 molecular weight circle of ctDNA is equivalent to one genome. Therefore, each *Olisthodiscus* cell has 650 identical copies of its chloroplast DNA. This is an intriguing observation, for the mechanism that maintains fidelity among the many different molecules through each replication cycle is not yet known. By random mutation, one would expect to find a greater genetic diversity.

To summarize these chloroplast DNA characterization studies, one might ask what constraints have limited the evolution of overt differences in circularity, size, polyploidal state, and molecular diversity in ctDNA among the nongreen and chlorophytic plant groups?

Chloroplast DNA packaging

Is chloroplast DNA in a conformationally constrained state?

It is known that all highly twisted DNA molecules found in nature are negatively supercoiled. Chloroplast DNA is no exception to this rule (Figure 11.2A). Studies by Kolodner and co-workers (Kolodner & Tewari 1975, Kolodner, Tewari & Warner 1976) have shown that the purified, supertwisted ctDNA molecules of spinach, pea, and lettuce are of intermediate superhelical density when compared to other closed circular DNA types, such as SV40, f factor, PM2, or rat mitochrondrial DNA.

An early model of prokaryotic DNA packaging proposed by Worcel and Burgi (1972) suggested that DNA is folded so that loops of DNA emanate from a common core. These loops would be considered to be independent domains. Each loop would limit a rotational event (i.e., the unwinding of a supercoil). If a break should occur in one domain, the superhelical twists of the remaining domains would maintain their torsional tension. The mode of maintaining these domains remains to be determined.

Rosette or floret structures in which DNA loops emanate from a common center have been isolated (Aldrich and Cattolico 1981) from *Olisthodiscus* chloroplasts. These structures (Figure 11.2B) contain an electron-opaque core of unknown composition, although these cores strongly resemble the catenáted knots within DNA preparations of tailless P2 bacteriophage capsids (Liu et al. 1981). The rosette structures of *Olisthodiscus* plastids are of different sizes, may be almost of unit genome length, and are frequently connected by fine skeins of DNA. Rosettes were found only in ctDNA preparations that were not highly purified. This observation suggests that ctDNA unfolding and ctDNA purification are simultaneous events. Moreover, rosette structures were recovered from the most dense portion of the gradient after CsCl–ethidium bromide centrifugation. This observation is consistent with a limited dye binding capacity imposed by a conformationally constrained molecule. Controls have been done to eliminate the possibility that these rosettes are artifacts of preparation (e.g., the ammonium acetate concentration in the hypophase and hyperphase was varied in DNA spreads, cytochrome *c* age and concentration was examined, and ϕX174 DNA that was co-spread with *Olisthodiscus* gave no rosette forms. Similar rosette structures have been seen (Herrmann, Kowallik, & Bohnert, 1974; Yoshida et al. 1978) in spinach, *Beta, Oenothera,* and *Artirrhinum* chloroplast DNA preparations.

As the ctDNA is further unfolded (Figure 11.2C), molecules that display distinct "organizational centers" from which loops of twisted or untwisted

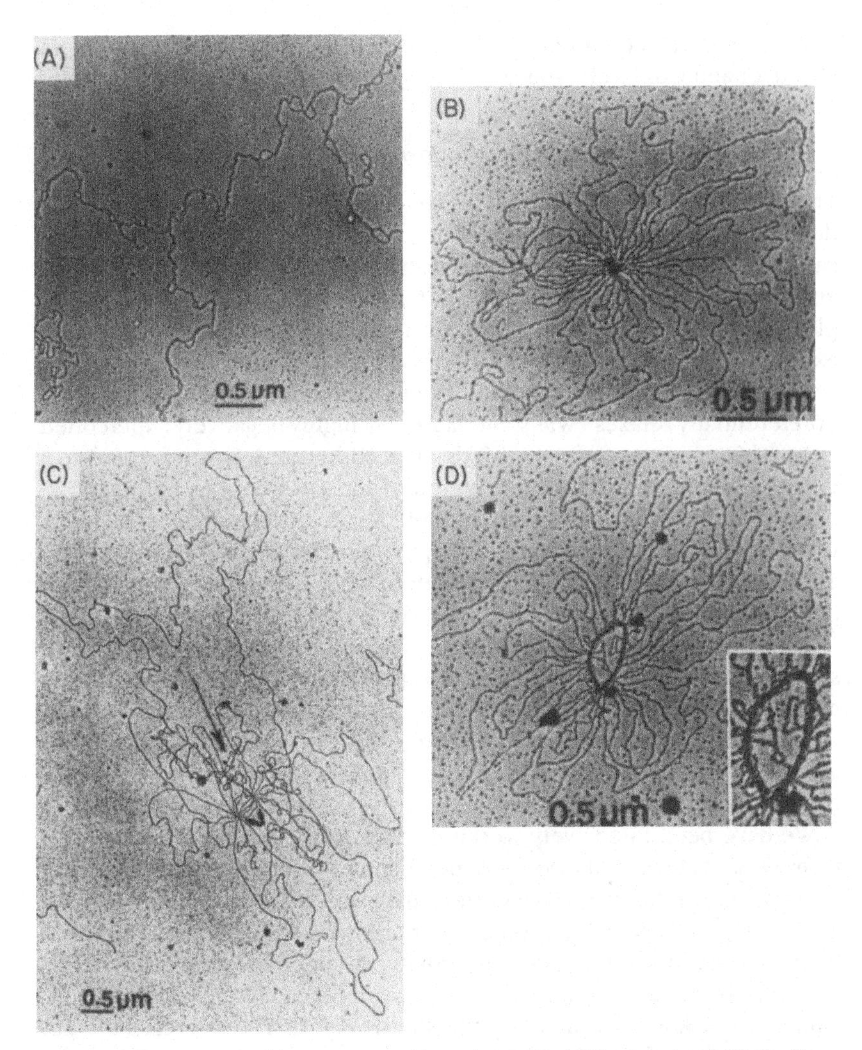

Fig. 11.2. Topoisomers of *Olisthodiscus* chloroplast DNA. (A) Supertwisted *Olisthodiscus* ctDNA molecule. (B) Rosette form with central core structure. (C) Maintenance of "organization" centers (arrow) within a partially unfolded ctDNA molecule. (D) Potentially catenated ctDNA displays a "key ring" structure (insert).

DNA emanate (forming independent domains) are seen to occur. This highly organized internal arrangement was maintained even though a double-stranded break occurred in the ctDNA molecule. These data suggest that the ctDNA molecule is folded so that only a restricted portion (a domain) can unwind within a restricted area.

Finally, on rare occasions, a ctDNA molecule with a "key ring" arrangement was seen (Figure 11.2D). The "key ring" terminology has been applied to similar structures observed in Col E1 plasmid catenated DNA (Kreuzer and Cozzarelli 1980).

What causes ctDNA to form these topoisomers?

The interconversion of DNA from one topological form to another is mediated by a group of enzymes known as topoisomerases. These enzymes can twist, knot, relax, catenate, and decatenate DNA. These are important enzymes, for they have been shown (Gellert 1981, Champoux 1978) to play an active role in DNA replication and recombination and in gene expression.

Two major categories of these enzymes are recognized. Topoisomerase I preferentially relaxes (Wang and Liu 1979) highly negatively supercoiled DNA. This enzyme covalently binds (Tse, Kirkegaard, & Wang 1980) to a DNA molecule, then nicks one strand of the DNA double helix, allows passage of the unaffected strand through the nick, then religates the break in the DNA strand. In prokaryotes (Depew, Liu, & Wang 1978) topoisomerase I binds to the DNA in the 5′ position, whereas in eukaryotes (Champoux, 1977) the enzyme binds in the 3′ position.

Topoisomerase II (Kornberg 1980, Liu, Liu, & Alberts 1980) can both supercoil relaxed DNA molecules or relax DNA that is in the supercoiled state. This is done by making a double-stranded break in DNA, passing an intact duplex through this break, then religating the disrupted duplex. This enzyme also has the capacity to catenate, decatenate, knot, and relax double-stranded DNA molecules.

We have been intensively working on the isolation of topoisomerases from *Olisthodiscus.* This work has been complicated by the fact that only three plants, cauliflower (Fukata and Fukasawa 1982), wheat germ (Dynan et al. 1981), and spinach (Siedlecki, Zimmerman, & Weissbach 1983) have been analyzed for topoisomerase activity. Recent data demonstrates that both topoisomerase I and II enzymes may occur in *Olisthodiscus* cells (S. Bressler and R.A. Cattolico, unpublished). Figure 11.3 represents a gel profile in which fractions from a hydroxyapatite column were tested for topoisomerase activity. It can be seen (Figure 11.3, inset) that a classic ladder of DNA relaxation occurs when the supertwisted plasmid pBR322 is used as the substrate. However, the eluted enzyme does not decatenate *Crithidia fasciculata* kinetoplast DNA (data not shown). These data lead us to conclude that isolated *Olisthodiscus* nuclei give a positive topoisomerase I signal. Most interesting are the preliminary data obtained in Figure 11.4. In this experiment, nucleoids that were isolated from the *Olisthodiscus* chloroplasts were dissociated and without further purification, tested for the presence of topoisomerase activity. pBR322 was totally relaxed (Figure 11.4A), and *Crithidia* DNA was decatenated by the nucleoidal preparation, suggesting that a topoisomerase II may be associated with

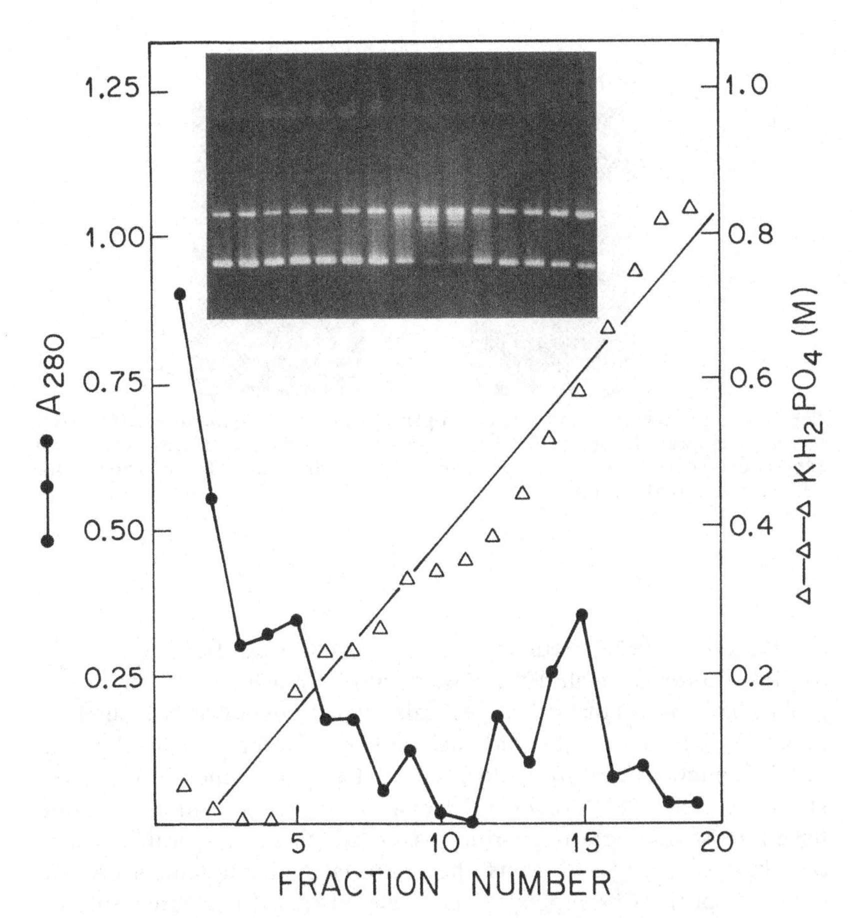

Fig. 11.3. Topoisomerase activity occurs in ***Olisthodiscus luteus*** nuclei. Elution profile of fractions recovered from a hydroxylapitite column. A subset of these fractions was assayed using pBR322 DNA (inset).

chloroplast DNA. Experiments are in progress to verify these observations (i.e., to show that an endonuclease is not responsible for this result) and to isolate and further characterize this enzyme.

How is chloroplast DNA packaged into the plastid?

The data presented above provide some clues to the mechanism of placing ctDNA into a supertwisted state, thus achieving one level of chloroplast DNA packaging, but how these supertwisted molecules are packaged into the plastid remains purely speculative.

Electron microscopic analysis (Bisalputra and Bisalputra 1969, Kowallik

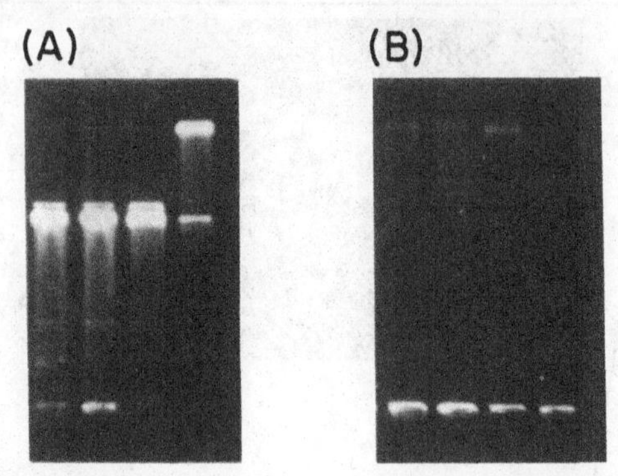

Fig. 11.4. Topoisomerase activity in chloroplasts of *Olisthodiscus luteus.* (A) Unpurified enzyme was assayed using pBR322 DNA and (B) *Crithidia faciculata* DNA. In both experiments control DNA samples were run in the far right gel lane. The direction of DNA migration is toward the top.

and Haberkorn 1971), Feulgen staining (Ris and Plaut 1962), plus recent studies (Kuroiwa et al. 1981, Coleman 1978) using fluorometric dye probes have shown that ctDNA is localized into nucleoidal structures. Although spherical, rod, or ovoid nucleoids of chlorophytic plants are scattered throughout the thylakoids, the nucleoids of some chromophytes (Kuroiwa et al. 1981, Coleman & Heywood 1981, R.A. Cattolico, unpublished) exist as a single macro-ring-shaped structure that is localized on the periphery of the plastid. Thus, the entire ctDNA complement of some members of the Chromophyta has a defined spatial position within the organelle.

The DNA in this nucleoidal ring structure is probably not naked, but is associated with histone-like proteins. Studies in spinach indicate that approximately seven acid-soluble proteins are associated with plastid DNA (Briat et al. 1982), and a similar low number probably occurs in the *Olisthodiscus* plastid (P. Rizzo, personal communication). Thus, we are confronted with the concept of the chloroplast "chromosome" – a supercoiled DNA molecule associated with histonelike proteins that is present in multiple copies. These "chromosomes" are localized within a defined position in the plastid.

Many questions still remain to be answered with respect to the packaging and folding of chloroplast DNA. For example, (1) are topoisomerases prokaryotic- or eukaryotic-like in their attachment to ctDNA and in their activity? (2) Are topoisomerases and histone-like proteins coded for by the chloroplast or the nuclear genome? (3) Are the histone-like proteins of the chloroplast as phylogenetically conserved as nuclear histones?

Chromophyte chloroplast DNA gene expression

What chloroplast proteins are synthesized in vivo?

A chloroplast molecule with a molecular weight of 90 to 100 $\times$ 10^6 has the capacity to code for approximately 100 proteins. Given the hypothesized evolutionary isolation of the chrysophyte chloroplast (Taylor 1979; Gibbs 1981, Whatley and Whatley 1981, Cavalier-Smith 1982), one may ask if the DNA of this organelle codes for a similar or different set of protein products than that of the chlorophytic plant.

The classic method to analyze protein products synthesized by the chloroplast was developed by Ellis and co-workers (Blair and Ellis 1973, Ellis 1977, Ellis, Highfield, & Silverthorne 1977). Isolated plastids are pulsed with a radioactive precursor, then the protein products are analyzed by gel electrophoresis. This method was used in our study of *Olisthodiscus* chloroplast protein synthesis. Cells are collected through a Percoll pad, resuspended in a buffer which contains sorbitol and BSA as osmotica, then subjected to disruption using a French pressure cell. This approach (Reith and Cattolico 1985a) results in a chloroplast preparation that is 60 to 70% refractile by phase microscopy. Using electron microscopic analysis, the plastids are seen to have a good morphological appearance. Thylakoid stacking is maintained and no blebbing of limiting membranes is seen to occur. There are few contaminating mitochondria, and no nuclei in the chloroplast preparation. However, these chloroplasts do not function as expected:

1. Protein synthesis is light independent. When [^{35}S]methionine, [^{3}H]leucine, or [^{3}H]arginine incorporation into protein product is measured (Reith and Cattolico 1984a), chloroplasts utilize these polypeptide precursors with equal efficiency whether plastids are maintained in the light or in the dark (Figure 11.5A). A series of control experiments (Reith and Cattolico 1985a) using chloramphenicol, cycloheximide, Triton X-100, carbonyl cyanide *m*-chlorophenylhydrazone (CCCP), 3-(3,4-dichlorophenyl)-1,1-dimethylurea (DCMU), oligomycin, and [^{3}H]sorbitol demonstrates that protein synthesis occurs on 70 S ribosomes and does not occur on cytoplasmic ribosomes or by contaminating bacteria, and that the energy used in protein synthesis does not originate from photosynthetic electron capture or phosphorylation events in either the chloroplasts or in contaminating mitochondria. Finally, the osmoticum does not serve as an energy source for protein synthesis.

2. Isolated plastids have a reduced photosynthetic capability. Though the isolated *Olisthodiscus* plastids retain Hill activity, they evolve O_2 and fix CO_2 for only a short time period (Figure 11.5B). Mixing experiments using spinach and *Olisthodiscus* chloroplasts demonstrate that this limited photosynthetic capacity is not due to a diffusable inhibitor (Martin and Bassham 1980) such as that seen to occur in pine chloroplast preparations. The

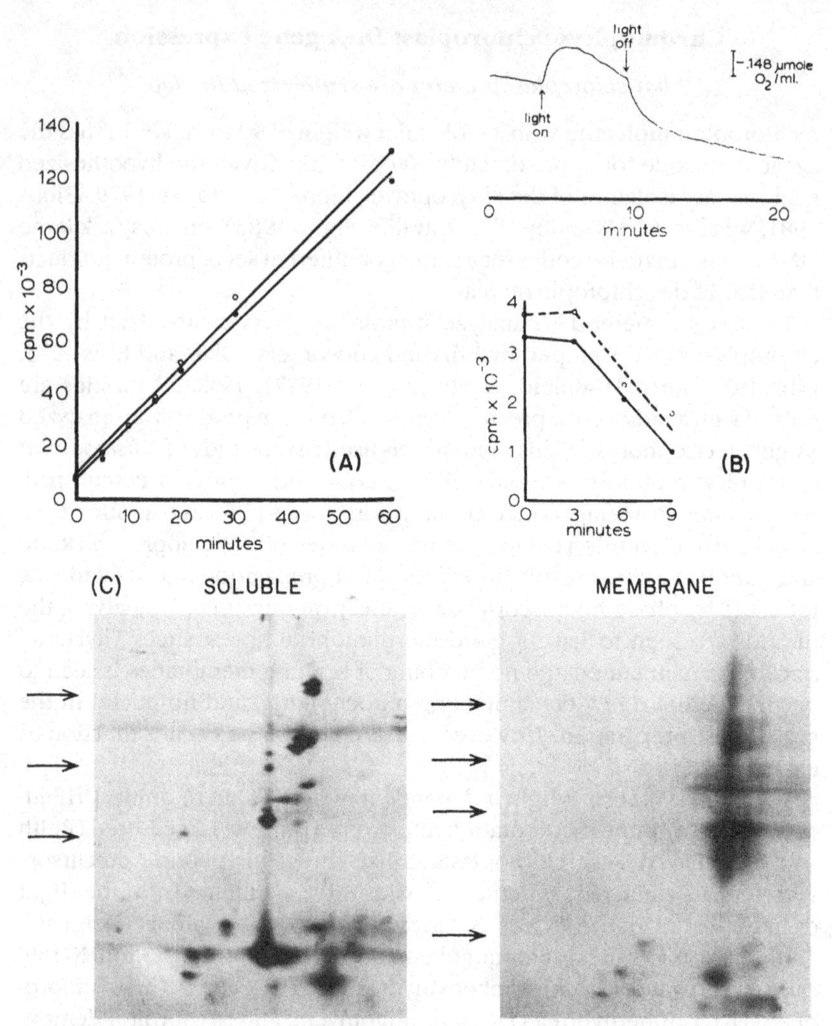

Fig. 11.5. (A) Incorporation of [^{35}S]methionine by isolated chloroplasts in the presence (o-o) and absence (•-•) of light. Each point is a mean of two replicates. Variation between replicates is no greater than 5%. (B) Phytosynthetic acticity of isolated chloroplasts. Oxygen evolution (top graph). Incorporation of $NaH^{14}CO_3$ (bottom graph): (o-o) light incorporation (•-•) dark incorporation. Variation between replicates is no greater than 7%. (C) Autoradiograph of in vivo labeled chloroplast proteins following molecular weight markers which were BSA (68,000), ovalbumin (45,000), carbonic anhydrase (29,000), and myoglobin (17,000).

addition of catalase to the isolated plastids (Robinson, Smith, & Gibbs 1980) has demonstrated that the contamination of the plastid preparation by H_2O_2 from broken chloroplasts also does not account for the reduced functionality of the organelle preparation.

3. All proteins are membrane associated. When the proteins made by isolated chloroplasts were separated into membrane and soluble compo-

nents, then analyzed by two-dimensional electrophoresis, it was noted that all newly synthesized proteins were membrane associated. This is a confusing result, for among the newly made proteins is the large subunit of ribulose-bisphosphate carboxylase, an enzyme that is known to be localized in the stroma. These data lead us to believe that some aspect of normal translation is deficient in our isolated plastids.

Why has such anomalous data with respect to protein synthesis been obtained with the *Olisthodiscus* system when controls with isolated spinach chloroplasts (Reith and Cattolico 1985a) have given expected results? The best answer that can be given at this time is the complexity of the organelle itself. It is well known (Heldt and Sauer 1971) that the two membranes that surround the chlorophytic plastid are differentially premeable to ions and macromolecules. Work with *Ochromonas* (Gibbs 1979) demonstrates that both the chloroplast endoplasmic reticulum (ER) and the chloroplast envelope membranes are involved in a transport function. The *Olisthodiscus* chloroplast is surrounded by a membrane set similar to that observed in the *Ochromonas* system. In chrysophytes, unlike chlorophytic plants, the chloroplast ER may be so intimately integrated with the cytosol ER (S.B. Barlow and R.A. Cattolico, unpublished) that removal of the plastid from its normal cellular position irreversibly damages the organelle membrane systems. Thus, even though refractility may be maintained in the isolated plastid, damage to one or more of the composite membrane sets will result in a loss of functionality. The problem of maintaining chloroplast functionality during isolation would probably be worse in systems such as *Ochromonas* or *Dinobryon* where the nuclear and chloroplast envelopes are contiguous.

What chloroplast proteins are synthesized in vitro?

An alternate approach was taken (Reith and Cattolico 1985b) to determine which proteins were coded by the *Olisthodiscus* chloroplast genome. Whole cells were radioactively labeled in the presence of either chloramphenicol or cycloheximide, the plastids were then isolated and an analysis of the protein products made by standard two-dimensional gel electrophoretic techniques. A spectrum of controls were run to demonstrate that the inhibitors chosen were functioning properly (Galling 1982).

When cells are labeled in the presence of cycloheximide, chloroplast protein production displays an expected biosynthetic response. Cell cycle experiments demonstrate that a majority of the organelle coded proteins are synthesized in the light but not in the dark, and that 35 soluble and 15 membrane-associated polypeptides are made (Figure 11.5C). The proteins that are synthesized range in size from 12,000 to greater than 100,000 in molecular weight, nearly all electrofocus in the range of pH 5 to 8, include no high molecular weight ($>$30,000) basic proteins, and, as expected, a small number display hydrophobic characteristics.

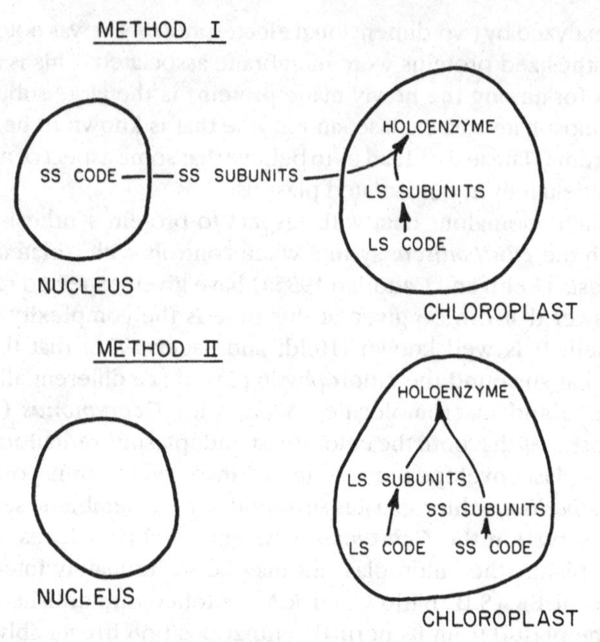

Fig. 11.6. Potential coding and assembly sites of ribulose-1,5-bisphosphate carboxylase subunits in holoenzyme construction.

To date only approximately ten proteins have been (see Bottomley and Bohnert 1982 for review) positively identified as being chloroplast DNA coded. We have begun a comparison analysis of *Olisthodiscus* plastid proteins with those reported for chlorophytic plant systems. Identification (Reith and Cattolico 1985b) of these proteins has lead to the following observations:

1. The photogene or 32,000 molecular weight protein is initially synthesized in *Olisthodiscus* as a slightly larger polypeptide that is then processed into the mature protein. This protein is important in electron transport. It exists in all chlorophytic plants studied to date and is also synthesized as a precursor molecule in these plant types.

2. The apoprotein of the chlorophyll–protein *P*700 complex has been identified. In spinach (Zeilinski and Price 1980), *Acetabularia* (Green 1980), and *Olisthodiscus*, the protein has an approximate molecular weight of 68,000.

3. Most interesting of the proteins synthesized by the chloroplast in *Olisthodiscus* is the photosynthetic enzyme, ribulose-1,5-bisphosphate carboxylase (RuBPcase). In chlorophytic plants, the enzyme is composed of two subunit types. The large subunit polypeptide contains the active site of the holoenzyme. This polypeptide is coded in the chloroplast DNA and is evolutionarily highly conserved (McFadden and Tabita 1974). The small

subunit is coded by the nuclear DNA. Its function is yet unknown but it is suggested that this polypeptide helps maintain the three-dimensional structure of the holoenzyme. The small subunit is extremely variable among all plant species analyzed (Wildman 1982).

Analysis of RuBPcase in *Olisthodiscus* using two different inhibitor sets (cycloheximide and chloramphenicol or anisomycin and erythromycin) presents strong preliminary evidence that *both* proteins of this enzyme are chloroplast coded (Figure 11.6). Recent studies (Reith and Cattolico, 1985c) have localized large and small RuBPcase genes to the chloroplast by identifying subunit protein products following expression of cloned ctDNA fragments in a linked transcription–translation system. RuBPcase large and small subunit genes are each present in two copies located on an inverted repeat that also contains the photogene plus the ribosomal RNA cistrons.

The unusual observation that the small subunit of *Olisthodiscus* may be ctDNA coded has led us (Newman and Cattolico 1985) to analyze more carefully the physical chemistry of the *Olisthodiscus* enzyme. To our knowledge, no studies have ever been published on RuBPcase from a nongreen plant system. As seen in Table 11.3, the large subunit of the *Olisthodiscus* enzyme has a molecular weight of 55,000, whereas the small subunit is 15,000 in size. The holoenzyme is composed of eight large and eight small polypeptide subunits. Holoenzyme size determined by Model E ultracentrifugation and by summation of subunit molecular weight after gel electrophoresis is 580,000 and 560,000, respectively. Similar data have been obtained for *Chaetoceros gracilis* and *Griffithsia pacifica* (Table 11.3).

Is there precedence for the possibility that both subunits of RuBP case may be coded in the chloroplast of Olisthodiscus?

A good deal of inferential data seem to support the possibility that organelles may vary in the coding site of holoenzyme components.

1. In yeast, a subunit of the mitochondrial ATPase complex (polypeptide 9) is coded by the mitochrondrion, while in *Neurospora,* the same polypeptide is coded by the nucleus (Fox 1981). This observation suggests that components of a holoenzyme can vary in coding location among composite organelles within a cell.

2. It has been suggested (Steinmuller, Kaling, & Zetsche 1983) that both subunits of RuBPcase may be coded by the chloroplast in *Cyanidium caldarium* and *Porphyridium aerugineum.* The authors conclusion is based on the fact that the small subunit of RuBPcase is synthesized on poly A− mRNA in these nongreen algal types whereas the small subunit is translated (Bottomley and Bohnert 1982) on poly A+ mRNA in chlorophytic plants.

Table 11.3. *Characterization of ribulose-1,5-bisphosphate carboxylase from nongreen plant systems*

Organism	Enzyme subunits molecular weight		Large/small subunit polypeptide ratio
	Large	Small	
Olisthodiscus luteus	55,000	15,000	8/8
Chaetoceros gracilis	54,000	15,000	8/8
Griffithsia pacifica	57,000	16,000	8/8

Organism	Holoenzyme size molecular weight
Olisthodiscus luteus	560,000 (17.9 S ± 0.21)[a]
	580,000[b]
Chaetoceros gracilis	552,000[b]
Griffithsia pacifica	584,000[b]

[a] Sedimentation analysis.
[b] Polyacrylamide gel analysis.

3. Recently, it has been reported (Heinhorst and Shively 1983) that both subunits of RuBPcase are synthesized on the cyanelle DNA of *Cyanophora paradoxca.*

Obviously, not all the data are in, and this story is certainly in its preliminary stages of development – thus extreme caution must be used. Whether taxonomic groups that have different ancestral chloroplasts will differ qualitatively and quantitatively in ctDNA coding profiles awaits further investigation. These data do demonstrate, however, that we are making progress using molecular analysis as a tool for the analysis of organelle relatedness.

Summary

At the onset of this discussion, we said that chloroplasts were semiautonomous organelles. Given this fact, the nucleus and chloroplast must work as a coevolutionary team (Taylor 1979). It will be most exciting to see the similarities and differences in maintenance strategies that have evolved among various phylogenetic teams, especially among the much neglected nongreen plant types.

References

Aldrich, J. & Cattolico, R.A. 1981. Isolation and characterization of chloroplast DNA from the marine chromophyte, *Olisthodiscus luteus:* Electron microscopic visualization of isomeric molecular forms. *Plant Physiol.* 68: 641–7.

Aldrich, J., Gelvin S., & Cattolico, R.A. 1982. Extranuclear DNA of a marine chromophytic alga: Restriction endonuclease analysis. *Plant Physiol.* 69: 1189–95.
Behn, W. & Herrmann, R.G. 1977. Circular DNA in the beta-satellite DNA of *Chlamydomonas reinhardtii. Mol. Gen. Genet.* 157: 25–30.
Bisalputra, T. & Bisalputra, A.A. 1969. The ultrastructure of chloroplast of a brown alga *Sphacelaria* sp. I. Plastid DNA configuration–the chloroplast genophore. *J. Ultrastruc. Res.* 29: 151–70.
Blair, G.E. & Ellis, R.J. 1973. Protein synthesis in chloroplasts. *Biochim. Biophys. Acta* 319: 223–34.
Bohnert, J.H., Crouse, E.J., & Schmitt, J.M. 1982. Organization and expression of plastid genomes. *In:* Parthier, B. & Boulter, D. [Eds.] *Nucleic Acids and Proteins in Plants,* Vol. II. Springer-Verlag, Berlin, pp. 475–530.
Bottomley, W. & Bohnert, H.J. 1982. The biosynthesis of chloroplast proteins. *In:* Parthier, B. & Boulter, D. [Eds.] *Nucleic Acids and Proteins in Plants,* Vol. II. Springer-Verlag, Berlin, pp. 531–96.
Bovenberg, W.A., Kool, A.J., & Nijkamp, H.J. 1981. Isolation, characterization and restriction endonuclease mapping of the *Petunia hybrida* chloroplast DNA. *Nuc. Acids. Res.* 9: 503–17.
Briat, J., Giqot, C., Lauhere, J., & Machu, R. 1982. Visualization of a spinach plastid transcriptionally active DNA–protein complex in a highly condensed structure. *Plant Physiol.* 69: 1205–11.
Cattolico, R.A. 1978. Variation in plastid number: Effect on chloroplast and nuclear DNA complement in the unicellular alga *Olisthodiscus luteus. Plant Physiol.* 62: 558–62.
Cattolico, R.A. & Gibbs, S.P. 1975. Rapid filter method for the microfluorometric analysis of DNA. *Anal. Biochem.* 69: 572–82.
Cattolico, R.A., Boothroyd, J., & Gibbs, S. 1976. Synchronous growth and plastid replication in the naturally wall-less alga, *Olisthodiscus luteus. Plant Physiol.* 57: 497–503.
Cavalier-Smith, T. 1982. The origins of plastids. *Biol. J. Lin. Soc.* 17: 289–306.
Champoux, J. 1977. Strand breakage by the DNA untwisting enzyme results in covalent attachment of the enzyme to DNA. *Proc. Natl. Acad. Sci. USA* 74: 3800–4.
–1978. Proteins that affect DNA conformation. *Ann. Rev. Biochem.* 47: 449–79.
Coleman, A.W. 1978. Visualization of chloroplast DNA with two fluorochromes. *Exp. Cell Res.* 114: 95–100.
–1979. Use of the fluorochrome 4′6,-diamidino-2-phenylindole in genetic and developmental studies of chloroplast DNA. *J. Cell Biol.* 82: 299–305.
Coleman, A.W. & Heywood, P. 1981. Structure of the chloroplast and its DNA in chloromonadophycean algae. *J. Cell Sci.* 49: 401–9.
Crouse, E.J., Schmitt, J.M., Bohnert, H.J., Driesel, G.K., & Herrmann, R.G. 1978. Intramolecular compositional heterogeneity of *Spinacia* and *Euglena* chloroplast DNAs. *In:* Akayunoglou, G. & Argyroudi-Akoyunoglou, J.H. [Eds.] *Chloroplast Development.* Elsevier, Amsterdam, pp. 1775–86.
Depew, R.E., Liu, L., & Wang, J.C. 1978. Interactions between DNA and *Escherichia coli* protein. *J. Biol. Chem.* 253: 511–18.
Dynan, W., Jendrisak, J., Haeger, P., & Burgess, R. 1981. Purification and characterization of wheat germ DNA topoisomerase I (Nicking-closing enzyme). *J. Biol. Chem.* 256: 5860–5.
Ellis, R.J. 1977. Protein synthesis by isolated chloroplasts. *Biochim. Biophys. Acta* 463: 185–215.
Ellis, R.J., Highfield, P.E., & Silverthorne, J. 1977. The synthesis of chloroplast

proteins by subcellular systems. *In:* Hall, D.O., Coombs, J., & Goodwin, T.W. [Eds.] *Proceedings of the Fourth International Congress on Photosynthesis.* The Biochimical Society, London, pp. 497–506.

Ersland, D. & Cattolico, R.A. 1981. Nuclear deoxyribonucleic acid characterization of the marine chromophyte *Olisthodiscus luteus. Biochim.* 20: 6886–92.

Ersland, D., Aldrich, J., & Cattolico, R.A. 1981. Kinetic complexity, homogeneity and copy number of chloroplast DNA from the marine alga *Olisthodiscus luteus. Plant Physiol.* 68: 1468–73.

Fox, T. 1981. More surprises from mitochondria. *Nature (London)* 292: 109–10.

Fukata, H. & Fukasawa, H. 1982. Isolation and partial purification of two distinct DNA topoisomerases from cauliflower inflorescence. *J. Biochim.* 91: 1337–42.

Galling, G. 1982. Use (and misuse) of inhibitors in gene expression. *In:* Parthier, P., & Boulter, D. [Eds.] *Nucleic Acids and Proteins in Plants,* Vol. II. Springer-Verlag, Berlin, pp. 663–77.

Gellert, M. 1981. DNA topoisomerases. *Ann. Rev. Biochim.* 50: 879–910.

Gibbs, S.P. 1979. Route of entry of cytoplasmically synthesized proteins into chloroplasts of algae possessing chloroplast ER. *J. Cell Sci.* 35: 253–66.

–1981. The chloroplasts of some algal groups may have evolved from endosymbiotic eukaryotic algae. *Ann. N.Y. Acad. Sci.* 361: 193–207.

Gibbs, S.P. & Poole, R.J. 1973. Autoradiographic evidence for many segregating DNA molecules in the chloroplast of *Ochromonas danica. J. Cell Biol.* 59: 318–28.

Gibbs, S.P., Chu, L.L., & Magnussen, C. 1980. Evidence that *Olisthodiscus luteus* is a member of the Chrysophyceae. *Phycologia.* 19: 173–7.

Green, B. & Padmanabhan, M. 1978. The kinetic complexity of *Acetabularia* chloroplast DNA. *Biochim. Biophys. Acta* 521: 67–73.

Green, B.R. 1980. Protein synthesis by isolated *Acetabularia* chloroplasts. *In vitro* synthesis of the apoprotein of the *P*700-chlorophyll alpha protein complex (CPI). *Biochim. Biophys. Acta.* 609: 107–20.

Hedberg, M.F., Huang, Y-S, & Hommersand, M.H. 1981. Size of the chloroplast genome in *Codium fragile. Science* 213: 445–7.

Heinhorst, S. & Shively, J.M. 1983. Encoding of both subunits of ribulose-1,5-bisphosphate carboxylase by organelle genome of *Cyanophora paradoxa. Nature (London)* 304: 373–4.

Heldt, H.W. & Sauer, F. 1971. The inner membrane of the chloroplast envelope as the site of specific metabolite transport. *Biochim. Biophys. Acta* 234: 83–91.

Herrmann, R.G., Kowallik, K.V., & Bohnert, H.J. 1974. Structural and functional aspects of the plastome I. The organization of the plastome. *Portug. Acta. Biol.* 14: 91–110.

Herrmann, R. Bohnert, H., Kowallik, K., & Schmitt, J. 1975. Size, conformation and purity of chloroplast DNA of some higher plants. *Biochim. Biophys. Acta* 378: 305–17.

Herrmann, R.G., Palta, H.K., & Kowallik, K.V. 1980. Chloroplast DNA from three archegoniates. *Planta* 148: 319–22.

Kolodner, R. & Tewari, K.K. 1975. The molecular size and conformation of the chloroplast DNA from higher plants. *Biochim. Biophys. Acta.* 402: 372–90.

Kolodner, R., Tewari, K.K., & Warner, R.C. 1976. Physical studies on the size and structure of the covalently closed circular chloroplast DNA from higher plants. *Biochim. Biophys. Acta* 447: 144–55.

Kornberg, A. 1980. *DNA Replication.* Freeman, San Francisco, pp. 307–15.

Kowallik, K. & Haberkorn, G. 1971. The DNA structures of the chloroplast of *Prorocentrum micans* (Dinophyceae). *Arch. Mikrobiol.* 80: 252–61.
Kreuzer, K. & Cozzarelli, N. 1980. Formation and resolution of DNA catenanes by DNA gyrase. *Cell* 20: 245–54.
Kuroiwa, T., Suzuki, T., Ogawa, K., & Kawano, S. 1981. The chloroplast nucleus: Distribution, number, size, and shape, and a model for the multiplication of the chloroplast genome during chloroplast development. *Plant and Cell Physiol.* 22: 381–96.
Lamppa, G.K., Elliot, L.V., & Bendich, A.J. 1980. Changes in chloroplast number during pea leaf development. An analysis of a protoplast population. *Planta* 148: 437–43.
Leadbeater, B.S.C. 1969. A fine structural study of *Olisthodiscus luteus* Carter. *Br. Phycol. J.* 4: 3–17.
Lewin, R.A. & Withers, N.W. 1975. Extraordinary pigment complement of a prokaryotic alga. *Nature (London)* 256: 735–7.
Liu, L., Liu, C., & Alberts, B. 1980. Type II topoisomerases: Enzymes that can unknot a topologically knotted DNA molecule via a reversible double-strand break. *Cell* 9: 697–707.
Liu, L., Perkocha, L., Calendar, R., & Wang, J. 1981. Knotted DNA from bacteriophage capsids. *Proc. Natl. Acad. Sci. USA* 78: 5498–502.
Loeblich, A.R. & Fine, K.E. 1977. Marine chloromonads: More widely distributed in neritic environments than previously thought. *Proc. Biol. Soc. Wash.* 90: 388–99.
Manning, J.E. & Richards, O.C. 1972. Isolation and molecular weight of circular chloroplast DNA from *Euglena gracilis. Biochim. Biophys. Acta* 259: 285–96.
Martin, D. & Bassham, J.A. 1980. Photosynthetic CO_2 fixation in spinach chloroplasts inhibited by pine chloroplasts or extracts of pine chloroplasts. *Physiol. Plant.* 48: 213–20.
McFadden, B.A. & Tabita, F.R. 1974. D-ribulose-1,5-diphosphate carboxylase and the evolution of autotrophy. *Bio. Systems* 6: 93–112.
McIntosh, L. & Cattolico, R.A. 1978. Preservation of algal and higher plant ribosomal RNA integrity during extraction and electrophoretic quantitation. *Anal. Biochim.* 91: 600–12.
Mucke, H., Loffelhardt, W., & Bohnert, H.J. 1980. Partial characterization of the genome of the "endosymbiotic" cyanelles from *Cyanophora paradoxa. FEBS Letters* 111: 347–52.
Newman, S. & Cattolico, R.A. 1985. Is RuBPcase in *Olisthodiscus luteus* an evolutionary divergent enzyme? *Plant Physiol.* (in review).
Palmer, J.D. 1982. Physical and gene mapping of chloroplast DNA from *Atriplex triangularis* and *Cucumis sativa. Nuc. Acids. Res.* 10: 1593–605.
Reith, M.E. & Cattolico, R.A. 1985a. *In vitro* polypeptide synthesis by chloroplasts of the chromophytic alga *Olisthodiscus luteus. Biochimistry.* 24: 2550–6.
–1985b. *In vivo* polypeptide synthesis by chloroplasts of the chromophytic alga *Olisthodiscus luteus. Biochimistry.* 24: 2556–61.
–1985c. The inverted repeat of *Olisthodiscus luteus* ctDNA contains the genes for both subunits of RuBPcase. *Proc. Nat. Acad. Sci.* (in press).
Ris, H. & Plaut, W. 1962. Ultrastructure of DNA-containing areas in the chloroplast of *Chylamydomonas. J. Cell Biol.* 13: 383–91.
Robinson, J.M., Smith, M.G., & Gibbs, M. 1980. Influence of hydrogen peroxide upon carbon dioxide photoassimiliation in the spinach chloroplast. *Plant Physiol.* 65: 755–9.

Schimper, A.F.W. 1883. Über die Entwickelung der Chlorophyllkörner und Farbkörper. *Bot.* 41: 105–14.
Scott, N.S. & Possingham, J.V. 1980. Chloroplast DNA in expanding spinach leaves. *J. Expt. Bot.* 31: 1081–92.
Siedlecki, J., Zimmerman, W., & Weissbach, A. 1983. Characterization of a prokaryotic topoisomerase I activity in chloroplast extracts from spinach. *Nucleic Acids Res.* 11: 1523–36.
Steinmuller, K., Kaling, M., & Zetsche, K. 1983. In vitro synthesis of phycobilins and ribulose-1,5-bisphosphate carboxylase by non-polyadenylated-RNA of *Cyanidium caldarium* and *Porphyridium aerugineum. Planta* 159: 308–13.
Taylor, F.J.R. 1979. Symbionticism revisited: A discussion of the evolutionary impact of intracellular symbioses. *Proc. Roy. Soc. Lond. B* 204: 267–86.
Thompson, J.A., Hansmann, P., Knoth, R., Link, G., & Falk, H. 1981. Electron microscopical localization of the 23 and 16 S rRNA genes with an inverted repeat for two chromoplast DNAs. *Curr. Genet.* 4: 25–8.
Tse, Y., Kirkegaard, K., & Wang, J. 1980. Covalent bonds between protein and DNA. *J. Biol. Chem.* 12: 5560–5.
Von Berg, K.H., Schmidt, M., Von Berg, G., Sturm, K., Hennig, A., & Kowallik, K.V. 1982. The chloroplast genome (plastome) from algae of different phylogenetic relationships. *British Phycol. J.* 77: 235.
Wang, J., & Liu, L. 1979. DNA topoisomerase: Enzyme that catalyzes the concerted breaking and rejoining of DNA backbone bonds. *In:* Taylor, J.H. [Ed.] *Molecular Genetics.* Academic Press, New York, pp. 65–88.
Whatley, J.M. 1983. Plastids – past, present and future. *Intern. Rev. Cytol.* 14: 329–73.
Whatley, J.M. & Whatley, F.R. 1981. Chloroplast evolution. *New Phytol.* 87: 233–47.
Wildman, S.G. 1982. Further aspects of fraction 1 protein evolution. *In:* Schiff, J. [Ed.] *The Origin of Chloroplasts.* Elsevier North Holland, New York, pp. 229–42.
Woodcock, C.L.F. & Bogorad, L. 1970. Evidence for variation in the quantity of DNA among plastids of *Acetabularia. J. Cell Biol.* 44: 361–75.
Worcel, A. & Burgi, E. 1972. On the structure of the folded chromosome of *Escherichia coli. J. Mol. Biol.* 71: 127–47.
Yoshida, Y., Laulhére, J.P., Rozier, C., & Mache, R. 1978. Visualization of folded chloroplast DNA from spinach. *Biol. Cell.* 32: 187–90.
Zielinski, R.E. & Price, C.A. 1980. Synthesis of thylakoid membrane proteins by chloroplasts isolated from spinach: Cytochrome *b*559 and *P*700 – chlorophyll a – protein. *J. Cell Biol.* 85: 435–45.

PART IV

Population dynamics

12

Phytoplankton dynamics in a chrysophycean lake

PETER A. SIVER
Department of Biology
Western Connecticut State University
Danbury, Connecticut, 06810 U.S.A.

AND

JAN S. CHOCK
East Woods School
31 Yellow Cote Road
Oyster Bay, New York, 11771, U.S.A.

Introduction

The Chrysophyceae are often found to be an important component of the phytoplankton in temperate, oligotrophic lakes (Hutchinson 1967). Taxa from genera such as *Dinobryon, Synura, Mallomonas,* and *Uroglena* are often the most important part of the flora (Hutchinson 1967, Kling & Holmgren 1972). Kristiansen and Takahashi (1983) pointed out that even though the Chrysophyceae are mentioned in many studies, ecological investigations are few. In addition, few studies have included vertical distributions on a seasonal basis. The Chrysophyceae have often been described as a cold-water group. However, many species have recently been identified in warm climates (Takahashi & Hayakawa 1979, Kristiansen & Takahashi 1983). Clearly, much more work needs to be done concerning the distribution and ecology of the Chrysophyceae.

The purpose of this study was to investigate the vertical and seasonal dynamics of the phytoplankton in a small oligotrophic lake, Lake Lacawac.

Lake Lacawac is a 13,000-year-old, 21 hectare ice scour lake located in the Pocono Mountains of northeast Pennsylvania (town of Lake Aerial). The lake, with a maximum depth of 13 meters, is situated in red Catskill sandstone, and the basin is lined with red clay. Lake Lacawac is part of a sanctuary owned by the Nature Conservatory and sits in an undisturbed watershed.

Materials and methods

Samples were taken on a monthly or bimonthly basis from November 1980 to November 1981. Sampling continued on a monthly basis through November 1982 but will only be briefly discussed in this paper. Water samples for phytoplankton, nutrient analysis, dissolved oxygen, pH, and optical density at 350 nm (a relative estimate of organic matter) were taken from the center of the lake with a van Dorn bottle at 0.5 meter or 1 meter intervals. In the field, underwater light was measured with a Licor LI-185 B quantum meter, temperature and conductivity with a YSI 33 SCT meter, and the Secchi disk depth was recorded.

In the laboratory pH, optical density at 350 nm, and dissolved oxygen were analyzed immediately. Samples for phytoplankton analysis were preserved with Lugol's solution. Live phytoplankton samples from several depths were observed to aid future identification. Dissolved oxygen was measured using the modified Winkler technique, while samples (unfiltered) for total phosphorus were acid persulfate digested and measured by the stannous chloride method (American Public Health Association 1980). Filtered samples were analyzed for nitrate by the cadmium reduction procedure (American Public Health Association 1980). Samples were frozen if not immediately processed.

Phytoplankton was prepared for quantification by centrifuging at 2000 rpm for 10 minutes, decanting the supernatant with a vacuum pump and resuspending the cells in a known volume. Four samples were enumerated from each site using a Palmer–Maloney chamber.

Samples with scaled chrysophytes were prepared for scanning (SEM) or transmission electron microscopic (TEM) observation by washing with distilled water at least five times. A concentrated drop was air-dried onto Formvar-coated copper grids and observed with a Hitachi HU-11A TEM or onto aluminum stubs, coated with gold and examined with a Coates and Welter HP550 SEM.

Results

Physicochemical data

Lake Lacawac, a dimictic lake, had a temperature range of 0° – 26°C during the study period. Fall (October to December) and spring (March to April) overturns were spaced between winter (January) and summer thermal stratification. Ice covered the lake from mid-December through most of March, reaching a maximum thickness of 40 cm in January to February while 30 cm of snow covered the lake during January. In February a warm spell melted the snow and warmed the surface waters to 3°C, creating density currents that mixed the water column. During summer stratification, which began in May and extended to mid-September, the hypolimnion reached a maximum of 9°C.

The extinction coefficient for light transmission, n, ranged from a low of 0.49 m^{-1} in December to a high of 1.01 m^{-1} in January; n was between 0.67 and 0.86 m^{-1} from February through October. The secchi disk depth ranged from 3.2 m in March to 6.8 m in September and was inversely proportioned to the phytoplankton concentration (Figure 12.1).

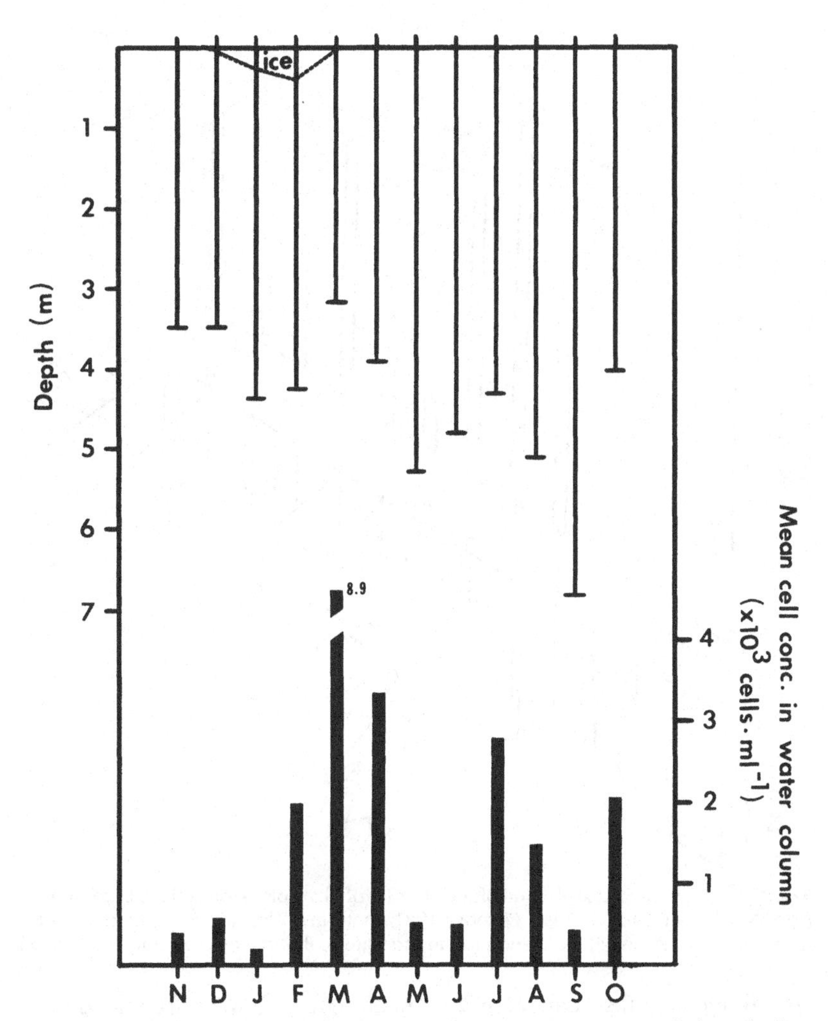

Fig. 12.1. The secchi disk depth (top) versus the mean concentration of phytoplankton cells (lower) in the water column from November 1980 through October 1981 for Lake Lacawac.

The water column was usually saturated with dissolved oxygen, except during summer stratification when the lower portion of the hypolimnion was anoxic. Nitrate and total phosphorus levels, analyzed in November, January, March, and June, were less than 0.01 mg NO_3 N liter^{-1} and 10 μg PO_4 P liter^{-1} (detection limits with a 1-cm cuvette), respectively, at all depths. Conductivity ranged from a minimum of 20 μS_{20} in January surface waters to a maximum of 44 μS_{20} in the hypolimnion during September. The mean conductivity for the water column was lowest in the winter months and gradually increased to a high in August during summer stratification. In general, bottom waters had greater conductivity then surface waters.

Lake Lacawac water was stained slightly yellow/brown in color. The

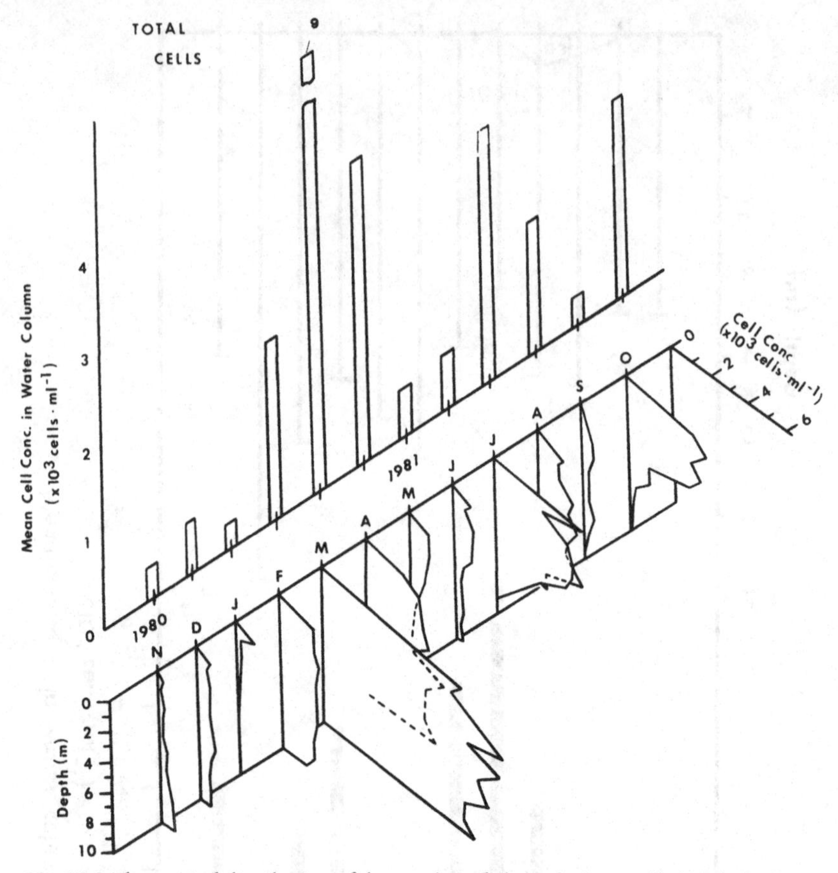

Fig. 12.2. The vertical distribution of the number of phytoplankton cells in Lake Lacawac from November 1980 through October 1981 (bottom). The top graph represents the mean phytoplankton cell concentration in the water column over the same time period.

optical density (absorbance) at 350 nm of unfiltered water from all depths was between 0.02 and 0.05. However, from June through August, when suspended and dissolved organic matter accumulated in the hypolimnion, the optical density increased from 0.02 at the surface to 1.1–1.3 at 10 meters in depth.

Although pH ranged from 5.25 to 6.6 during the study, it was usually between 5.4 and 6.2. The pH decreased with depth during summer stratification.

Distribution and composition of phytoplankton

Phytoplankton concentrations were low in Lake Lacawac, ranging from 2.3×10^2 to 9.0×10^3 cells ml^{-1} (Figure 12.2). Cell concentrations were low (less than 10^3 cells ml^{-1}) during the winter months, began increasing

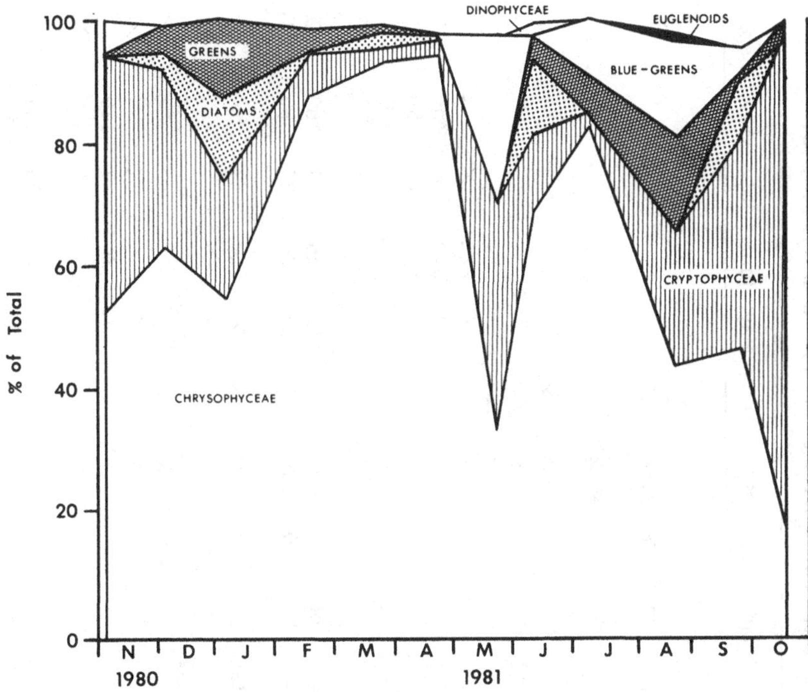

Fig. 12.3. The percentage of the total number of phytoplankton cells in each algal class from November 1980 through October 1981.

in February, and reached a seasonal high with ice out in March at 4° to 5°C. The March peak declined as water temperatures increased, reaching a second minimum (less than 10^3 cells ml^{-1}) during May and June. A smaller peak occurred during July (26°C), but concentrations declined during August to a third minimum (less than 10^3 cells ml^{-1}) in September. Concentrations increased again during fall overturn.

Phytoplankton were evenly distributed in the water column during the fall (November to December) and spring (March to April) overturns with no more than 12% of the total occurring at any one depth (Figure 12.2). In February, despite an ice cover, density currents mixed the water column and caused an even distribution of phytoplankton. During January, when ice and snow levels were greatest, over 66% of the phytoplankton were within the top 2 meters; 90% of the phytoplankton above 2 meters had flagella, while 54% of the community below 2 meters [*Asterionella formosa* and *Schroederia setigera*] lacked flagella. During summer stratification, the majority of phytoplankton were found in the epilimnion, however, by late August several population strata had developed in the hypolimnion.

The Chrysophyceae were the most important algal group, accounting for over 78% of all cells (Figure 12.3). In terms of importance on a percentage monthly basis the Chrysophyceae accounted for over 62% of the total phy-

Table 12.1. *Phytoplankton species found in Lake Lacawac between November 1980 and October 1981*[a]

Taxon	Nov.	Dec.	Jan.	Feb.	Mar.	Apr.	May	June	July	Aug.	Sept.	Oct.
Chrysophyceae												
Mallomonas acaroides Perty	130	×						63			16	21
Dinobryon cylindricum Imhof	×	126	37	78	2154	2018						
Synura petersenii Korshikov v. *glabra* (Korsh.) H. Pestalozzi	7	278	90	1620	4932							
Synura spinosa Korshikov		×	×	×	×							
Kephyriopsis ovum Pascher et Ruttner					1166	×						
Uroglena sp.					179							
Uroglenopsis sp.						1125						63
Kephyrion sp.							×	22	×	×		
Stichogloea doederleinii (Schm.) Wille							173	309	2264	336	91	
Pseudokephyrion sp.									27	96		42
Mallomonos akrokomos Ruttner in Pascher											×	188
Synura uvella Ehrenb.											38	63
Cryptophyceae												
Rhodomonas lacustris Pascher et Ruttner	112	145	28	98	90	33	112	40	27	160	82	1625
Cryptomonas erosa Ehrenb. and *C. ovata* Ehrenb.	40	38	12	39	90	33	66	29	27	80	41	42
Chlorophyceae												
Schroederia setigera (Schroed.) Lemm.	×	38	23	20	×	×	×	×	×	×	×	42
Elakatothrix gelatinosa Wille		×	×				×	×				×
Quadrigula chodatii (Tanner-Fullman) Smith	×	×	×									
Chlamydomonas spp.	×	×	×	×	×	×				×	×	
Arthrodesmus incus (Breb.) Hassall	×	×	×				×	×	×	×	×	×
Oocystis lacustris Chodat							×	23	27	146		
Pediastrum tetras (Ehrenb.) Rolfs.								×	×	×	×	
Crucigenia tetrapedia (Kirchr.) W. and G. West										×	×	×
Tetraedron caudatum (Corda) Hansgirg									55	114		

Golenkinia sp.	×							×				×
Cyanophyceae												
Nostoc planktonicum Drouet	18	×				×	142					×
Merismopedia tenuissima Lemm.									304	195		
Microcystis incerta Lemm.										65	13	84
Bacillariophyceae												
Asterionella formosa Hassall		×	35	20	359	66	×	74				
Tabellaria fenestrata (Lyng.) Kützing	×	2									13	
Euglenophyceae												
Trachelomonas volvocina Ehrenb.	×		×			×			×	16	×	×
Euglena gracilis Kelbs										16		
Other												
Chloromonad cf.					×		33			32	×	×

[a] Only those taxa present on at least three different dates or that form greater than 1% of the community on one date are listed. The numbers refer to the actual concentrations (cells per milliliter). An × represents less than 1% of the population.

toplankton. The Cryptophyceae was the second most important group accounting for 24% and 14% of the phytoplankton on a monthly and total cell basis, respectively. The importance of the Chrysophyceae, as a percentage of the total number of phytoplankton, fell below 45% during only May and October (Figure 12.3) when the Cryptophyceae was most abundant. The Cyanophyceae, Chlorophyceae, and Bacillariophyceae were of much less importance, accounting for 6% of the total number of cells during the study.

Although over 100 species of phytoplankton were found during the study, only 31 were present on at least 25% of the collection dates or formed greater than 1% of the community on a single date (Table 12.1). Eleven and ten of these taxa belonged to the Chrysophyceae and Chlorophyceae, respectively. Even though the Cryptophyceae were the second most important class, in terms of cell numbers, it was composed of only two (or three) taxa (Table 12.1). *Synura petersenii* v. *glabra* (Figure 12.11), *Dinobryon cylindricum, Stichogloea doederleinii,* and *Rhodomonas lacustris* accounted for over 70% of all phytoplankton cells. Although *S. petersenii* v. *glabra* was three times more abundant than *R. lacustris* on a total cell basis, they both accounted for 19% on a monthly basis.

Seasonal succession of the major taxa

Synura petersenii v. *glabra, Dinobryon cylindricum, Kephyriopsis ovum, Uroglenopsis* sp., *Uroglena* sp., and *Mallomonas acaroides* formed a cold-water chrysophyte group that dominated the phytoplankton from November through April (Figures 12.4–12.8). Of this group all taxa, except *Mallomonas acaroides,* were always found at temperatures less than or equal to 12°C. *Mallomonas acaroides* (Figures 12.4, 12.8 and 12.12) had a temperature range from 7° to 20°C. On four occasions, October to November 1980, June 1981, September to October 1981, and May 1982, when the temperature cooled or warmed to between 15° to 20°C, *M. acaroides* grew and formed an important component of the phytoplankton community (Figure 12.8). During November 1980 and June 1981, *M. acaroides* accounted for 36% (1.3×10^2 cells ml^{-1}) and 11% (6.4×10 cells ml^{-1}) of the phytoplankton communities, respectively. *Synura petersenii* v. *glabra* and *Dinobryon cylindricum* were the most abundant and important taxa over the 4-month period from December 1980 through March 1981 (Figures 12.5 and 12.6). Both taxa had similar distributional patterns, first appearing in November and reaching maxima at 4.5°C during late March, with ice out. *Synura petersenii* v. *glabra* was equally distributed in the water column except in January when over 80% were above 1 meter. The organism reached a maximum cell concentration of 4.9×10^3 cells ml^{-1} in

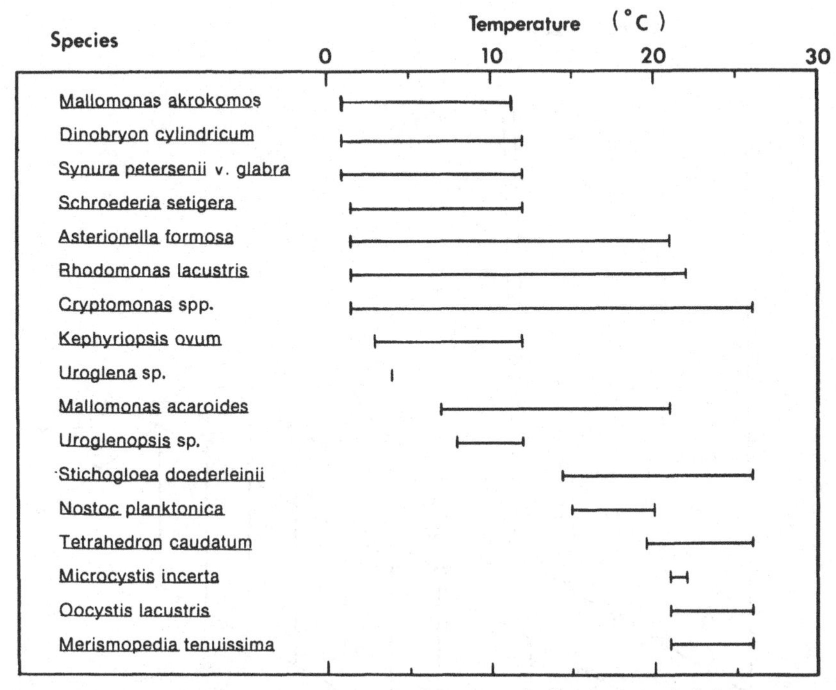

Fig. 12.4. The temperature range for each of 17 phytoplankton species in Lake Lacawac. The ranges represent those in which the organism was found.

March and was gone from the water column by April (Figure 12.5). Although *D. cylindricum* also had a peak at 1 meter in January, it was more sporadically distributed in the water column during the other months. *Dinobryon cylindricum* reached a maximum concentration of 2.1×10^3 cells ml^{-1} in March and began lowering in April when most of the population had settled to the bottom (Figure 12.6).

Mallomonas akrokomos (Figure 12.13), rare during the 1980–1981 winter, appeared in the fall of 1981 when the water temperature lowered to 15°C, and became a codominant with *S. petersenii* v. *glabra* and *D. cylindricum* during the 1981–1982 winter. *Asterionella formosa* and *Kephyriopsis ovum* (Figure 12.7) were also important constituents of the winter flora, reaching concentration maxima of 3.2×10^2 and 1.2×10^3 cells ml^{-1} in March. Both taxa were gone by the end of April; however, *A. formosa* reformed a small surface bloom in June. Although of much less importance, a cold-water group of Chlorophyceae consisting of *Schroederia setigera, Elakatothrix gelatinosa, Quadrigula chodatii,* and *Arthrodesmus incus* had population maxima at temperatures less than 12°C. These cold-water greens were restricted to deep, cooler waters during warmer months.

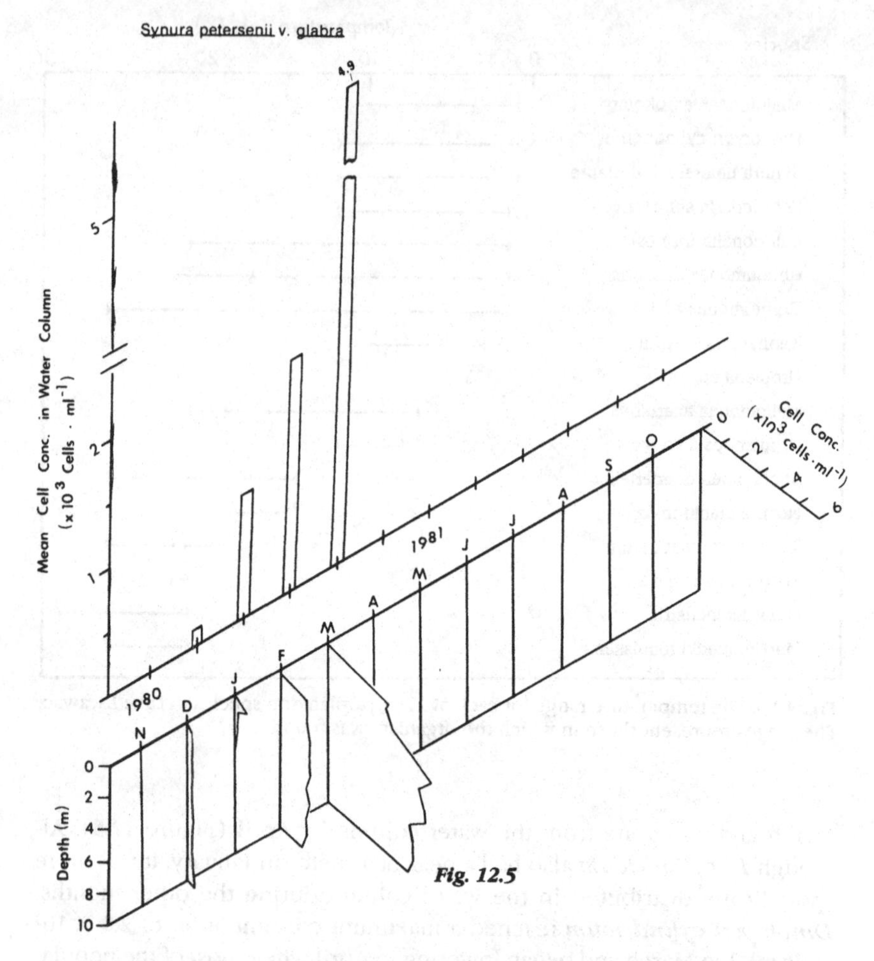

Fig. 12.5–12.10. The vertical distribution of the number of phytoplankton cells for a given species in Lake Lacawac from November 1980 through October 1981 (bottom). The top graph represents the mean cell concentration of each taxon in the water column over the same time period. ***Fig. 12.5.*** ***Synura petersenii*** v. ***glabra. Fig. 12.6. Dinobryon cylindricum. Fig. 12.7. Kephyriopsis ovum. Fig. 12.8. Mallomonas acaroides. Fig. 12.9. Stichogloea doederleinii. Fig. 12.10. Rhodomonas lacustris.***

The most abundant and dominant phytoplankton from May through September 1981, was *Stichogloea doederleinii* (Figure 12.9). This taxon began growth in May at 15°C and formed its largest concentration of 4.2×10^3 cells ml^{-1} in surface waters during July when the temperature ranged from 23° to 26°C. A warm-water Chlorophyceae group, consisting of *Oocystis lacustris, Pediastrum tetras, Crucigenia tetrapedia*, and *Tetraedron caudatum* were mainly found in surface waters between 20° to 26°C (Table 12.1). The major Cyanophyceae, *Nostoc planktonicum, Merismopedia*

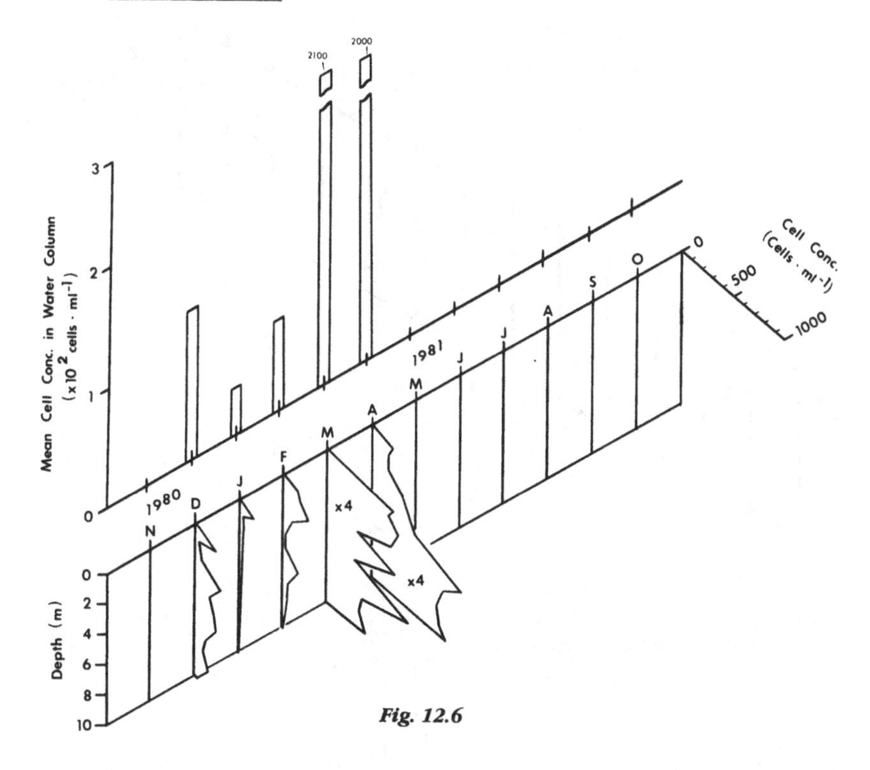

Fig. 12.6

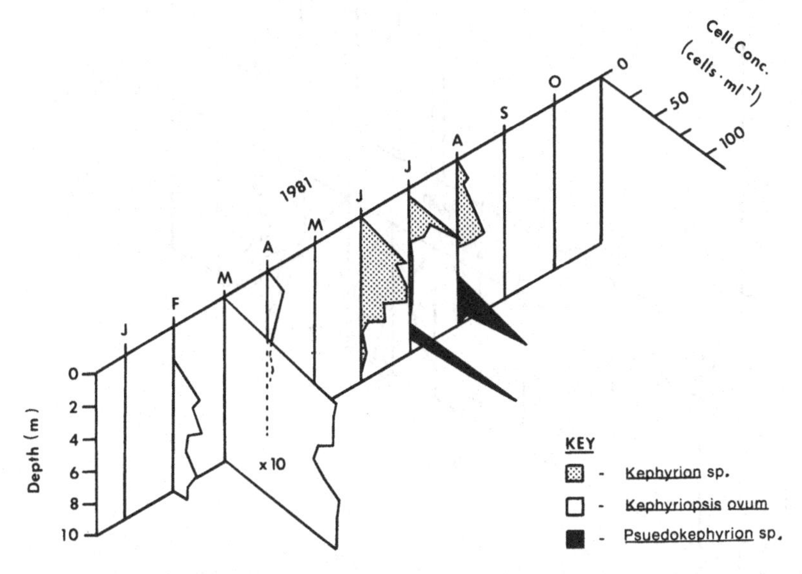

Fig. 12.7

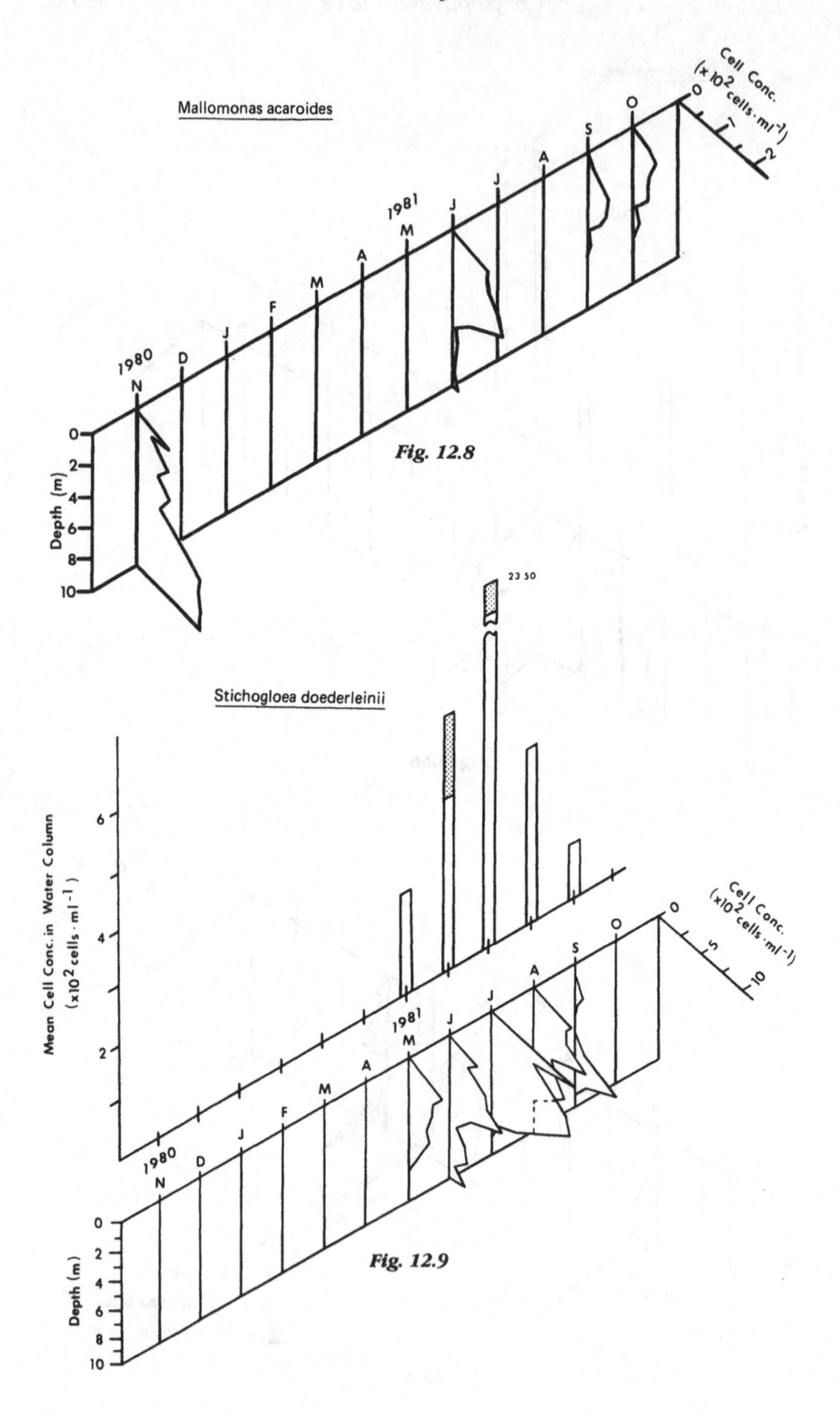

Fig. 12.8

Fig. 12.9

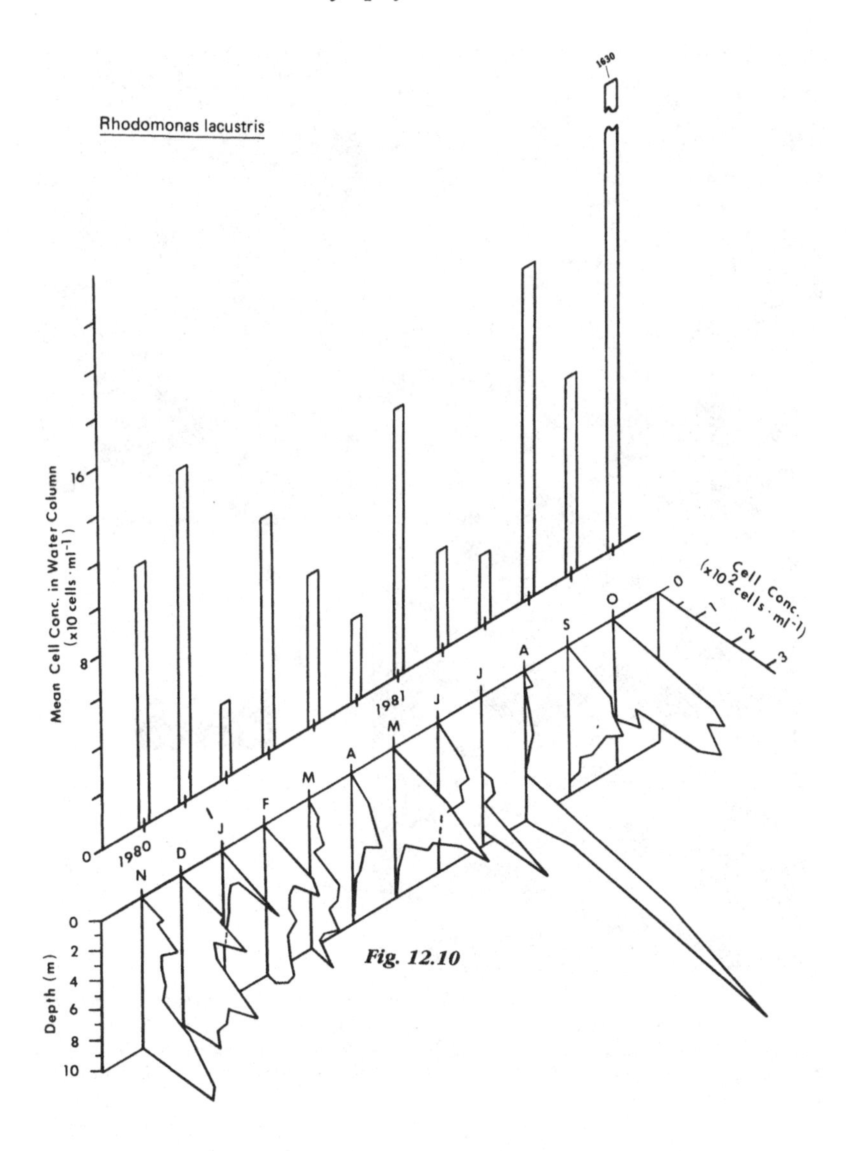

Fig. 12.10

tenuissima, and *Microcystis incerta* were warm-water summer forms also found at the surface.

The cryptomonads, *Rhodomonas lacustris* and *Cryptomonas* spp. (*C. ovata* and *C. erosa*), present throughout the entire year, had identical vertical and seasonal distributions (Figure 12.10). During January, April to May, and September to October both genera formed surface maxima after having had deep water peaks in the preceeding months (Figure 12.10). In August, several organisms (*Euglena gracilis* and *Pseudokephyrion* sp.) besides the cryptomonads developed deep-water strata.

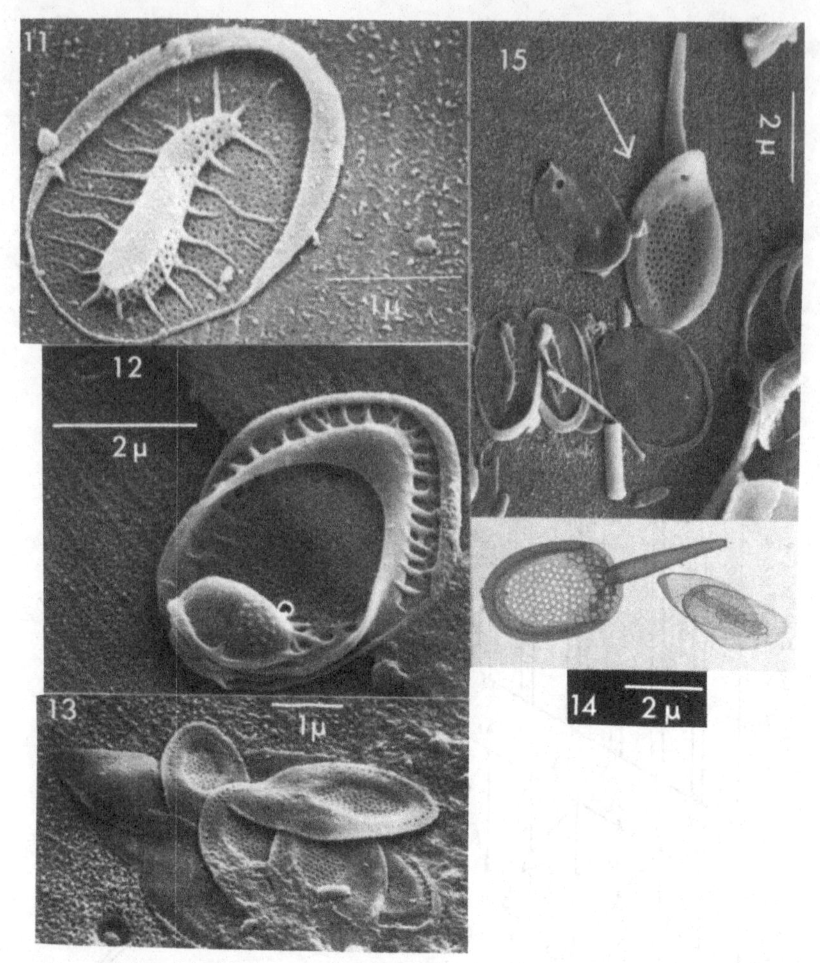

Figs. 12.11 – 12.15. Electron micrographs of scaled chrysophytes from Lake Lacawac. ***Fig. 12.11.*** Scale of *Synura petersenii* v. *glabra* (×24,000). ***Fig. 12.12.*** Scale of *Mallomonas acaroides* (×13,700). ***Fig. 12.13.*** Body scales of *Mallomonas akrokomos* (×13,500). ***Fig. 12.14 and 12.15.*** Scales of *Synura spinosa.* ×8000 in Figure 12.14. ×8230 in Figure 12.15.

Discussion

Many features are often used to describe the trophic condition of a lake (Wetzel 1983). Based on nutrient concentration, phytoplankton flora, and phytoplankton biomass, we feel that Lake Lacawac is best described as an oligotrophic lake. The fact that the total phosphorus and nitrate levels were below detectable limits places Lacawac as a nonpolluted, ultraoligotrophic lake (Vollenweider 1968, Wetzel 1983). Low phytoplankton concentrations and the insignificance of the Cyanophyceae and Chlorophyceae (collectively less than 4% of the total cell number) were also indicative of an oligotrophic condition.

The phytoplankton communities of many oligotrophic lakes are consistently dominated by algae belonging to the Chrysophyceae, especially the genera *Dinobryon, Synura, Mallomonas, Uroglena,* and *Chrysosphaerella* (Hutchinson 1967, Kling & Holmgren 1972, DeNoyelles & O'Brien 1978). Other groups of algae, such as the Cryptophyceae, are often important, but less significant than the Chrysophyceae in such lakes (Willen 1969, Schnidler & Nighswander 1970, Roijackers 1981).

The phytoplankton community in Lake Lacawac was dominated by Chrysophyceae (78%) and Cryptophyceae (13%) and, therefore, was very characteristic of an oligotrophic lake. *Synura petersenii* v. *glabra, Dinobryon cylindricum, Stichogloea doederleinii, Rhodomonas lacustris, Uroglenopsis* sp., and *Mallomonas* spp. accounted for 81% of the total number of phytoplankton on an integrated depth/seasonal basis. The phytoplankton community in Lake Lacawac closely resembled lakes in the Canadian shield region [class A lakes (Kling & Holmgren 1972)] and southern Sweden (Willen 1969) where the Chrysophyceae and the Cryptophyceae were the most abundant groups.

Lake Lacawac is unique in the Pocono Mountains of Pennsylvania because its watershed is totally undeveloped (Reif & Frock 1981). In several studies where similar oligotrophic bodies of water were enriched with nitrogen and phosphorus, there was a major shift in phytoplankton composition from the Chrysophyceae to the Chlorophyceae and/or Cyanophyceae (Schindler et al. 1971, DeNoyelles & O'Brien 1978, Findlay 1978). In addition, there were large increases in biomass. Generally, species of Chlorophyceae and Cyanophyceae present in the lakes prior to enrichment, but low in numbers, increased and bloomed following enrichment (DeNoyelles & O'Brien 1978). Probably, as the levels of nitrogen and phosphorus became nonlimiting, taxa with the fastest growth rates (greens and blue-greens) outcompeted and dominated the slower-growing Chrysophyceae (Moss 1973, DeNoyelles & O'Brien 1978). In each study, once nutrient enrichment stopped, there was a shift back to chrysophyte domination.

During the summer, Lake Lacawac had species of blue-greens that caused major blooms in nearby lakes but were outcompeted by chrysophytes within the lake. Because phosphorus and nitrogen levels were low (and probably limiting), we feel that increased nutrient loading would cause a shift to a Chlorophyceae and/or Cyanophyceae dominated lake as was found in the above referenced enrichment study. Since green algae usually outcompete blue-green algae under nutrient enrichment at low pH (Shapiro 1973), they may become dominant in Lake Lacawac.

The cryptomonads, *Rhodomonas lacustris* and *Cryptomonas* spp., had similar vertical and seasonal distribution and were holoplanktonic (perennial), found living throughout the year under a wide range of environmental conditions. During the year the cryptomonads alternated population maxima between the epilimnion and hypolimnion. This pattern may be related to zooplankton activity, especially *Daphnia catawba* and *Holopedium gibberum* during the summer months (Tessier 1983).

The seasonal distribution of phytoplankton (total numbers) followed a pattern common to oligotrophic lakes in temperate regions (Wetzel 1983). Low winter concentrations were followed by a short-lived spring maximum (March) directly after ice-out, then a population crash (May–June), followed by smaller summer (July) and fall (October) peaks. Low winter concentrations in the lake (10^2 cells ml^{-1}) were probably the result of reduced light levels and slower growth rates of the cold-adapted taxa.

In many temperate oligotrophic lakes, winter flora are dominated by small motile forms, especially *Rhodomonas, Cryptomonas, Dinobryon, Synura,* and *Mallomonas* (Rodhe 1955, Maeda & Schimura 1973, Wetzel 1983). During January, 75% of the phytoplankton in the water column and 90% of those above 2 meters belonged to these flagellated genera. With the onset of ice-out and spring overturn in March, concentrations increased from 2.3×10^2 cells ml^{-1} in January to the seasonal maximum of 9.0×10^3 cells ml^{-1}. The March peak was the result of increased growth of winter-adapted species after light levels were increased. We felt, as Wetzel (1983) concluded, the spring maximum was not controlled by temperature because it was still low (4°C).

In many temperate lakes, nutrient depletion has been found the cause of the decline in the spring peak. Since nitrate and total phosphorus levels were always below 0.01 mg NO_3 N $liter^{-1}$ and 10 μg PO_4 P $liter^{-1}$ we feel these nutrients did not cause the decline. In addition, since populations of *Mallomonas acaroides* and *Asterionella formosa* formed soon after the decline, silica depletion was probably not the cause of the spring crash. Perhaps another nutrient was involved. Zooplankton activity was also not believed to have caused the spring decline (A. Tessier, personal communication).

We believe an increase in temperature was the major cause of the spring decline. The cold-water flora consisting of *Synura petersenii* v. *glabra, Dinobryon cylindricum, Kephyriopsis ovum, Uroglena* sp., *Uroglenopsis* sp., and *Mallomonas akrokomos* declined rapidly as temperatures increased above the 12° to 15°C range. The importance of this range, in terms of being where the switch from a cold- to a warm-water flora occurs was, in part, supported by the literature. *Synura petersenii* v. *glabra* (=*S. glabra*) (Figure 12.11) has been found at 2° to 20°C (Takahashi 1978), 14°–16°C (Roijackers 1981), and 0°–10°C (Kristiansen 1975). *Synura spinosa* Korsh (Figures 12.14 and 12.15) was found in Lake Lacawac to have a similar distribution as *S. petersenii* v. *glabra,* although with a much smaller concentration. In past studies, forms of *S. spinosa* were found to occur below 15°C (Kristiansen 1975, 1978, Takahashi 1978). *Mallomonas akrokomos* has been reported at temperatures of 12°C (Kristiansen 1978), 0.3° and 8.6°C (Asmund & Hilliard 1961), and 0° to 13°C (Asmund 1956). During the winter of 1981–1982 and the following spring, essentially the same cold-water phytoplankton community consisting of *S. petersenii* v. *glabra, S. spinosa, Dinobryon cylindricum, M. akrokomos,* and *Uroglenopsis* sp. reformed. However, ice-out was much later in the season and the water temperature rose above the 12° to 15°C range rapidly, perhaps not allowing for as large a spring peak as in 1980–1981. This led us to conclude that temperature and light were the major factors controlling species composition and biomass, respectively, in Lake Lacawac.

Temperature also appeared to control the occurrence of *Mallomonas acaroides* since it consistently appeared each time the water cooled or warmed to within the 15° to 20°C range. This does not mean that *M. acaroides* cannot survive outside of this range. The taxon always appeared when the water temperature cooled to 20°C and was found as low as 7°C. However, when temperatures increased, it was not found until about 15°C. Thus, it probably survives but may not reproduce at temperatures lower than 15°C. The temperature requirements of *M. acaroides* seem consistent with previous studies where it was found between 8° to 19°C (Asmund 1959, Takahashi 1978, Gretz, Sommerfeld & Wujek 1979).

Although many chrysophytes are true cold-water forms, the generalization that the group prefers cold northern water may simply be due to the lack of studies in warmer waters (Kristiansen 1975, 1980, Kristiansen & Takahashi 1983). In the present study, although no scaled chrysophytes were found above 20°C, the alga *Stichogloea doederleinii* dominated the summer flora.

In summary, the phytoplankton of Lake Lacawac was dominated throughout the year by the Chrysophyceae and secondarily by the Cryptophyceae. We feel the nutrient condition of the lake controlled the type of flora that existed in it (Chrysophyceae instead of Cyanophyceae and/or Chlorophyceae), while temperature and light levels were more important in determining the succession of taxa and biomass, respectively. Future work on the interaction of environmental parameters on chrysophyte ecology, especially involving culture studies, is needed.

References

American Public Health Association 1980. *Standard Methods for the Examination of Water and Wastewater,* 15th ed. American Public Health Association, New York.

Asmund, B. 1956. Electron microscope observations on *Mallomonas* species and remarks on their occurrence in some Danish ponds. II. *Bot. Tidsskr.* 53: 75–85.

– 1959. Electron microscope observations on *Mallomonas* species and remarks on their occurrence in some Danish ponds and lakes III. *Dansk. Bot. Ark.* 18: 1-50.

Asmund, B. & Hilliard, K. 1961. Studies on Chrysophyceae from some ponds and lakes in Alaska. I. *Mallomonas* species examined with the electron microscope. *Hydrobiol.* 17: 237-58.

DeNoyelles F. & O'Brien, W. 1978. Phytoplankton succession in nutrient enriched experimental ponds as related to changing carbon, nitrogen and phosphorus conditions. *Arch. Hydrobiol.* 84: 137-65.

Findlay, D. 1978. Seasonal succession of phytoplankton in seven lake basins in the experimental lakes area, Northwestern Ontario, following artificial eutrophication. Data from 1974 to 1976. Fishery and Marine Service Report no. 1466.

Gretz, M., Sommerfeld, M., & Wujek, D. 1979. Scaled Chrysophyceae of Arizona - A preliminary survey. *J. Arizona Nevada Acad.* 14: 75-80.

Hutchinson, G.E. 1967. *A Treatise of Limnology,* Vol. II. *Introduction to Lake Biology and the Limnoplankton.* Wiley, New York.

Kling, H. & Holmgren, S. 1972. Species composition and seasonal distribution of phytoplankton in the Experimental Lakes Area, northwestern Ontario. Canadian Fishery Marine Service Technicial Report No. 337.

Kristiansen, J. 1975. On the occurrence of the species of *Synura* (Chrysophyceae). *Verb. Internat. Ver. Limnol.* 19: 2709-15.

- 1978. Studies on the Chrysophyceae of Bornholm II. *Bot. Tidsskrift.* 73: 71-85.

- 1980. Chrysophyceae from some Greek lakes. *Nova Hedwigia* 33: 167-94.

Kristiansen, J. & Takahashi, E. 1983. Chrysophyceae: introduction and bibliography. *In:* Rosowski, J. & Parker, B. [Eds.] *Selected Papers in Phycology* Vol. II. Phycological Society of America, Lawrence, Kansas, pp. 698-704.

Maeda, O. and S. Ichimura. 1973. On the high density of a phytoplankton population found in a lake under ice. *Int. Rev. Ges. Hydrobiol.* 58: 673-85.

Moss, B. 1973. The influence of environmental factors on the distribution of freshwater algae: An experimental study, II. The role of pH and the carbon dioxide-bicarbonate system. *J. Ecol.* 61: 157-77.

Reif, C. & Frock, R. 1981. Planktonic algae of Lacawanna County, Pennsylvania. *Proc. PA. Acad. Sci.* 54: 48-60.

Rodhe, W. 1955. Can plankton production proceed during winter darkness in subarctic lakes? *Verb. Internat. Ver. Limnol.* 12: 117-22.

Roijackers, R.M.M. 1981. Chrysophyceae from freshwater localities near Nijmegen, The Netherlands. *Hydrobiol.* 76: 179-89.

Schindler, D. & Nighwander, J. 1970. Nutrient supply and primary production in Clear Lake, Eastern Ontario. *J. Fish. Res. Bd. Canada* 27: 2009-35.

Schindler, D., Arnstrong, F., Holmgren, S., & Brunskill, G. 1971. Eutrophication of lake 227, Experimental Lakes Area, northwestern Ontario, by addition of phosphate and nitrate. *J. Fish. Res. Bd. Canada.* 28: 1763-82.

Shapiro, J. 1973. Blue-green algae: why they become dominant. *Science,* 179: 382-4.

Takahashi, E. 1978. *Electron Microscopical Studies of the Synuraceae (Chrysophyceae) in Japan.* Tokai University Press, Tokyo.

Takahashi, E. & Hayakawa, T. 1979. The Synuraceae (Chrysophyceae) in Bangledesh. *Phykos* 18: 129-47.

Tessier, A. 1983. Coherence and horizontal movements of patches of *Holopedium gibberum* (Cladocera). *Oecologia,* 60: 71-5.

Vollenweider, R.A. 1968. Scientific fundamentals of the eutrophication of lakes

and flowing waters, with particular reference to nitrogen and phosphorus as factors in eutrophication. Paris, Report to the Organization for Economic Cooperation and Development, No. DAS/CSI/ 68.27.
Wetzel, R.G. 1983. *Limnology.* Saunders, New York.
Willen, T. 1969. Phytoplankton from Swedish lakes II. Lake Assjön. 1961–62. *Oikos,* 20: 67–77.

13

Blooms of *Uroglena americana* in relation to concentrations of B group vitamins

AKIRA KURATA
Lake Biwa Research Institute
1-10, Uchide-hama, Otsu, Shiga Prefecture, 520 Japan

Introduction

Eutrophication has proceeded extensively in Lake Biwa recently. Since the first occurrence in 1977, the blooms of the flagellate *Uroglena americana* Calkins have occurred annually from April to June in almost the whole area of the lake and have caused serious troubles for people receiving their water supply from this lake. It has been reported that various kinds of marine and freshwater phytoplankton require B group vitamins, especially thiamine, biotin, and vitamin B_{12} for their growth in axenic cultures (Provasoli 1958, Provasoli & Carlucci 1974, Swift 1980). However, the relationship between the concentrations of B group vitamins and the occurrence of blooms of *U. americana* in the water of Lake Biwa has not yet been made clear.

Thus, the production of these vitamins by bacteria closely associated with the *Uroglena* colonies was examined and discussed, in relation to the occurrence of *Uroglena* blooms.

Materials and methods

Sampling was made twice a week in spring and once a month in the other season from 1978 to 1980 at the designated stations in Lake Biwa (Figure 13.1).

Determination of thiamine, biotin, and vitamin B_{12} was carried out according to the method described previously (Kurata et al. 1976).

Bacterial strains isolated from the colony of *U. americana* were kindly placed at our disposal by Ishida and Nakahara (Ishida & Nakahara 1981). These strains had been isolated from the *Uroglena* colonies by washing about ten times with sterile distilled water.

This study was financially supported by the Environment Conservation Bureau of the Prefectural Government of Shiga. The author is grateful to the staff of Shiga Prefectural Institute of Public Health and Environmental Science for their kind assistance in sampling program during the investigation period.

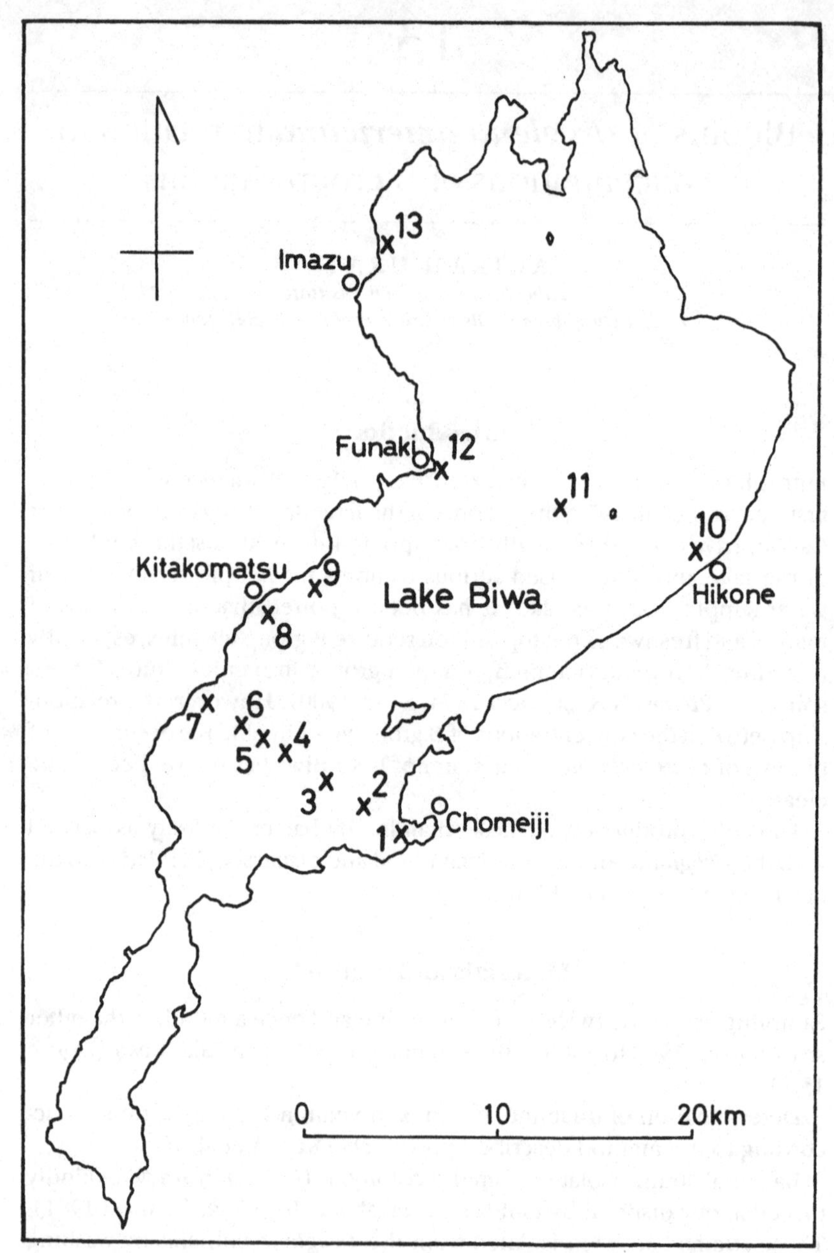

Fig. 13.1. Location of the sampling stations set up in Lake Biwa.

The bacterial strains were incubated for 14 days at 20°C in an oligotrophic medium that contains only 1 mg of casamino acids as an organic carbon source in 1 liter of distilled water. The composition of the low nutrients medium is as follows: $FeSO_4 \cdot 7H_2O$, 0.05 g; KH_2PO_4, 0.05 g; casamino acids (vitamin-free), 1.0 mg; 1 liter distilled water, pH 7.2. The composition of ZoBell 2216 freshwater medium is as follows: $FeSO_4 \cdot 7H_2O$, 0.05 g; KH_2PO_4, 0.05 g; 1 liter distilled water, pH 7.2. Immediately after collection of water samples and without any fixation, the number of *Uroglena* colonies was counted directly on a square scaled glass slide using a microscope.

Results and discussion

The conspicuous blooms of *U. americana* have occurred annually in almost the whole area of Lake Biwa since their first occurrence in the spring of 1977, and most frequently off Kitakomatsu and the island of Okinoshima near Chomeiji. However, one of the most difficult problems in the investigation of blooms is to predict when and where the blooms will occur. Therefore, there have been few reports until now concerned with the relationship between various kinds of nutritional parameters, especially B group vitamins, and the organisms involved.

The concentrations of B group vitamins, one of the most important growth factors for phytoplankton, were investigated from the spring of 1978 to the spring of 1980. The results obtained in the spring of 1978 are shown in Figure 13.2. The bloom of *U. americana* was observed five times at five stations in the investigation area from May 4 to June 27 during 1978. The concentration of biotin ranged from 0.24 to 3.25 ng/liter at each station. Wherever the bloom of the flagellate occurred, the concentration of biotin was always high. However, a close relationship could not be found with the concentrations of the other kinds of vitamins. Fluctuations of the concentrations of thiamine and vitamin B_{12} in 3-meter deep water at each station from May to June of 1978 are shown in Figure 13.3.

Generally, the seasonal changes in the concentrations of B group vitamins have an inverse relationship to those of population densities of phytoplankton in lakes and reservoirs owing to the active utilization of these vitamins by different kinds of phytoplankton (Kurata et al. 1976). It appears that this is the first report on the relation between the concentration of biotin and *Uroglena* blooms.

From April 12 to May 21 in 1979, the bloom of *U. americana* occurred more often at each sampling station than during the same period of 1978. The concentration of biotin ranged from 0.15 to 3.71 ng/liter at these stations, and the correlation between the concentration of biotin and the occurrence of the bloom was ascertained more clearly (Figure 13.4).

The colony number of *U. americana* was counted at the depths of 0, 3, 5,

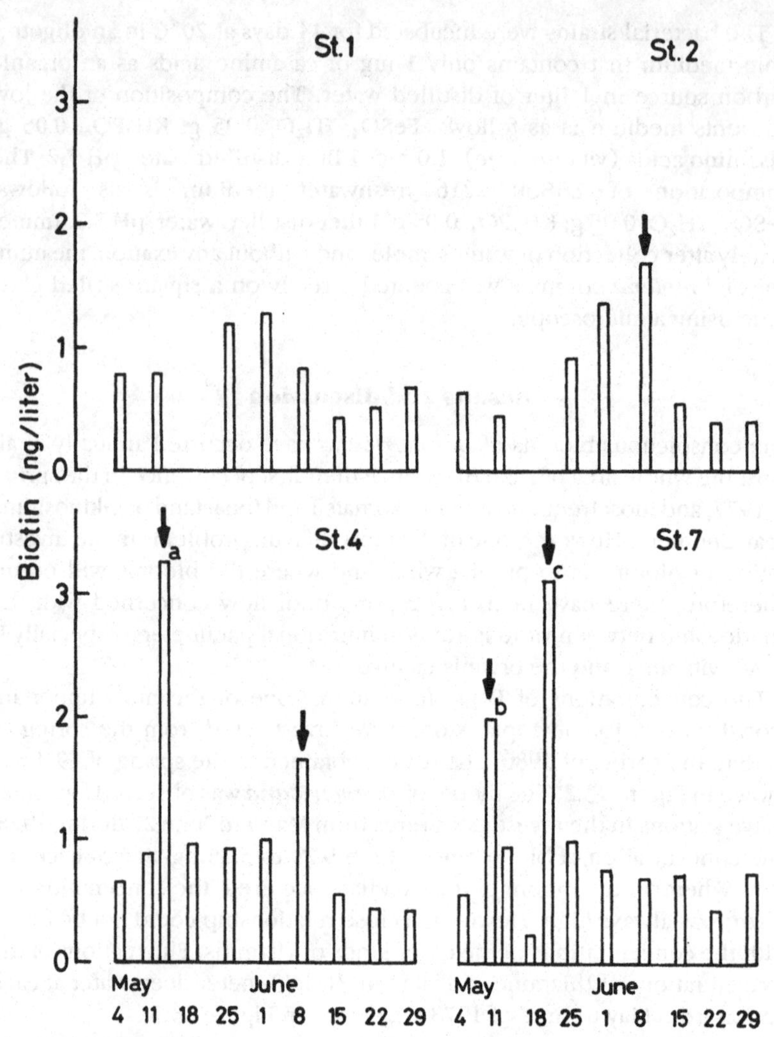

Fig. 13.2. Variation of the concentration of biotin in water at each station (St.) in the north basin of Lake Biwa from May 4 to June 29, 1978. (a) St. 5, (b) St. 6, and (c) St. 9. ↓, bloom of *U. americana* (approximately > 100 colonies per ml of water) observed. The lake water assume a faintly brown color with approximately 50 colonies of the flagellate per ml of water.

10, 20, and 30 meters at the stations 4 and 8 from the beginning of the investigation in year 1979 in order to clarify the relationship between the concentration of biotin and the colony number of *U. americana* (Figures 13.5 and 13.6). The correlation coefficients were 0.503 and 0.636, respectively.

In the spring of 1980, blooms were less developed and observed only on

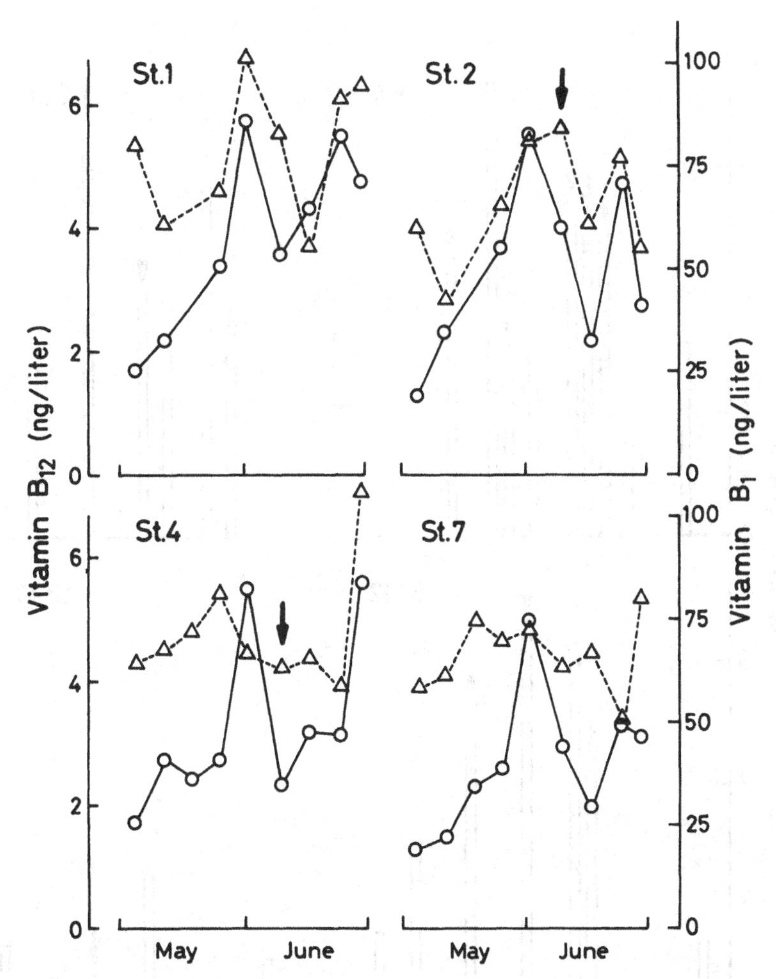

Fig. 13.3. Fluctuation of the concentrations of vitamins B_1 and B_{12} in 3 meter deep water at each station in the north basin of Lake Biwa during May and June of 1978. △- -, B_1; ○—, B_{12}. ↓ , bloom of *U. americana* observed.

two occasions at the end of May and at a few stations. Therefore, it was not possible to collect additional information regarding the concentration of biotin and the population density of the organism. The results obtained are shown in Figure 13.7.

It may be concluded that the occurrence of high concentrations of biotin and high population densities of *U. americana* must be due to production of vitamins by the flagellate itself or by the microorganisms closely associated with it. Carlucci and Bowes (1970) reported on the production of thiamine, biotin, and vitamin B_{12} by a few species of diatoms in axenic cultures. It has been reported that a number of periphytic and epiphytic

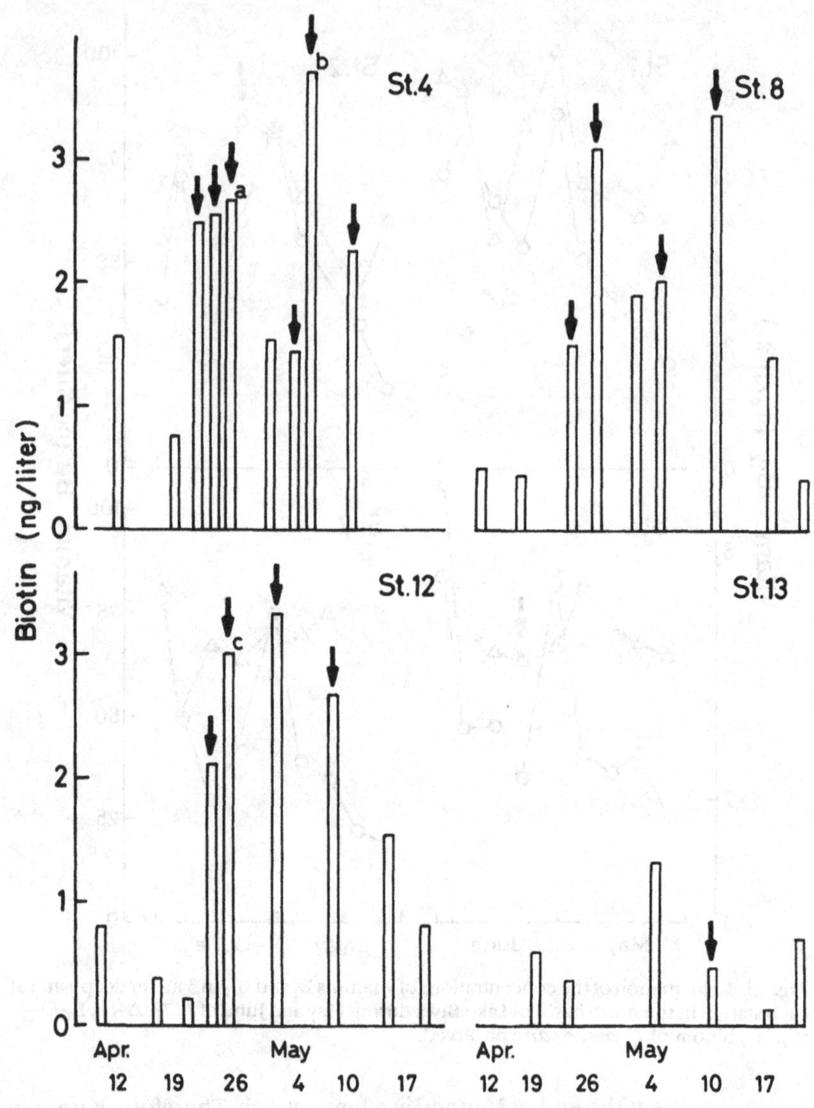

Fig. 13.4. Variation of the concentration of biotin in water at each station (St.) in the north basin of Lake Biwa from April 12 to May 21, 1979. (a) St. 2, (b) St. 5, and (c) St. 10. ↓, bloom of *U. americana* observed.

microorganisms, especially bacteria, actively produce B group vitamins in aquatic environments of Lake Biwa (Kurata 1981). A great deal of gelatinous matter is excreted from the *Uroglena* cells, and it forms a favorable substratum for periphytic biotin-producing bacteria. The nutritional requirements of *U. americana*, particularly its vitamin requirements, have not yet been made clear because of the difficulty to obtain an axenic cul-

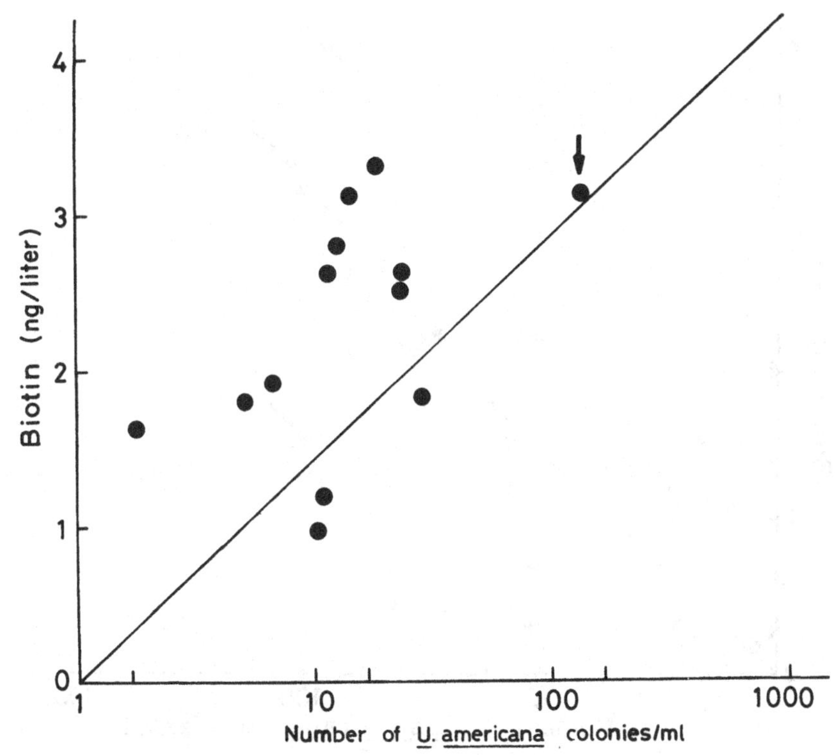

Fig. 13.5. Relationship between the concentration of biotin and the colony number of *U. americana* in water at station 4 during the investigation period. The correlation coefficient was 0.503. ↓ , bloom of *U. americana* observed.

ture. However, recently Ishida and Nakahara (1981) reported on the thiamine and vitamin B_{12} requirements of the flagellate using a monoaxenic culture. Accordingly, the production of vitamins by bacteria associated with the colony of *U. americana* was examined using the ZoBell 2216 freshwater medium, and the results are shown in Table 13.1.

All bacterial strains produced vitamins actively, and two strains produced three kinds of vitamins simultaneously. Therefore, the production of vitamins by these bacterial strains was examined further in the extreme oligotrophic medium (Table 13.2), similar to the lake water. The cell density of bacteria in the medium is shown in Table 13.3. Most of strains grew well in this medium and actively produced vitamins. The results of the vitamin production by bacteria are shown in Table 13.4. Two strains produced all kinds of vitamins simultaneously even in this oligotrophic medium. All bacterial strains produced biotin and 69.4–90.5% of the vitamin produced was actively excreted to the medium. Vitamin B_{12} was produced slightly by two strains and was not excreted in the medium at all.

Accordingly, it is suggested that *U. americana* must be supplied with the

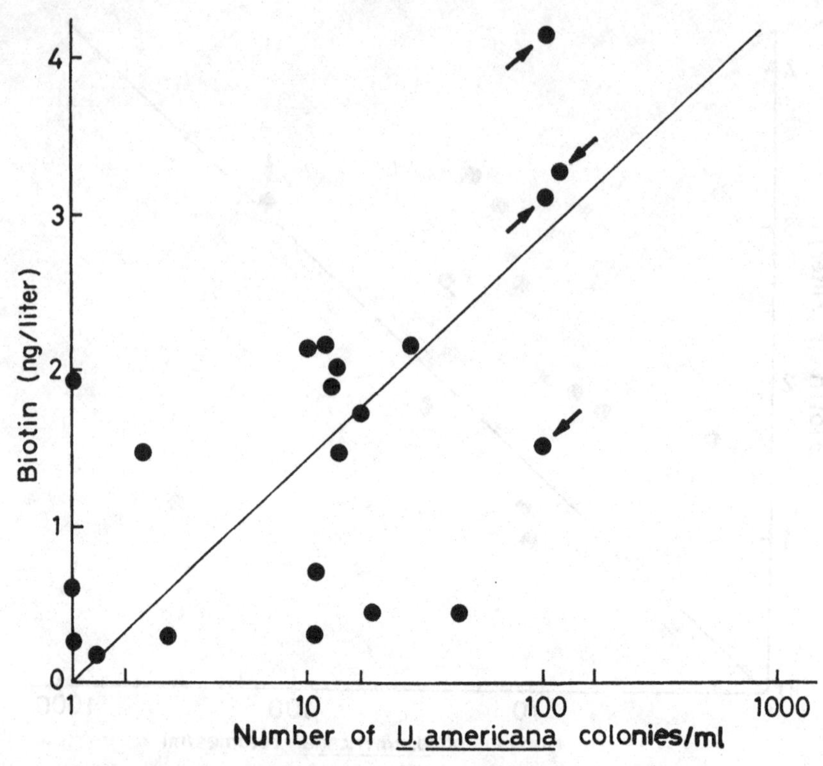

Fig. 13.6. Relationship between the concentration of biotin and the colony number of *U. americana* in water at station 8 during the investigation period. The correlation coefficient was 0.636. ↓ , bloom of *U. americana* observed.

Table 13.1. ***Production of thiamine, biotin, and vitamin B_{12} by bacterial strains associated with*** U. americana[a]

Bacterial strain	Vitamin		
	Thiamine	Biotin	Vitamin B_{12}
I_1	+	+++	−
I_2	+	+	−
I_3	−	+	−
NIA	+++	+++	++
NbA-1	+	+	+
NbA-2	++	++	−
NbB	++	+	−
BA	+	+	−
U_4	+	+	−

[a] ZoBell 2216 freshwater medium.

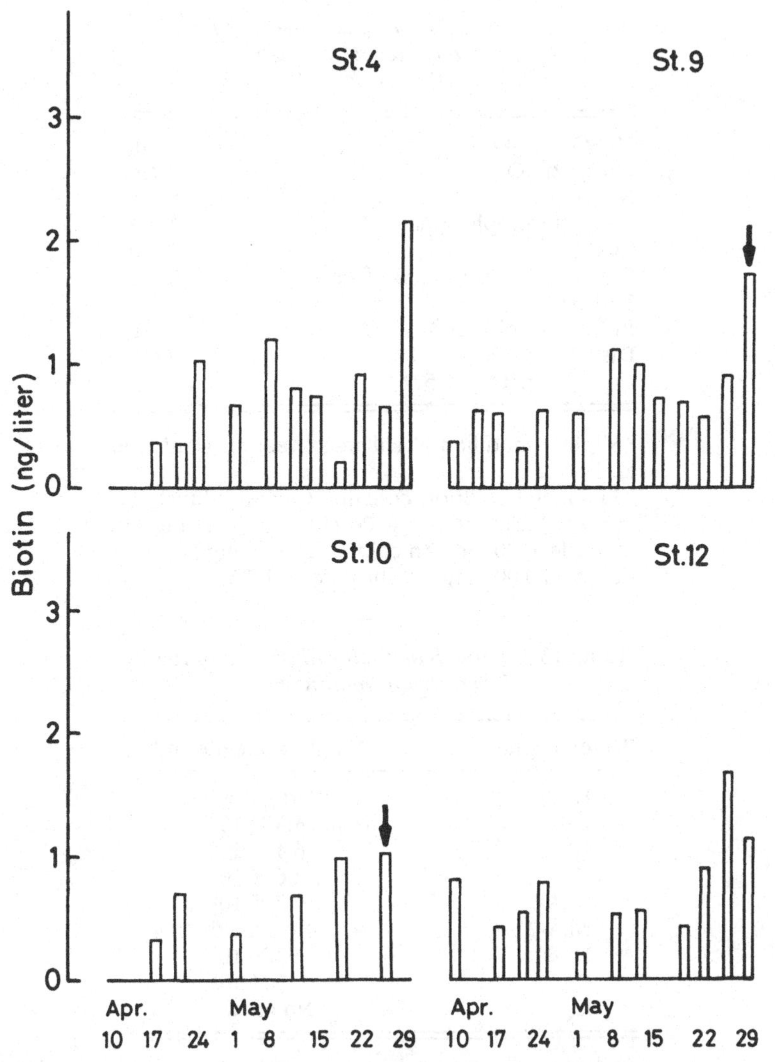

Fig. 13.7. Variation of the concentration of biotin in water at each station in the north basin of Lake Biwa from April 10 to May 29, 1980. ↓ , bloom of *U. americana* observed.

vitamins by bacteria associated with the colony and that the bacteria may utilize organic matter excreted by the flagellate; thus, there is a symbiosis or a certain mutualism between the alga and the bacteria. It must be attributable to the production of vitamins by the associated bacteria that *U. americana* required essentially thiamine and vitamin B_{12} and did not require biotin for its growth in the monoaxenic culture as reported by Ishida

Table 13.2. *Composition of the medium for bacteria associated with* U. americana[a]

$MgSO_4 \cdot 7H_2O$	10 mg
$CaCl_2 \cdot 2H_2O$	10 mg
NH_4NO_3	5 mg
β-Na_2 glycerophosphate $5H_2O$	4 mg
KCl	1 mg
Casamino acids (vitamin-free)	1 mg
Fe-EDTA	0.5 mg
PIV metal solution[b]	1 ml
Distilled water	999 ml
pH 7.5	

[a] Slightly modified Nakahara's medium (Ishida et al. 1978).
[b] PIV metal solution contains (per milliliter): 1.0 mg Na_2 EDTA, 0.04 mg Fe chloride, 0.01 mg Mn chloride, 0.005 mg Zn chloride, 0.005 mg Mo chloride, and 0.001 mg Co chloride; pH 7.5.

Table 13.3. *Growth of bacterial strains in the low nutrient medium*[a]

Bacterial strain	Number of cells/ml[b]
I_1	No growth
I_2	3.4×10^6
I_3[c]	6.4×10^6
NIA	2.6×10^6
NbA-1	9.2×10^5
NbA-2	No growth
NbB	8.3×10^6
BA	1.5×10^5
U_4	No growth

[a] Incubated at 20°C for 14 days.
[b] Averaged values of duplicate.
[c] Oligocarbophilic.

and Nakahara (1981). Recently, Swift and Guillard (1978) stressed the ecological significance of vitamin B_{12} even though its requirement for growth by diatoms is not absolute. In the case of *U. americana* blooms in Lake Biwa, it is probable that the growth of the flagellate is increased by biotin produced by the bacteria associated with the flagellate.

Table 13.4. *Production of thiamine, biotin, and vitamin B_{12} by bacterial strains associated with* U. americana *in the low nutrients medium*

Bacterial strain	Thiamine		Biotin		Vitamin B_{12}	
	Extracellular (ng/liter)	Intracellular (ng/liter of culture)	Extracellular (ng/liter)	Intracellular (ng/liter of culture)	Extracellular (ng/liter)	Intracellular (ng/liter of culture)
I_2	11 (35.5)[a]	20 (64.5)	2.5 (69.4)	1.1 (30.6)	ND	ND
I_3	13 (43.3)	17 (56.7)	2.0 (66.7)	1.0 (33.3)	ND	ND
NIA	14 (51.9)	13 (48.1)	1.4 (77.8)	0.4 (22.2)	ND	0.9 (100)
NbA-1	13 (72.2)	5 (27.8)	1.9 (90.5)	0.2 (9.4)	ND	0.5 (100)
NbB	18 (45.0)	22 (55.0)	1.3 (72.2)	0.5 (27.8)	ND	ND
BA	7 (58.3)	5 (41.7)	1.1 (78.6)	0.3 (21.4)	ND	ND

[a] Numbers in parentheses are percentages.

References

Carlucci, A.F., & Bowes, P.M. 1970. Vitamin production and utilization by phytoplankton in mixed culture. *J. Phycol.* 6: 393–400.

Ishida, Y. & Nakahara, H. 1981. Biwako ni okeru plankton no ijo hassei kiko ni kansuru chosa kenkyu hokoku (In Japanese). *In:* Kadota, H. [Ed.] *Biwako plankton ijo hassei chosa dan,* Kyoto, pp.120–31.

Kurata, A. 1981. The production of B group vitamins by epiphytic microorganisms on macrophytes in Lake Biwa. *Verh. Internat. Verein. Limnol.* 21: 596–9.

Kurata, A., Saraceni, C., Ruggiu, D., Nakanishi, M., Melchiorri-Santolini, U., & Kadota, H. 1976. Relationship between B group vitamins and primary production and phytoplankton population in Lake Mergozzo (Northern Italy). *Mem. Ist. Ital. Idrobiol.* 33: 257–84.

Provasoli, L. 1958. Nutrition and ecology of protozoa and algae. *Ann. Rev. Microbiol.* 12: 279–308.

Provasoli, L. & Carlucci, A.F. 1974. Vitamins and growth regulators. *In:* Stewart, W.D.P. [Ed.] *Algal Physiology and Biochemistry.* Blackwell, Oxford, pp. 741–89.

Swift, D.G. 1980. Vitamins and phytoplankton growth. *In:* Morris, I. [Ed.] *The Physiological Ecology of Phytoplankton.* Blackwell, Oxford, pp. 329–68.

Swift, D.C. & Guillard, R.R.L. 1978. Unexpected response to vitamin B_{12} of dominant centric diatoms from the spring bloom in the Gulf of Maine (Northeast Atlantic Ocean). *J. Phycol.* 14: 377–86.

14

Light quality effects on carbon and sulfur uptake of a metalimnetic population of the colonial chrysophyte *Chrysosphaerella longispina*

FRANCES R. PICK[1]
Department of Biology
Trent University
Peterborough, Ontario, K9J 7B8, Canada

AND

R. L. CUHEL
Department of Biology and Living Resources
University of Miami
Miami, Florida, 33149, U.S.A.

Introduction

Chrysophytes, particularly colonial forms, have been observed to form subthermocline chlorophyll maxima in many northern lakes (Brook, Baker, & Klemer 1971, Fee 1976, Pick, Nalewajko & Lean 1984). These peaks can be easily missed because they often extend less than 0.25 meter. The chlorophyll concentrations can be up to 30 times surface values, but it is not clear if this biomass is important to the total primary production of these lakes (Fee 1976, Pick et al. 1984).

Fee (1976) reported high rates of production from the hypolimnion of several lakes of the Experimental Lakes Area (ELA) in northwestern Ontario and suggested that up to 50% of the primary production in ELA lakes occurred in the chrysophyte layers found in the hypolimnia of these lakes. However, on further analysis, the model used to generate these numbers was found to overestimate hypolimnetic but not epilimnetic production (Fee 1978). Because an incubator was used it was thought that the light quality at depth was not being properly simulated and that this was the likely cause for the error (Fee 1978, 1980). This hypothesis is tested here on a subthermocline population of *Chrysosphaerella longispina* Lauterborn emend. Nicholls, by comparing photosynthesis under "white" light with that under "green" light, which predominates at these peaks (Fee 1978). Sulfur uptake and the fraction of sulfur incorporated into protein is

[1] Present address: Department of Biology, York University, 4700 Keele Street, Toronto, Ontario M3J 1P3, Canada.

also examined. This measurement approximates protein synthesis rates (Cuhel, Taylor, & Jannasch 1982a,b) and provides a better estimate of growth when combined with carbon fixation (Bates 1981, Cuhel, Ortner, & Lean 1984).

Methods and materials

The chrysophyte sample was collected from Williams Bay, Jacks Lake (44° 41′20″ N latitude, 78° 02′54″ W longitude) where a population of *Chrysosphaerella longispina* has been observed to form a band for the past 6 years (Pick et al. 1984). The identity of this species was confirmed by scanning electron microscopy according to the recent reassessment of the genus by Nicholls (1980). The peak was located using in vivo fluorescence (Pick et al. 1984) at a depth of 8.5 m at dawn on 18 August 1982. The peak population migrates upward in the morning over 1.5 meters then begins to descend in the afternoon and through the night reaching a maximum depth before dawn (Pick and Lean 1984). A large carboy protected from light was used to collect the sample directly from the outflow of the fluorometer.

The carboy was subsampled to fill 12 replicate BOD bottles of 300 ml capacity. Half were spiked with [^{14}C]bicarbonate (to a final activity of 2×10^5 dpm ml^{-1}) and the other half with [^{35}S]sulfate (to a final activity of 4×10^6 dpm ml^{-1}). The bottles were then placed in an incubator at five light levels and maintained at 10°C, which corresponds to the in situ temperature at 8.5 meters. These were exposed to the normal "white" light produced by a high-pressure, quartz halogen, street light. After 1.5 hours, the bottles were retrieved and duplicate 25 ml subsamples filtered separately through 0.45 μm Sartorius and 12 μm Nuclepore filters. Volumes of 200 ml were passed through a 30 μm Nitex screen exclusively to retain colonies of *C. longispina,* which were then backwashed onto GF/F filters and frozen for subsequent analysis of subcellular components as described by Cuhel et al. (1982a). The filtrate passed through the 30 μm Nitex net was collected directly on glass fiber (Whatman GF/F) filters and frozen for the same analysis. The latter filtrations were also conducted on the bottles spiked with ^{35}S. The entire procedure as above was repeated with a "green" screen to simulate more closely conditions of light quality in situ.

Phytoplankton were enumerated by epifluorescence microscopy on samples preserved with 10% glutaraldehyde and 0.1 M sodium cacodylate. Details of this method are given in Caron, Pick, & Lean (1985).

Measurements of quantum irradiance in situ and in the incubator were made with a Li-Cor 185 Quantum/Radiometer/Photometer equipped with a cosine-corrected Li-192 S quantum underwater sensor that measures photosynthetically active radiation in the 400–700 nm wave band. Relative spectral values of quantum irradiance were made with a Lambda Instruments spectroradiometer.

Results

The light quality spectrum found under surface-incident light and at different depths in Williams Bay is compared to that of the incubator under white light and with a green screen (Figure 14.1). Under white light, the blue, red, and far-red portions of the spectrum are more important components of the total than under green light and light at depth in situ. However, the "white" light used here approximates more closely the underwater light regime in the epilimnion with less blue and far-red when compared to the surface-incident spectrum. The latter spectrum is closer to the spectrum used by other workers using incubators (Fee 1978).

The photosynthesis versus quantum irradiance curve generated under white light conditions is very similar to that obtained under green light for whole lake water (WLW), the >12 μm (not shown) and >30 μm fractions (Figure 14.2A). The >30 μm fraction at P_{max} (the maximum rate of photosynthesis at saturating quantum irradiances) represented approximately 40% of the total uptake, and the fraction greater than 12 μm represented 57% of the total. The former fraction contained 90% of the *C. longispina* cells, and the total concentration of cells in the sample was 4.3×10^3 cells ml^{-1}, representing 83% of the total algal biomass. The phytoplankton in the below 12 μm fraction were predominantly small cyanobacteria (1 μm^3) and unidentified chrysophytes (probably *Ochromonas* sp.). The former were abundant (1.16×10^5) but represented only 11% of the total algal biomass.

Total sulfur uptake was significantly enhanced under green light conditions compared to white light for both fractions, but the >30 μm fraction represented only 25–30% and 32–46% of the total uptake under white and green light, respectively (Figure 14.2). Photoinhibition of photosynthesis occurred in the first two chambers for both fractions but photoinhibition of S uptake only occurred in WLW.

The subcellular extraction procedure yields sequential fractions of low molecular weight (LMW) organic compounds, alcohol-soluble protein, lipids, hot trichloroacetic acid-soluble (HTCA) material (nucleic acids, carbohydrate, and polyphosphates), and residue proteins. The overall error of the technique in natural waters is about 7% (Cuhel et al. 1983). The more important fractions are summarized in Table 14.1 for relative carbon incorporation. The main difference between the greater than 30 μm fraction and WLW is the higher percentage of label in the LMW carbon fraction, with proportional decrease in lipid, both under low (25–35 μeinsteins m^{-2} sec^{-1}) and high light (250 μeinsteins m^{-2} sec^{-1}). This remains true under green light. The HTCA fraction (which contains largely carbohydrates) is similar under both light treatments, comprising about 20% of the total label. The alcohol-soluble protein represented only 1–1.9% of the total and is not given in Table 14.1.

The difference between the green and white light treatments lies in the higher percentage of protein carbon labeled in both fractions under low

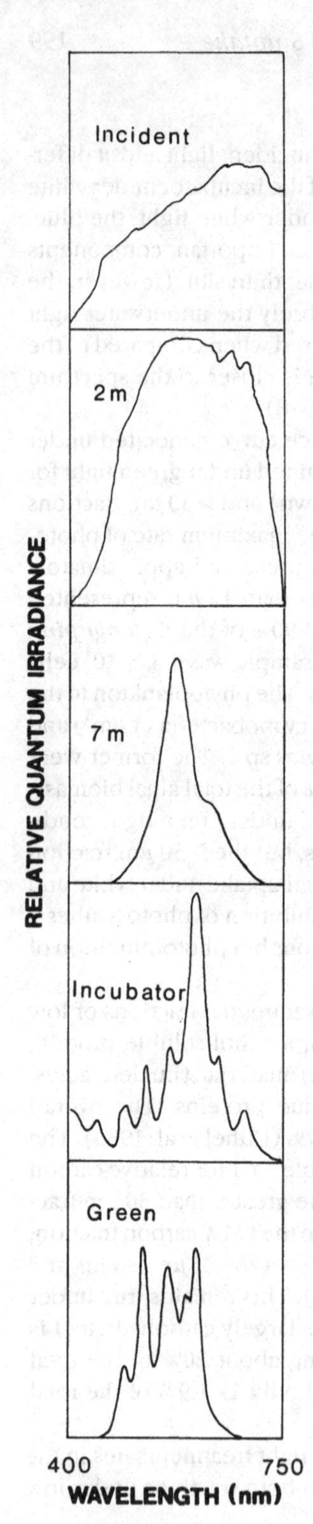

Fig. 14.1. From top to bottom, spectral distributions of quantum irradiance measured under surface-incident radiation, underwater radiation at 2 meters, 7 meters, and under the "white" and the "green" light of an incubator. Relative units.

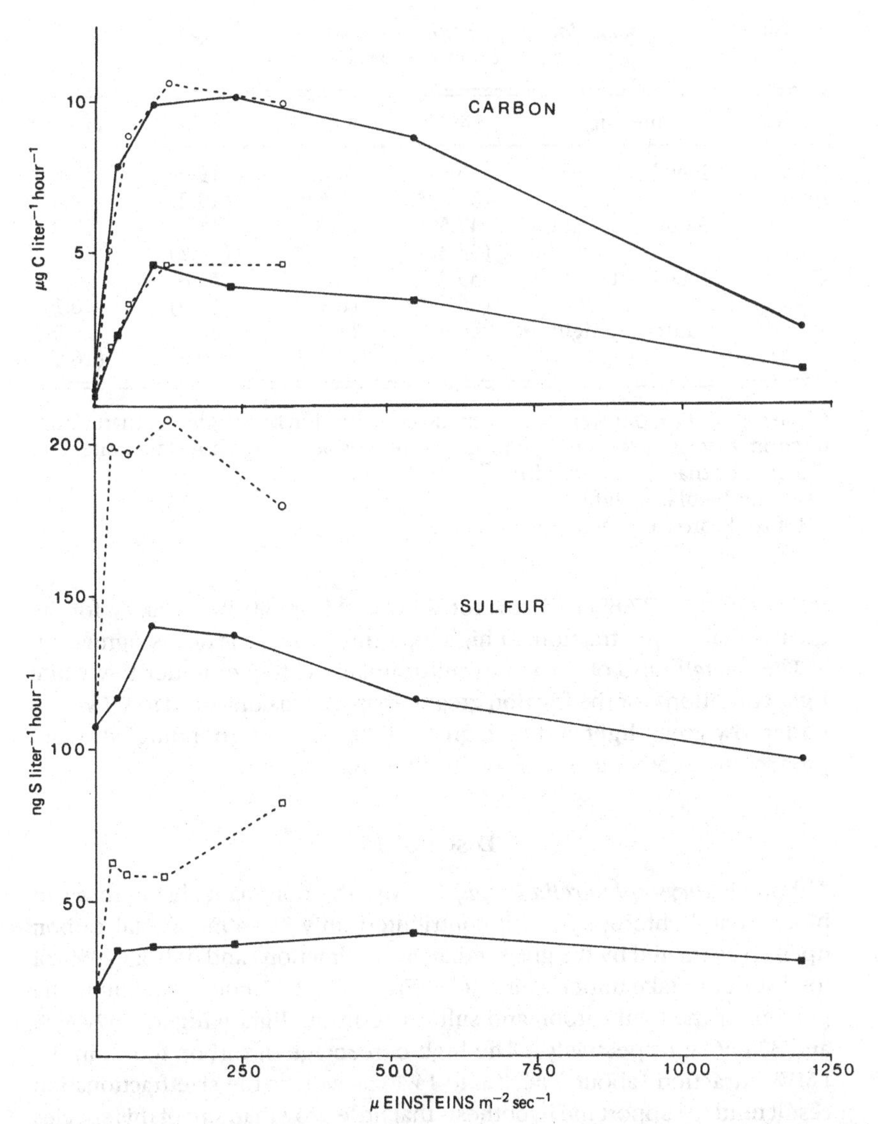

Fig. 14.2. Upper graph: photosynthesis versus quantum irradiance for whole lake water (•) and the greater than 30 μm fraction (■) under white light (solid line) and under green light of an incubator (dashed line). Lower graph: sulfate uptake versus quantum irradiance for the same sample.

green. Low green light best simulates actual conditions in situ at 7–8 m, where light levels rarely exceed 35 μeinsteins m^{-2} sec^{-1}. Absolute rates of protein-C formed are shown in Figure 14.3. Similarly, absolute rates of protein-S formed are significantly higher under green light for both size fractions (Figure 14.3). At low green light, the percentage of protein-S

Table 14.1. *Subcellular distribution of ^{14}C as a percentage of the total radioactivity incorporated*[a]

Light	Intensity	LMW[b]	Lipid	HTCA[c]	Protein
White light	Low light	45.1	14.8	19.0	19.6
		(52.4)*	(7.5)*	(19.2)	(19.8)
	Saturating light	47.5	11.6	18.8	17.9
		(54.3)*	(7.1)*	(20.2)	(17.4)
Green light	Low light	35.7	15.7	18.8	27.9
		(46.5)*	(8.1)*	(18.0)	(26.1)
	Saturating light	46.6	11.2	20.3	20.5
		(56.9)*	(6.3)*	(19.0)	(16.9)*

[a] The numbers in brackets represent the contribution by the greater than 30 μm fraction. Asterisks denote significant differences between whole lake water and the greater than 30 μm fraction.
[b] Low molecular weight.
[c] Hot trichloroacetic acid-soluble.

increased from 27.8% to 37.8% for WLW and from 30.4% to 40.9% for the greater than 30 μm fraction. At high light this increase is not as significant.

The overall ratio of carbon to sulfur uptake is higher under low white light conditions for the fraction greater than 30 μm compared to WLW, but under low green light and high green light, the corresponding ratios approach one another at a level of 50–55 (Figure 14.4).

Discussion

Although *Chrysosphaerella longispina* was the dominant alga in terms of biomass and chlorophyll *a*, it contributed only 30–40% of total carbon uptake (estimated by the greater than 30 μm fraction) and only 25–35% of total sulfur uptake under white light (Figure 14.2). Under green light, the percent of the total carbon and sulfur fixed was slightly higher (37–47% and 32–46%, respectively). The high percentage of carbon found in the LMW extraction (about 50%, Table 14.1) as well as the size fractionation result tend to support the hypothesis that little growth in situ of this species occurs below the thermocline (Pick et al. 1984). The *C. longispina* fraction appeared to be fixing much carbon, which was not being translated into protein synthesis. Under both low light treatments the >30 μm fraction fixed 30–35% of the total protein-C and total protein-S. The <30 μm fraction was the more metabolically active size fraction.

Contrary to the conclusions of Fee (1978), no difference was found in total carbon uptake under the two light quality treatments (Figure 14.2). However, the spectrum of the incubator used in this study resembled more closely the light regime at 2 meters rather than that of surface incident light

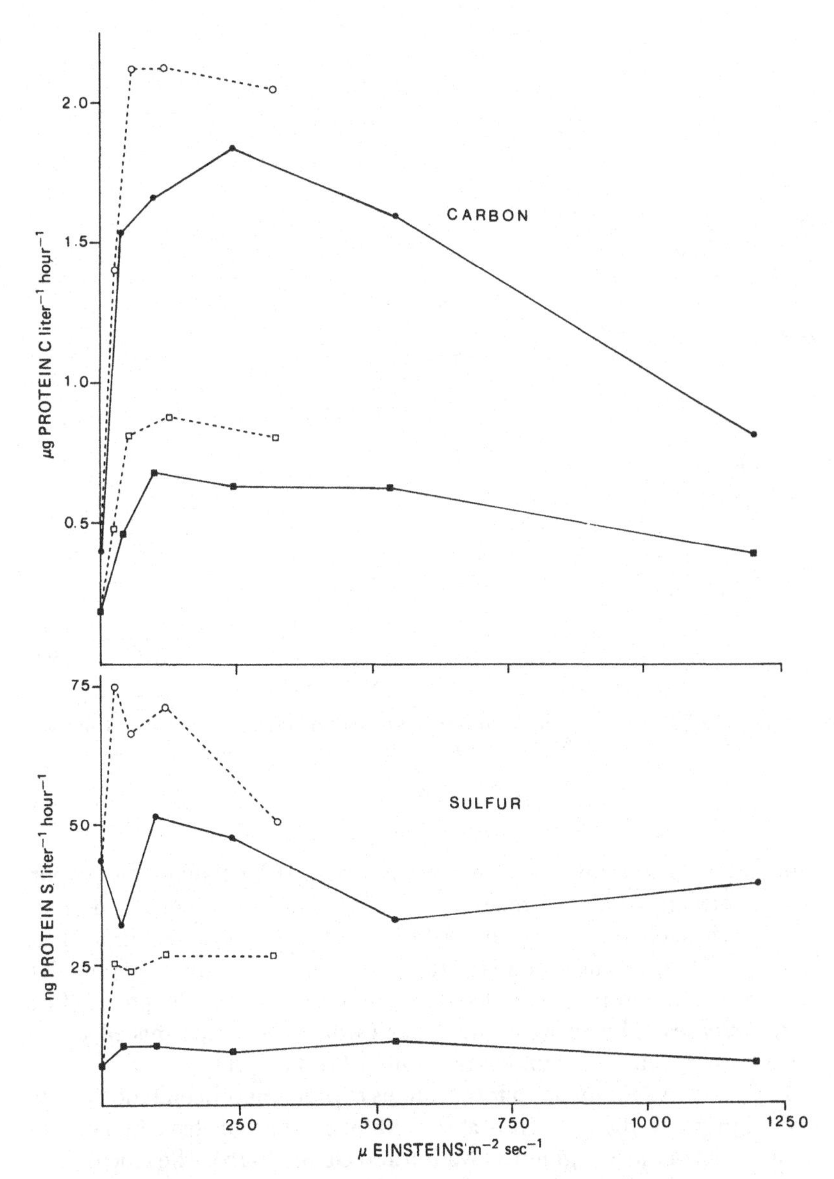

Fig. 14.3. Upper graph: protein-C synthesized versus quantum irradiance for the whole lake water (•) and the greater than 30 μm fraction (■) under white (solid line) and green light of an incubator (dashed line). Lower graph: protein-S synthesized versus quantum irradiance for the same sample.

as with Fee (1978). The proportion of carbon allocated to protein was higher under green light, and this was confirmed by sulfur uptake (Figure 14.3). Wallen and Geen (1971a) showed that protein constituted a greater fraction of the photosynthate in blue and green light than in white light at

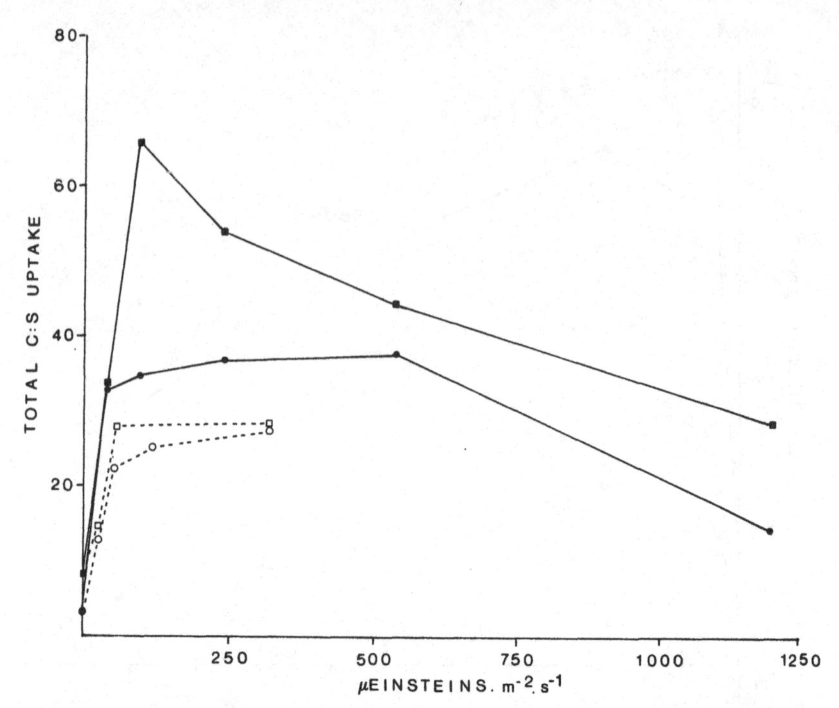

Fig. 14.4. Carbon to sulfur uptake ratios versus quantum irradiance white (solid line) and green (dashed line) light for whole lake water (•) and the greater than 30 μm fraction (■).

similar levels of irradiance in two marine algae. Here, though, the proportion of protein produced relative to other fractions was enhanced under green light at the expense of the LMW fraction rather than the carbohydrate fraction of Wallen and Geen (1971a) (their ethanol-soluble compounds). In general, the proportion of fixed ^{14}C incorporated into the protein fraction increases at lower light intensities (Morris 1981) and this effect was even more pronounced under green light (Table 14.1).

Light quality is known to affect changes in pigment concentrations. Cells grown in green light which is absorbed by carotenoids have higher concentrations of these pigments (Wallen and Geen 1971b). The ratio of carotenoid to chlorophyll *a* is high in subthermocline algal peaks as is the ratio of chlorophyll *c* to chlorophyll *a* (Pick 1982). Most Chrysophyceae contain fucoxanthin, which in vivo has absorption extended to 590 nm and is very efficient, unlike most carotenoids, in transferring energy to chlorophyll *a* (Govindjee and Zilinskas Braun 1974). Although most plants use red light in photosynthesis this metalimnetic chrysophyte population is adapted to using light in the green part of the spectrum, even though this energy seems translated mainly into maintenance rather than growth.

Summary

At similar levels of irradiance, no difference was observed on total C uptake by a metalimnetic population of *Chrysosphaerella longispina* under white light or green light. However, green light did increase the proportion of the label found in protein and enhanced total sulfur uptake. Even though *C. longispina* constituted 83% of the total phytoplankton biomass, its contribution to photosynthesis under low green light which occurs in situ below the thermocline was only 37% and to protein-S synthesis only 34%. These data support the hypothesis that growth in situ of this species is minimal below the thermocline.

References

Bates, S.S. 1981. Determination of the physiological state of marine phytoplankton by use of radiosulfate incorporation. *J. Exp. Mar. Biol. Ecol.* 51: 219–39.

Brook, A.J., Baker, A.L., & Klemer, A.R. 1971. The use of turbidimetry in studies of the population dynamics of phytoplankton populations with special reference to *Oscillatoria agardhii. Mitt. Internat. Verein. Limnol.* 19: 244–52.

Caron, D., Pick, F.R. & Lean, D.R.S. 1985. Chroococcoid cyanobacteria in Lake Ontario: vertical and seasonal distributions during 1982. *J. Phycol.* 21: 171–5.

Cuhel, R.L., Taylor, C.D. & Jannasch, H.W. 1982a. Assimilatory sulfur metabolism in marine microorganisms: sulfur metabolism, growth, and protein synthesis of *Pseudomonas halodurans* and *Alteromonas luteo-violaceus* during sulfate limitation. *Arch. Microbiol.* 130: 1–7.

– 1982b. Assimilatory sulfur metabolism in marine microorganisms: considerations for the application of sulfate incorporation into protein as a measurement of natural population protein synthesis. *Appl. Environ. Microbiol.* 43: 160–8.

Cuhel, R.L., Jannasch, H.W., Taylor, C.D. & Lean, D.R.S. 1983. Macromolecular growth and synthesis in the northwestern Atlantic Ocean. *Limnol. Oceanogr.* 28: 1–18.

Cuhel, R.L., Ortner, P.B. & Lean, D.R.S. 1984. Night synthesis of protein by algae. *Limnol. Oceanogr.* 29: 731–44.

Fee, E.J. 1976. The vertical and seasonal distribution of chlorophyll in lakes of the Experimental Lakes Area, northwestern Ontario: Implications for primary productivity. *Limnol. Oceanogr.* 21: 767–83.

– 1978. A procedure for improving estimates of in situ primary production at low irradiances with an incubator technique. *Verh. Int. Verein. Theor. Angew. Limnol.* 20: 59–67.

– 1980. Important factors for estimating annual phytoplankton production in the Experimental Lakes Area. *Can. J. Fish. Aquat. Sci.* 37: 513–22.

Govindjee & Zilinskas Braun, B. 1974. Light absorption, emission and photosynthesis. *In:* Stewart, W.P.D. [Ed.] *Algal Physiology and Biochemistry.* Blackwell, Oxford, pp. 346–90.

Morris, I. 1981. Photosynthetic products, physiological state and phytoplankton growth. *Can. Bull. Fish. Aquat. Sci.* 210: 83–102.

Nicholls, K.H. 1980. A reassessment of *Chrysosphaerella longispina* and *C.*

multispina, and a revised key to related genera in the Synuraceae (Chrysophyceae). *Pl. Syst. Evol.* 135: 95–106.
Pick, F.R. 1982. Vertical stratification of phytoplankton in lakewater. Ph.D. thesis. University of Toronto.
Pick, F.R. & Lean, D.R.S. 1984. Diurnal movements of metalimnetic phytoplankton. *J. Phycol.* 20: 430–6.
Pick, F.R., Nalewajko, C. & Lean, D.R.S. 1984. The origin of a metalimnetic chrysophyte peak. *Limnol. Oceanogr.* 29: 125–34.
Wallen, D.G. & Geen, G.H. 1971a. Light quality in relation to growth, photosynthetic rates and carbon metabolism in two species of marine plankton algae. *Mar. Biol.* 10: 34–43.
– 1971b. Light quality and concentration of proteins, RNA, DNA and photosynthetic pigments in two species of marine plankton algae. *Mar. Biol.* 10: 44–51.

15

Effects of environmental temperature on the vegetative growth and sexual life history of *Dinobryon cylindricum* Imhof

CRAIG D. SANDGREN[1]
Department of Biology
The University of Texas at Arlington
Arlington, Texas 76019 U.S.A.

Introduction

The response of phytoplankton growth to a gradient of environmental temperatures is known from a large number of studies (Aruga 1965, Smayda 1969, Eppley 1972, Moss 1973, Goldman & Carpenter 1974, Goldman & Ryther 1976, Goldman 1977, Senft, Hurchberger, & Roberts 1981, Spencer 1983). Growth rates typically describe a negatively skewed "normal" distribution with rates being particularly sensitive to temperature changes above the optimal range. Most species show broad optimal ranges between 15° and 25°C, although this frequently fails to correspond to the temperatures at which these forms are typically found in nature (Smayda 1969, Durbin 1974). Interactions between temperature and either light intensity or nutrient supply (Jitts et al. 1964, Yoder, 1979, Goldman & Mann 1980, Tilman, Mattson, & Langer 1981, Mechling & Kilham 1982) are known to be important in influencing phytoplankton growth rates, yields, photosynthetic rates, and competitive fitness. Temperature alone or temperature in combination with other physical and chemical aspects of the planktonic habitat have long been considered key factors in regulating the patterns of seasonal phytoplankton succession in lakes and the ocean (Findenegg 1943, Rodhe 1948, Fogg 1965, Lund 1965, Hutchinson 1967, Smayda 1980). Temperature's importance in regulating the life history patterns of phytoplankton species has also been demonstrated frequently, particularly with regard to the survivorship and germination requirements of benthic resting cysts (Holmes 1966, Roelofs & Oglesby 1970, von Stosch 1973, Durbin 1978; von Stosch and Fecher 1979, Spencer, Volpp, & Lembi 1980, Hollibaugh, Seibert, & Thomas 1981, Dale 1983, Heaney, Chapman, & Morison 1983, O'Kelley 1983).

[1]Present address: Department of Biological Sciences, University of Wisconsin at Milwaukee, P. O. Box 413, Milwaukee, Wisconsin 53201 U.S.A.

Temperature has been considered particularly important in determining the geographical and seasonal distributions of planktonic chrysophycean algae. Because of the acknowledged difficulty in culturing chrysophytes under defined conditions, our understanding of the temperature tolerances (and other environmental tolerances as well) of these species has come primarily from correlations to existing ambient conditions when chrysophytes are common in the plankton. On the strength of these observations, it would appear that many chrysophytes are stenothermal in comparison to the thermal tolerances of other planktonic microalgae. Typically, most species are considered to be geographically or seasonally restricted to cool waters (as reviewed in Kristiansen and Takahashi 1982) and to exhibit population maxima during the winter or as bimodal peaks following diatom blooms in the spring and autumn (Ruttner 1930, Findenegg 1943, Hutchinson 1944, 1967, Kristiansen 1965, 1975). Increasing temperature is often proposed to trigger resting cyst (statospore) production by natural chrysophyte populations (Ruttner 1930, Fott 1959, 1964, Willén 1963, Kristiansen 1965; Sheath, Hellebust, & Sawa 1975), the inference being that encystment is cued by the onset of conditions outside the tolerance limits of the vegetative cells.

It has recently been possible to establish many clonal populations of the common and widely distributed chrysophyte *Dinobryon cylindricum* Imhof and to trigger sexual statospore production for this species under defined laboratory conditions (Sandgren 1981). This culture system was used in the present study to evaluate the thermal tolerances of this species over an ecologically significant range and to examine the effect of environmental temperature on statospore production and survivorship. The role of temperature in determining the seasonal periodicity of this species is discussed.

Materials and methods

The experiments presented here utilize three clones of *Dinobryon cylindricum* Imhof that constitute a portion of a previously documented bipolar, heterothallic mating group (Sandgren 1981). Clones 1 and 5 are male, while clone 13 is female. These clones were chosen for the present series of experiments in order to test the consistency among genetically distant clonal populations of any results suggesting that aspects of the *Dinobryon* life history are temperature dependent. All three clones were isolated from mesotrophic, north-temperate lakes during the spring of the year when ambient water temperatures ranged from 10° to 20°C. Sexual copulations in *D. cylindricum* are initiated within several days of mixing compatible clones in culture; nonloricate cells of the male clone fuse anteriorly to loricate female cells to produce binucleate zygotic statospores.

All experiments were performed in controlled environmental cham-

under standard cool white fluorescent lighting of 200 μeinstein $m^{-2} \cdot hour^{-1}$ on a 16-hour light–8-hour dark (16L:8D) cycle. Three experi-perimental temperatures were employed: 10°, 15°, and 20°C. These encompass the range that populations of this species would encounter during their typical vernal growth episodes in the lakes that were the sources of the clones. All clonal populations were acclimated to the experimental temperatures for at least 30 days by making weekly dilutions of batch cultures to keep cells in exponential growth. At the previously estimated growth rates for these clones (Sandgren 1978, 1981), this period of acclimation would represent a minimum of 15 cell cycles under the experimental temperature conditions prior to the initiation of experimental manipulations. DY III, a defined artificial medium designed specifically for *Dinobryon* (Lehman 1976), was used throughout these studies. Phosphorus is the primary limiting nutrient.

Clonal growth experiments were performed in triplicate 30-ml culture tubes. Growth dynamics and final cell yields were determined from tube cultures started with an inoculum of 5000 cells/ml from exponentially growing parent cultures. These cultures were sampled at 4-day intervals, and cell densities were estimated from eight replicate hemocytometer counts. Interval growth rates were calculated on the basis of the exponential growth model. Colony size frequency distributions for each clone were determined by counting the number of cells per colony for 93 colonies sampled from exponentially growing cultures at each experimental temperature.

Sexual encystment dynamics and final statospore yields were determined from triplicate tube cultures started with 2500 cells/ml of both male and female clones. Two crosses, clones 1 × 13 and 5 × 13, were monitored at each experimental temperature for 16 days. This period constitutes essentially the entire exponential and linear growth phases of *Dinobryon* cultures under the defined standard conditions. Samples were taken at 4-day intervals and enumerated as above with both vegetative cells and statospores being simultaneously counted in hemocytometer fields.

Studies of statospore survivorship and germination were conducted on populations of clones 5 × 13 cysts collected on day 16 from 500-ml batch cultures grown under the same standard lighting at each temperature. These statospore populations were stored in opaque polyethylene bottles with loose caps at 4°C in a refrigerator or at 15°C in a growth chamber. At intervals after the initial 40-day period required to kill the remaining vegetative cells in the collections, subsamples were centrifuged, decanted, and then placed in tubes of fresh DY III medium at 15°C under standard lighting. Survivorship was determined following 7 days of incubation by estimating the ratio of living to dead (i.e., "empty") statospores in triplicate populations of 100 cysts. Germination potential was estimated from these same subsamples by making weekly observations over a 28-day period.

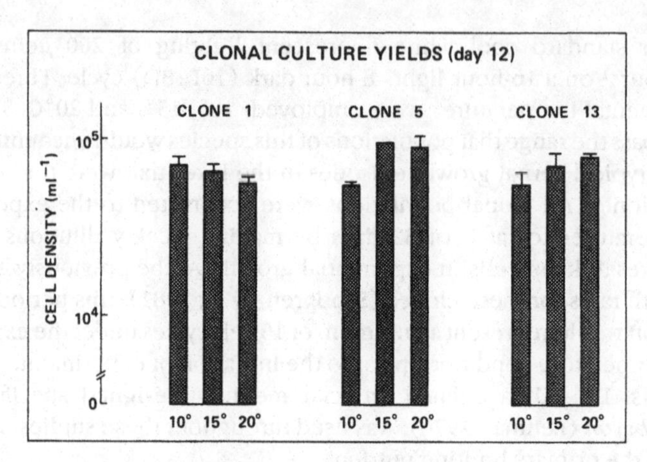

Fig. 15.1. Yield of *Dinobryon* cells in clonal cultures after 12 days of growth under standard conditions with an initial inoculum of 5000 cells/ml. Cell densities are mean values from three replicate cultures ±SD.

Successful germination was signaled by both the observation of excysting cells and the establishment of new vegetative cell populations within this period of time. It must be emphasized that these experimental procedures test the import of the *encystment* temperature regime on statospore survivorship and germination; they do not adequately test the role of ambient environmental temperature in altering survivorship or in cuing statospore excystment.

Results

Effects of environmental temperature on Dinobryon *vegetative growth*

Both the final yield of cells and the population growth dynamics may be altered by environmental factors such as temperature. The final cell densities of the three *Dinobryon* clones after 12 days of growth from initial inocula of 5000 cells/ml showed few temperature-related differences (Figure 15.1). There was no statistically significant effect of temperature on yields of clones 1 and 13 (one way ANOVA, $\alpha = 0.05$). Clone 5 exhibited a reduced yield at 10°C compared to that at the higher temperatures (Student-Newmann-Keuls multiple comparison test, $\alpha = 0.05$). These rather uninteresting yield results, however, masked some important effects of environmental temperature on the growth dynamics of these clones (Figure 15.2). Calculations of interval growth rates clearly indicate that all three clones experienced a prolonged lag phase in growth at 10°C that was not apparent at higher temperatures. This phenomenon was most marked for clone 5. A result of these temperature-related growth discrepancies was

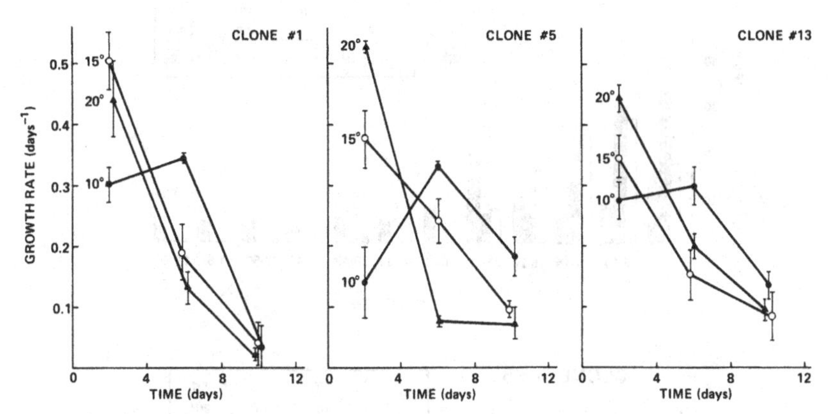

Fig. 15.2. Population dynamics of *Dinobryon* clones. Data from same cultures as in Figure 15.1. Growth rate estimates are mean values ±SD over 4-day periods of growth and are plotted at the midpoints of each time interval for all three experimental temperatures. Open circles, 15°C; closed circles, 10°C; closed triangles, 20°C.

that clonal populations at 10°C exhibit higher growth rates during days 4–8 in batch culture than during days 0–4.

This pattern was in contrast to that exhibited for growth at higher temperatures, where growth rates were highest during the initial 4-day period and cultures had entered the linear phase of growth by days 4–8. Initial growth of both male clones 1 and 5 was very rapid at 20°C so that nutrients (particularly phosphorus) were quickly depleted and growth rates were reduced during days 8–12 relative to those of the lower temperatures. The rates of zygotic statospore production in culture (and by analogy, in nature) should be particularly sensitive to the growth dynamics of the male clones because sexual encystment apparently requires the differentiation of male gametes by recent cell divisions in *Dinobryon* (Sandgren 1981).

An additional aspect of vegetative growth that was tested for temperature dependence is the size of the arborescent *Dinobryon* colonies (Figure 15.3). Apart from any temperature dependence, the three clones investigated here have very different mean colony size during exponential growth. The variance in colony size was directly related to mean colony size among these clones, with clone 1 being the most variable and clone 13 the least. These morphological distinctions are presumably genetically based, but vary with the physiological state of the population (C.D. Sandgren, unpublished observations). Temperature had a marked influence on colony size for clone 1 as demonstrated by an increase in both the mean and median colony size at 20°C (one way ANOVA and SNK multiple comparison test, $\alpha = 0.05$). There was a small but significant increase in mean colony size for clone 13 at 15°C relative to both the higher and lower

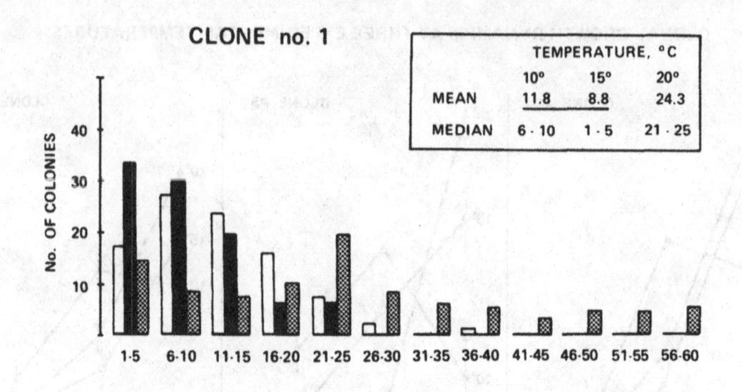

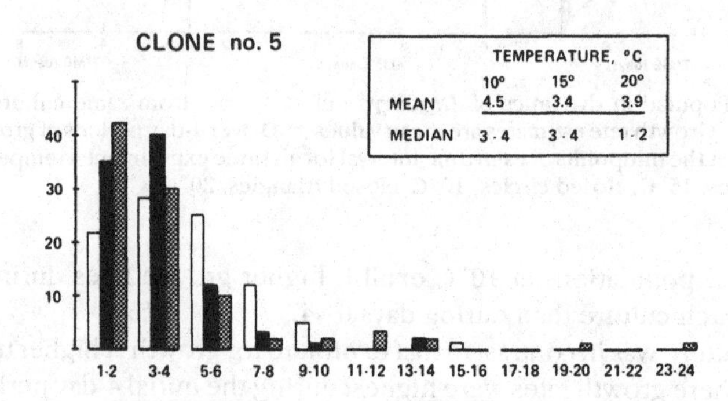

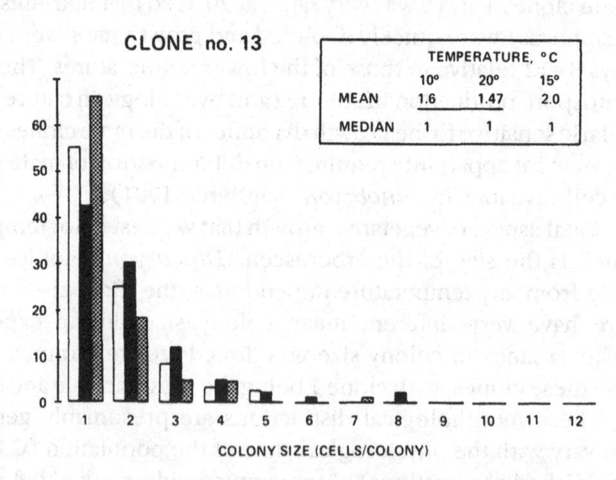

Fig. 15.3. Colony size distributions for three *Dinobryon* clones at three experimental temperatures. All measurements made from exponentially growing populations using acclimated cells. Dark horizontal bars beneath mean colony size estimates denote statistical similarities. White histograms, 10°C; black histograms, 15°C; dotted histograms, 20°C.

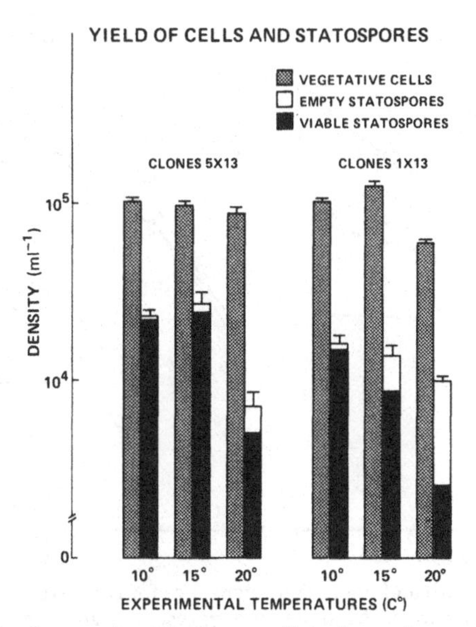

Fig. 15.4. Yields of vegetative *Dinobryon* cells and zygotic statospores following 16 days of growth under standard conditions at each temperature. Values plotted are means ±SD from three replicate crossed cultures inoculated with 2500 cells/ml of each mating type.

temperatures. Clone 5 demonstrated no significant temperature dependency on mean colony size. There was no obvious correlation between mean colony size and estimated clonal growth rates for any of these three clones that could be used to infer either a direct or an indirect relationship between these two variables.

Effects of environmental temperature on Dinobryon *sexual statospore production*

As with vegetative cell growth, temperature may effect both the final statospore yield and the dynamic pattern of sexual reproduction. Yields of vegetative cells and zygotic statospores after 16 days of growth starting with inocula of 2500 cells/ml of each mating type are shown in two sexual crosses (Figure 15.4). Final vegetative cell densities in mixed cultures (see Figure 15.4) exhibited no significant temperature dependence in the cross of clones 5×13, but the 1×13 cross resulted in significantly different final cell yields at all three temperatures (one way ANOVA and SNK multiple comparison test, $\alpha = 0.05$). Total statospore yields (black and white histogram bars in Figure 15.4) in both crosses were significantly depressed at 20°C relative to the lower temperatures. This result is particularly dramatic for clones 5×13 where final statospore yield at 20°C is less than

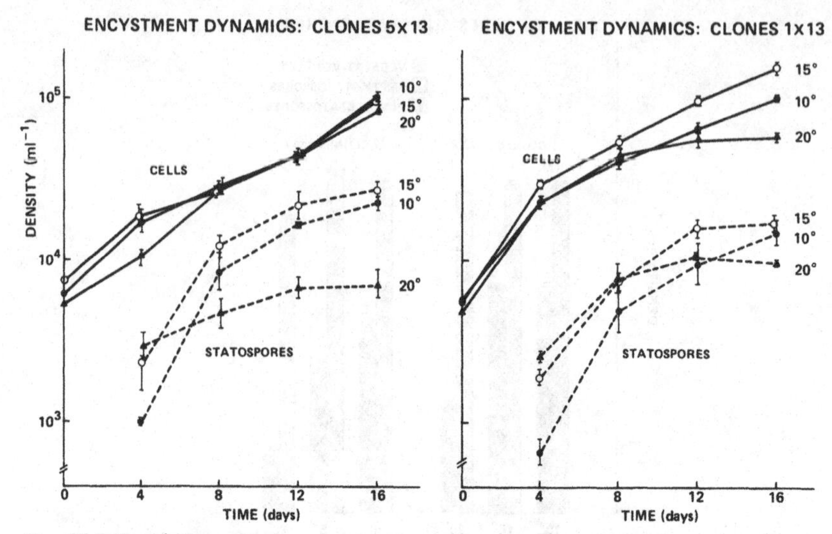

Fig. 15.5. Population growth and encystment dynamics for two sexual crosses of *Dinobryon cylindricum.* Results for each temperature represent means ±SD from three replicate cultures initiated at approximately 2500 cells/ml of each clone. Both cell and cyst densities are cumulative estimates over the entire course of the experiment. Closed circles, 10°C; open circles, 15°C; closed triangles, 20°C.

one-third the yield at 10° and 15°C, even though final vegetative cell densities are similar at all three temperatures.

Not only did increasing temperature effect statospore yields, but an increasing proportion of the statospores produced at higher temperatures were aborted prior to maturation and so appeared as empty cyst walls in the samples (represented by the white portion of the statospore histogram bars). For clones 5 × 13, the average percentage of aborted statospores for 10°, 15°, and 20°C cultures was 3.1, 11.9, and 29.8%, respectively. The values for clone cross 1 × 13 were 6.8, 37.2, and 72.5, respectively. These results suggest a profound effect of environmental temperature on the ability of statospores to mature in batch cultures that is independent of temperature effects on sexual copulations and statospore initiation.

Some explanation for the observed temperature dependence of statospore yields in these tube cultures can be deduced by an examination of the encystment dynamics (Figure 15.5) with reference to the dynamics of the individual clones under the same standard growth conditions (Figure 15.2). Statospore production during the initial 4-day period was comparatively low at 10°C for both sexual crosses. This could have resulted from the prolonged lag phase of growth demonstrated by all three clones at 10°C, or it could reflect a requirement for a minimal cell density to initiate sexual encystment. Apart from this initial difference, however, the encystment dynamics and, therefore, the final statospore yields of the 10° and 15°C cultures were very similar. The dynamics of encystment at 20°C for both crosses was distinctive, exhibiting a higher initial encystment rate (e.g., cysts produced per vegetative cell) that then declined rapidly after day 8 relative to the encystment patterns expressed at the lower temperatures.

The explanation for this distinctive pattern of encystment at 20°C must lie in the dynamics of the individual clones growing in the mixed culture. In monoculture, the growth of the male clone 5 was reduced to very low rates by day 8 (Figure 15.2). Therefore, few male gametes would be differentiated after this time, and sexual encystment in 5 × 13 crosses should be greatly depressed, as was observed. Most of the vegetative cells produced during the latter stages of growth for these crossed cultures would be predicted to be of clone 13 because the low expected clone 5 growth rates in combination with the loss of vegetative cells to sexual reproduction should greatly reduce the standing crop of the male clone. The decline in growth rate through time in batch cultures of clone 1 was less dramatic than for clone 5 (Figure 15.2). Therefore, encystment should continue for a longer period in mixed sexual cultures, and this was observed. By days 8–12, however, the growth of clone 1 was reduced to almost zero, and statospore production in crosses declined to zero as expected.

The cost of sexual reproduction for these *Dinobryon* populations can be estimated by comparing the actual vegetative cell densities in crossed cultures to the predicted densities that would occur if clones grew at the same rates as exhibited in monoclonal cultures. Day 8 data were selected for this comparison because significant encystment occurred by this time, and the cultures had only just entered the period of nutrient-limited linear growth (Table 15.1). When calculated in this way, the loss of cells from vegetative populations as a result of sexual reproduction was severe in most cases. Cultures of clones 5 × 13 experienced losses in the range of 44–64% of the potential cell density, while 1 × 13 cultures suffered losses in the range of 15–38%.

Effects of encystment temperature on statospore survivorship and germination potential

Survivorship as a function of storage time for populations of statospores collected from batch cultures and then stored at 4°C in the dark are shown (Figure 15.6). The values on day 0 of storage reflect the temperature dependence of initial cyst maturation as has already been described (Figure 15.4). The slopes of these three curves are very similar; the difference among these populations in cyst survivorship values on any date is primarily related to the initial values and not the slopes. Ambient temperature during encystment episodes would thus appear to have had little impact on cyst survivorship, and the number of potentially viable cysts was primarily a product of the number of mature statospores initially produced. All three

Table 15.1. *Cost of sexual reproduction in batch cultures of* Dinobryon cylindricum, *day 8*[a]

Cross	Temperature (°C)	Estimated cell yields without sexual reproduction (cells/ml)	Actual mean cell yields in crossed cultures (cells/ml)	Percentage decline in density
1 × 13	10	58,460	36,070	38.2
	15	59,171	50,070	15.4
	20	57,140	46,780	18.1
5 × 13	10	41,385	22,861	44.8
	15	48,523	17,278	64.4
	20	60,602	26,935	55.6

[a] Data represent a comparison of predicted potential and actually observed vegetative cell densities in crossed cultures on day 8 of growth. Estimated cell yields without sexual reproduction were derived from growth rates in clonal cultures and are corrected for an inoculum density of 2500 cells/ml. Actual mean cell yields were derived from the growth rates observed in the crossed cultures undergoing sexual encystment and are also corrected for inoculum densities of 2500 cells/ml of each mating clone.

populations contained viable statospores after 1 year of dark, cold storage. Based on the day 16 statospore yields already presented (Figure 15.4) and the survivorship values (Figure 15.6), the size of viable statospore populations after 1 year of storage for 10°, 15°, and 20°C encystment were estimated to be 3690, 3050, and 285 viable statospores/ml, respectively, under the standard conditions employed.

The potential of individual statospores in the above samples to germinate and produce new vegetative cell populations appeared independent of the encystment temperature (Table 15.2). All three statospore populations stored at 4°C exhibited an apparent initial dormancy period of 2–3 months before germination could be induced. One possible method of overcoming this dormancy appeared to be storage of samples at ambient temperatures (e.g., 15°C) rather than 4°C. At least the 15° and 20°C statospores could be germinated up to 12 months after initial storage regardless of which storage temperature regime was utilized. In all cases, the percentage of germinating statospores constituted a small proportion (10–20%) of the apparently viable spores in the sample.

These studies are the first documentation of *Dinobryon* statospore germination in culture, and they are the first demonstration that the entire sexual life history of a chrysophyte can be reproduced in vitro. A single ameboid cell emerged from germinating statospores and immediately secreted a cellulosic encystment vesicle such as was observed for *Dinobryon divergens* by Sheath and co-workers (1975). A complete account of events

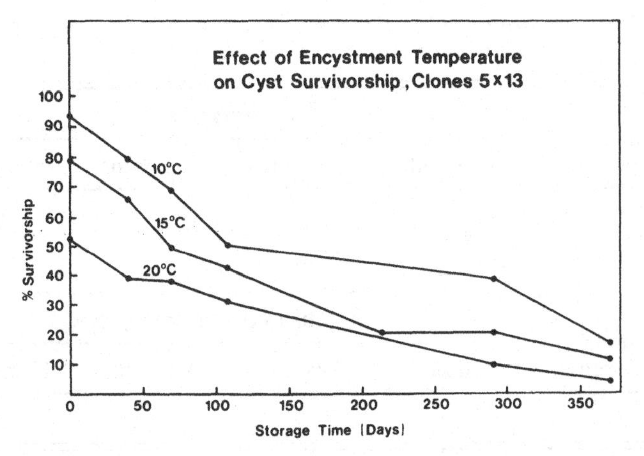

Fig. 15.6. Zygotic statospore survivorship curves for statospore populations produced under three encystment temperatures. Survivorship estimated as the percentage of living versus empty statospores in each sample after returning stored samples to light and fresh medium. Storage conditions were 4°C and dark for all three samples. Values plotted are mean values for three replicate counts of 100 statospores.

during germination of *D. cylindricum* sexual statospores will be published elsewhere.

Discussion

Temperature and vegetative growth

The results presented in this study suggest that *D. cylindricum* has the potential to produce vegetative cell populations of comparable densities over the temperature range of 10° to 20°C given adequate light and nutrients. Although all three clones tested showed a marked lag phase in growth at 10°C that was not evident at 15° and 20°C, the dynamics of population increase were such that final cell densities were similar at all temperatures when these batch cultures were allowed to develop into the linear growth phase. Comparisons of the growth rates for *D. cylindricum* over the temperatures tested here indicate that this species has a temperature tolerance range of the same breadth as many other north-temperature phytoplankton species that have been tested (Aruga 1965, Eppley, 1972, Moss 1973, Senft et al. 1981) although growth above 20°C has not yet been evaluated. It would, therefore, be incorrect to consider *D. cylindricum* a stenothermal species. Because the ambient epilimnetic water temperature in the lakes from which these clones were isolated rarely deviates from the 10°–20°C experimental temperature range (Munch 1972, Sandgren 1978, Lehman 1979), it would seem unlikely that the general pattern of seasonal temperature changes in these lakes would be the sole factor responsible for the observed vernal and autumnal population maxima of this species.

Table 15.2. *Germination potential of* Dinobryon cylindricum *zygotic statospores, clones 5 × 13*[a]

Encystment temperature regime	Dark storage at 4°C			
	2 months	3 months	4 months	12 months
10°C	0	+	+	0
15°C	0	+	+	+
20°C	0	+	+	+

Encystment temperature regime	Months in dark storage, 15°C			
	2 months	3 months	4 months	12 months
15°C	+	+	+	+

[a] Statospore populations were collected from clones 5 × 13 cross cultures on day 16 and stored in the dark at 4° or 15°C. Germination potential was determined by returning samples to the light at 15°C in fresh medium and observing encystment and vegetative population development during the subsequent 28-day interval.

These results do not of course discredit the significance of environmental temperature cycles or species-specific thermal tolerance ranges in influencing chrysophyte seasonal distribution patterns. The general preference of temperate chrysophytes for cool or moderate thermal regimes is too pervasive an observation (Rodhe 1948, Lund 1965, Hutchinson 1967) to be discarded as unimportant. This generalization does not, however, necessarily indicate a direct relationship between absolute thermal tolerance limits and ambient water temperatures. The potential synergistic interaction of temperature in modifying a species' light requirements (Jitts et al. 1964, Yoder 1979) and nutrient sequestering abilities (Goldman and Mann 1980, Tilman et al. 1981, Mechling and Kilham 1982) must be considered as contributing factors that may alter both the environmental tolerances and the competitive fitness of chrysophycean species in limnetic habitats. Temperature changes well within thermal tolerance ranges could indirectly influence chrysophyte seasonal growth patterns. Growth characteristics of chrysophytes above their thermal optima would obviously be most critical for understanding the absence of many species from warm summer conditions; unfortunately such temperatures were not included in the present experimental design. The growth response of *Dinobryon cylindricum* above 20°C might be particularly instructive in view of the obvious thermal stress signs exhibited by the sexual reproduction process at this temperature (see section on temperature and sexual encystment). Short-term thermal instability, as evidenced by sudden stochastic temperature changes, is a significant characteristic of the vernal epilimnion of most north-temperature lakes that has been little studied for its role in modifying growth rates and nutrient uptake characteristics of phytoplankton species. The capacity of species to adapt to such unpredictable environmental changes may provide a competitive advantage to some types of cells and may also be important for explaining the observed spring dominance by diatoms and chrysophytes in many lakes.

The interesting observations by Tilman and co-workers (1976) of colony size dependence upon specific growth rate and the species of limiting nutrient in *Asterionella formosa* Hass. prompted an analysis here of *D. cylindricum* colony size dependence upon temperature. Two of the three clones tested did show a significant shift in mean colony size along the temperature gradient of 10°, 15°, and 20°C. However, the great variability in colony size within each clonal population and the widely different mean colony sizes of these three clones suggest that it would be very difficult to document temperature-dependent colony size shifts in heterogeneous natural assemblages of *Dinobryon*. The general trend is toward larger colonies at higher temperatures.

Temperature and sexual encystment

Elevated temperatures (20°C) during encystment in batch cultures results in a marked decline in the yield of zygotic *Dinobryon* statospores. It is likely that this result in mixed populations of heterothallic mating clones is related to the temperature influence upon growth dynamics of the individual clones. This conclusion is based upon the assumption that potential gametes must be differentiated through a gametogenic cell cycle, and it is supported by the close correlation between crossed-culture growth rates and cyst production rates during the first 4-day interval (Figure 15.5). Subsequent encystment dynamics become difficult to interpret in batch cultures because each clone experiences different growth dynamics while suffering a varying loss of cells through sexual copulations. It is probable that the vegetative cell populations during the later stages of culture development contain a very disproportionate ratio of cells from the two mating types, and this would certainly influence cyst production rates. A direct effect of temperature on the process of gametogenesis itself cannot be ruled out by these experimental methods. A definitive experimental examination of the separate effects of specific clonal growth rates and environmental temperature on *Dinobryon* sexual encystment can only be accomplished in an open system under steady state conditions using compatible clones with identical growth rates. This is not currently possible.

It is clear from these studies and from previous examinations of *D. cylindricum* sexual reproduction in vitro (Sandgren 1981) that initiation of sexual reproduction is not directly determined by environmental temperature. Rather, sexual encystment appears to be initiated whenever sufficient cell densities of both heterothallic mating types are present; this seems independent of environmental temperature over an ecologically meaningful range. There are several previous studies that draw correlations between sexual (and asexual) encystment and the relatively elevated temperature present during the later stages of spring chrysophytc population maxima (Ruttner 1930, Fott 1959, 1964, Willén 1963, Kristiansen 1965, Sheath et al. 1975). It is my supposition that the significant correlation in these observations may be between encystment and the development of sufficiently dense populations to initiate sexual recognition and copulations rather than between encystment and elevated temperature. Environmental temperature does influence the size of statospore populations produced by *Dinobryon* through its effects on vegetative cell growth, but it is not a trigger for encystment initiation. Temperature may be important in initiating the development of some types of diatom resting cells (Holmes 1966, Hargraves and French 1983); its role in initiating dinoflagellate encystment is equivocal (Dale 1983). Temperature has a marked effect on *Chlorococcum* zygospore production (O'Kelley 1983) in culture, and Spencer et al. (1980) demonstrated a significant negative correlation between seasonal water temperature and the abundance of *Pithophora* akinetes in a small pond.

The effect of encystment temperature on the capacity of statospores to mature completely is an important phenomenon documented in this study. The frequency of aborted, empty cysts increases directly with encystment temperature over the range tested. Since clonal growth rates also generally increase with increasing temperature, this observation may reflect the available energy reserves in the cells acting as gametes, with more rapidly growing cells having less stored photosynthate to be used in carrying through the elaborate encystment process (Hibberd 1977, Sandgren 1980a,b, 1983a). It may also reflect a temperature dependence of the mechanism of silica deposition during cyst wall formation because a gradient of decreasing definition in cyst ornamentation with increasing temperature has been previously documented for these same sexual statospores (Sandgren 1983b). Azam, Hemmingsen, and Volcani (1974) suggested that silica transport into diatom cells is temperature sensitive, particularly near the upper physiological tolerance limits. It is unlikely that these observed morphological and physiological trends in *Dinobryon* statospores are the result of nutrient stress in batch cultures because they can be documented in day 4 cyst populations collected while the cultures are actively growing.

No significant effect of encystment temperature has been documented here on either statospore survivorship over a 1-year laboratory storage period or on the potential of statospores to germinate and produce new vegetative cell populations. However, because of the marked effect of temperature on cyst yields as well as on the percentage of cysts that fail to mature fully, the size of the potentially viable cyst population after any period of storage is inversely related to encystment temperature. Thus encystment temperature can have a dramatic impact on potential recruitment from benthic statospore populations. The response of chrysophycean statospores to gradients of temperature during storage and during germination have not yet been investigated, but such factors are obviously of great potential importance for understanding seasonal periodicity and perennation strategies for these species. Temperature and light and physiological dormancy requirements are the primary factors shown to influence survivorship and germination in the resting stages of other planktonic microalgae (Lewin 1949, Lund 1965, von Stosch 1965, 1973, Hargraves & French 1975, 1983, Yamamoto 1976, Rother & Fay 1977, Anderson & Morel 1979, Anderson 1980, Cain 1980, Davis et al. 1980, Dale 1983, Heaney et al. 1983).

Despite the varying size of viable statospore populations initially produced under the three experimental encystment temperature regimes, a sufficiently large number of viable cysts existed after 1 year of cold dark storage to regenerate successfully a vegetative cell population. Only the 10°C sample failed to yield vegetative cells, and this result must be confirmed before any ecological significance can be ascribed. If benthic survival for 1 year's time is a minimum requirement for successful perennation in seasonally restricted natural *Dinobryon* populations, then these zygotic statospores are a sufficient means of survival. Green algal zygotes and akinetes (Coleman 1983), akinetes of cyanobacteria (Roelfs & Oglesby 1970, Reynolds 1975), diatom resting spores (Hollibaugh et al. 1981) and dinoflagellate hypnozygotes (Huber & Nipkow 1923, Anderson 1980, Heaney et al. 1983) have all been demonstrated to persist for 1 year or longer in laboratory storage or in natural sediments.

Conclusions

These experimental studies do not support the perception of common chrysophytes such as *Dinobryon cylindricum* as cool stenothermal phytoplankton. This species can produce comparable yields of vegetative cells over the entire temperature range of 10° to 20°C, with highest specific growth rates occurring at 15° – 20°C. Thermal tolerances alone should not, therefore, be responsible for the commonly observed absence of this species from summer plankton assemblages in temperate lakes. The typical winter or spring – autumn bimodal seasonal distribution of *D. cylindricum* must result from a more complex interaction of physical, physiological, and biological interactions that present unfavorable combinations during the warmest season of the year. The production of sexual resting cysts occurs during the period of active vegetative population growth, and this process may, therefore, constitute a severe loss factor, certainly of a magnitude comparable to losses from cell sinking or zooplankton grazing experienced by natural *Dinobryon* populations. Temperature has a great effect on yields of sexual cysts, with the number of potentially viable cysts being much reduced at 20°C as compared to 10° and 15°C. The survivorship of viable cyst populations produced at all three temperatures, however, was the same under laboratory conditions intended to mimic natural sedimentary environments, and all three cyst populations were able to regenerate vegetative cells after 6–12 months of storage. Such cysts can thus be considered a sufficient mechanism for recruitment in this seasonally restricted phytoplankton. Both vegetative growth and sexual dynamic patterns exhibited significant interclonal variability, and this could potentially expand the thermal tolerance limits of the species. At elevated temperatures, the high costs of sexual reproduction in regards to vegetative cell losses together with low yields of viable cysts may mitigate against the evolution of warm water-adapted clones of this species and may thus stabilize the observed cool-temperature seasonal distribution.

References

Anderson, D.M. 1980. Effects of temperature conditioning on development and germination of *Gonyaulax tamarensis* (Dinophyceae) hypnozygotes. *J. Phycol.* 16: 166–72.

Anderson, D.M. & Morel, F.M.M. 1979. The seeding of two red tide blooms by the germination of benthic *Gonyaulax tamarensis* hypnocysts. *Estuarine and Coast Mar. Sci.* 8: 279–93.

Aruga, Y. 1965. Ecological studies of photosynthesis and matter production of phytoplankton. II. Photosynthesis of algae in relation to light intensity and temperature. *Bot. Mag., Tokyo* 78: 360–5.

Azam, F., Hemmingsen, B.B., Volcani, B.E. 1974. Role of silicon in diatom metabolism. VI. Silicic acid transport and metabolism in the heterotrophic diatom *Nitzschia alba*. *Arch. Microbiol.* 97: 103–14.

Cain, J.R. 1980. Inhibition of zygote germination in *Chlamydomonas moewusii* (Chlorophyceae, Volvocales) by N deficiency and sodium citrate. *Phycologia* 19: 184–9.

Coleman, A.W. 1983. The roles of resting spores and akinetes in chlorophyte survival. *In:* Fryxell, G. [Ed.] *Survival Strategies of the Algae*. Cambridge University Press, pp. 1–22.

Dale, B. 1983. Dinoflagellate resting cysts: "benthic plankton." *In:* Fryxell, G. [Ed.] *Survival Strategies of the Algae*. Cambridge University Press, pp. 69–136.

Davis, C.O., Hollibaugh, J.T., Seibert, D.L.R., Thomas, W.H. & Harrison, P.J. 1980. Formation of resting spores by *Leptocylindrus danicus* (Bacillariophyceae) in a controlled experimental ecosystem. *J. Phycol.* 16: 296–302.

Durbin, E.G. 1974. Studies on the autecology of the marine diatom *Thalassiosira nordenskioeldii* Cleve. I. The influence of daylength, light intensity and temperature on growth. *J. Phycol.* 10: 220–5.

– 1978. Aspects of the biology of resting spores of *Thalassiosira nordenskioeldii* and *Detonula confervacea*. *Mar. Biol.* 45: 31–7.

Eppley, R.W. 1972. Temperature and phytoplankton growth in the sea. *Fish. Bull.* 70: 1063–85.
Findenegg, I. 1943. Untersuchungen über die Ökologie und die Produktions - Verhaltnisse des Planktons in Kärntner Seengebiet. *Int. Rev. Hydrobiol.* 43: 366–429.
Fogg, G.E. 1965. *Algal Cultures and Phytoplankton Ecology.* University of Wisconsin Press, Madison.
Fott, B. 1959. Zur Frage der Sexualität bei den Chrysomonaden. *Nova Hedwigia* 1: 115–30.
– 1964. Hologamic and agamic cyst formation in loricate chrysomonads. *Phykos* 3: 15–18.
Goldman, J.C. 1977. Temperature effects on phytoplankton growth in continuous culture. *Limnol. Oceanogr.* 22: 932–6.
Goldman, J.C. & Carpenter, E.J. 1974. A kinetic approach to the effect of temperature on algal growth. *Limnol. Oceanogr.* 19: 756–66.
Goldman, J.C. & Mann, R. 1980. Temperature influenced variations in speciation and chemical composition of marine phytoplankton in outdoor mass cultures. *J. Exp. Mar. Biol. Ecol.* 46: 29–39.
Goldman, J.C. & Ryther, J.H. 1976. Temperature-species competition in mass cultures of marine influenced phytoplankton. *Biotechnol. Bioengng.* 18: 1125–44.
Hargraves, P.E., and French, F. 1975. Observations on the survival of diatom resting spores. *Nova Hedwigia Beihefte* 53: 229–38.
– 1983. Diatom resting spores: Significance and strategies. *In:* Fryxell, G. [Ed.] *Survival Strategies of the Algae.* Cambridge University Press, pp. 49–68.
Heaney, S.I., Chapman, D.V., & Morison, H.R. 1983. The role of the cyst stage in the seasonal growth of the dinoflagellate *Ceratium hirundinella* within a small productive lake. *Br. Phycol. J.* 18: 47–59.
Hibberd, David J. 1977. Ultrastructure of cyst formation in *Ochromonas tuberculata* (Chrysophyceae). *J. Phycol.* 13: 309–20.
Hollibaugh, J.T., Seibert, D.L.R., & Thomas, W.H. 1981. Observations on the survival and germination of resting spores of three *Chaetoceros* (Bacillariophyceae) species. *J. Phycol.* 17: 1–9.
Holmes, R.W. 1966. Short-term temperature and light conditions associated with auxospore formation in the marine centric diatom *Coscinodiscus concinnus* W. Smith. *Nature (London)* 209: 217–18.
Huber, G. and Nipkow, F. 1923. Experimentelle Untersuchungen über die Entwicklung und Formbildung von *Ceratium hirundinella* O. F. M. *Flora (Jena)* 116: 114–215.
Hutchinson, G.E. 1944. Limnological studies in Connecticut. VII. A critical examination of the supposed relationship between phytoplankton periodicity and chemical changes in lake water. *Ecology* 25:3–26.
– 1967. *A Treatise on Limnology.* Vol. 2. *Introduction to Lake Biology and the Limnoplankton.* Wiley, New York.
Jitts, H.R., McAllister, C.D., Stephens, K., & Strickland, J.D.H. 1964. The cell division rates of some marine phytoplankters as a function of light and temperature. *J. Fish. Res. Bd. Canada* 21: 139–57.
Kristiansen, J. 1965. Occurrence and ecology of *Chrysolykos planctonicus,* a chrysomonad with sexual reproduction. *Bot. Tidsskr.* 61: 98–105.
– 1975. On the occurrence of species of *Synura* (Chrysophyceae). *Verh. Internat. Verein. Limnol.* 19: 2709–15.
Kristiansen, J. & Takahashi, E. 1982. Chrysophyceae: Introduction and bibliog-

raphy. *In:* Rosowski, J.R. & Parker, B.C. [Eds.] *Selected Papers in Phycology* Vol. II. Phycological Society of America, Lawrence, Kan., pp. 698–711.

Lehman, J.T. Ecological and nutritional studies on *Dinobryon* Ehr.: seasonal periodicity and the phosphate toxicity problem. *Limnol. Oceanogr.* 21: 646–58.

– 1979. Physical and chemical factors affecting the seasonal abundance of *Asterionella formosa* Hass. in a small temperate lake. *Arch. Hydrobiol.* 87(3): 274–303.

Lewin, R.A. 1949. Germination of zygospores in *Chlamydomonas. Nature (London)* 165: 543–4.

Lund, J.W.G. 1965. The ecology of freshwater phytoplankton. *Biol. Rev.* 40: 231–93.

Mechling, J.A. and Kilham, S.S. 1982. Temperature effects on silicon limited growth of the Lake Michigan diatom *Stephanodiscus minutus* (Bacillariophyceae). *J. Phycol.* 18: 199–205.

Moss, B. 1973. The influence of environmental factors on the distribution of freshwater algae: an experimental study. III. Effects of temperature, vitamin requirements and inorganic nitrogen compounds on growth. *J. Ecol.* 61: 179–92.

Munch, C.S. 1972. An ecological study of the planktonic chrysophytes of Hall Lake, Washington. Ph.D. dissertation, University of Washington, Seattle.

O'Kelley, J.C. 1983. Environmental factors and sexual expression in *Chlorococcum echinozygotium* (Chlorophyceae). *J. Phycol.* 19: 57–64.

Reynolds, C.S. 1975. Interrelations of photosynthetic behavior and buoyancy regulation in a natural population of a blue-green alga. *Freshwater Biol.* 5: 323–38.

Rodhe, W. 1948. Environmental requirements of fresh-water plankton algae; experimental studies in the ecology of phytoplankton. *Symb. Bot. Upsal.* 10: 1–149.

Roelofs, T.D. and Oglesby, R.T. 1970. Ecological observations on the planktonic cyanophyte *Gloeotrichia echinulata. Limnol. Oceanogr.* 15: 224–9.

Rother, J.A., and Fay, P. 1977. Sporulation and the development of planktonic bluegreen algae in two Salopian meres. *Proc. Roy. Soc. Lond. Ser. B* 196: 317–32.

Ruttner, F. 1930. Das Plankton des Lunzer Untersees; seine Verteilung in Raum und Zeit wahrend der Jahre 1908-1911. *Intern. Rev. Ges. Hydrobiol. Hydrogr.* 23: 1–287.

Sandgren, C.D. 1978. Resting cysts of the Chrysophyceae: Their induction, development, and strategic significance in the life histories of planktonic species. Ph.D. Dissertation, University of Washington, Seattle.

– 1980a. An ultrastructural investigation of resting cyst formation in *Dinobryon cylindricum* Imhof (Chrysophyceae, Chrysophycota). *Protistologica* 16: 259–75.

– 1980b. Resting cyst formation in selected chrysophyte flagellates: An ultrastructural survey including a proposal for the phylogenetic significance of interspecific variations in the encystment process. *Protistologica* 16: 289–303.

– 1981. Characteristics of sexual and asexual resting cyst (statospore) formation in *Dinobryon cylindricum* Imhof. *J. Phycol.* 17: 199–210.

– 1983a. Survival strategies in chrysophycean flagellates: Reproduction and the formation of resistant resting cysts. *In:* Fryxell, G. [Ed.], *Survival Strategies in the Algae.* Cambridge University Press, New York, pp. 23–48.

– 1983b. Morphological variability in populations of chrysophycean resting cysts. I. Genetic (interclonal) and encystment temperature effects on morphology. *J. Phycol.* 19: 64–70.

Senft, W.H., II, Hunchberger, R.A., & Roberts, K.E. 1981. Temperature dependence of growth and phosphorous uptake in two species of *Volvox* (Volvocales, Chlorophyta). *J. Phycol.* 17: 323–9.

Sheath, R.G., Hellebust, J.A., & Sawa, T. 1975. The statospore of *Dinobryon divergens* Imhof: formation and germination in a subarctic lake. *J. Phycol.* 11: 131–8.

Smayda, T.J. 1969. Experimental observations on the influence of temperature, light, and salinity on cell division of the marine diatom, *Detonula confervaceae* (Cleve) Gran. *J. Phycol.* 5: 172–194.

– 1980. Phytoplankton species sucession. *In:* Morris, I. [Ed.], *The Physiological Ecology of Phytoplankton.* University California Press, Berkeley, pp. 493–570.

Spencer, D.F. 1983. Temperature and the growth of *Pithophora oedogonia* Witt. (Chlorophyta). *Phycologia* 22: 202–5.

Spencer, D.F., Volpp, T.R., & Lembi, C.A. 1980. Environmental control of *Pithophora oedogonia* (Chlorophyceae) akinete germination. *J. Phycol.* 16: 424–7.

Tilman, D., Kilham, S.S., & Kilham, P. 1976. Morphometric changes in *Asterionella formosa* colonies under phosphate and silicate limitation. *Limnol. Oceanogr.* 21: 883–6.

Tilman, D., Mattson, M., & Langer, S. 1981. Competition and nutrient kinetics along a temperature gradient: an experimental test of a mechanistic approach to niche theory. *Limnol. Oceanogr.* 26: 1020–33.

von Stosch, H.A. 1965. Sexualität bei *Ceratium cornutum* (Dinophyta). *Naturwiss.* 52: 112–13.

– 1973. Observations on vegetative reproduction and sexual life cycles of two freshwater dinoflagellates, *Gymnodinium pseudopalustre* and *Woloszynskia apiculata* sp. nov. *Br. Phycol. J.* 8: 105–34.

von Stosch, H.A. & Fecher, K. 1979. "Internal thecae" of *Eunotia soleirolii* (Bacillariophyceae): development, structure and function as resting spores. *J. Phycol.* 15: 233–43.

Willén, T. 1963. Notes on Swedish plankton algae. *Nova Hedwigia* 5: 39d56.

Yamamoto, Y. 1976. Effect of some physical and chemical factors on the germination of akinetes of *Anabaena cylindrica. J. Gen. Appl. Microbiol.* 22: 311–23.

Yoder, J.A. 1979. Effect of temperature on light-limited growth and chemical composition of *Skeletonema costatum* (Bacillariophyceae). *J. Phycol.* 15: 362–70.

PART V

Ecology and biogeography

16

Identification, ecology, and distribution of silica-scale-bearing Chrysophyceae, a critical approach

JØRGEN KRISTIANSEN[1]
Institute of Plant Anatomy and Cytology
University of Copenhagen
1307 Copenhagen K, Denmark

Introduction

The silica-scale-bearing Chrysophyceae (the families Mallomonadaceae and Paraphysomonadaceae) present special problems in identification and consequently also in the study of their ecology and distribution. In most cases, electron microscopy of their silica structures is necessary for reliable identifications. Only now, after such studies have been made on so many species, in so diverse localities, and from so many parts of the world, can ecological types be recognized, and occurrence and distribution patterns begin to appear. On the other hand, because of the increasingly sophisticated methods, still finer morphological details are used for taxonomy, and the problems are accentuated with regards to the relevant species criteria that are so crucial in ecological and biogeographical investigations.

Any synthesis of ecology and distribution of scale-bearing Chrysophyceae must be based on many individual investigations. It is evident that the results from such investigations must be both reliable and exhaustive in order to be useful.

However, many contributions, even quite recent ones, are very defective in these respects. Several species of Chrysophyceae described in the literature and still occurring in phytoplankton lists are impossible to identify (e.g., *Mallomonas producta, M. apochromatica,* and *M. minima*), and they must be considered dubious (Asmund & Kristiansen 1986). Several names (e.g., *Synura uvella* Ehrenberg) include more than one species. Often there is no proof of correct identification of valid species such as *Mallomonas acaroides,* and the species lists are most often incomplete, as species of *Paraphysomonas* or *Spiniferomonas* (= *Chromophysomonas*)

The author is indebted to many colleagues for discussions of these problems and to Kirsten Pedersen for typing the manuscript.

[1]Present address: Institut for Sporeplanter, University of Copenhagan, Øster Farimagsgade 2D, 1353 Copenhagen K, Denmark.

are rarely mentioned – even if they certainly must have been present, but they have been overlooked because of their small size.

From a starting point of this kind, the survey presented here must concentrate on the actual possibilities for making such a synthesis, on the identification of taxa, and on principles and viewpoints for investigations on ecology and distribution.

Identification and taxonomy

Any investigation on phytoplankton ecology and distribution must be based upon taxonomical units. This means that the basic problem is definition and identification.

The silica-scaled Chrysophyceae have traditionally in modern literature been termed Synuraceae, although as recently pointed out by Silva (1980) Mallomonadaceae has priority. This was a heterogeneous assemblage, even in quite fundamental cytological details, and it has been shown that the family must be divided into two distinct families (Preisig & Hibberd 1983, and Chapter 5, see also Andersen 1985).

The best known genera are *Synura, Mallomonas,* and *Chrysosphaerella,* but a few more exist. Some of these, *Paraphysomonas* and *Spiniferomonas* (also called *Chromophysomonas*), have recently been demonstrated to be very rich in species. Some other genera (e.g., *Conradiella* and *Catenochrysis*) are only rarely reported and they still seem somewhat enigmatic.

The genera are distinguished on light microscopical (LM) characters such as morphology of cells or colonies, flagella number, etc. However, within these genera, taxonomy is based on the silica scales covering the cell. This principle was established by Iwanov (1899) and Korshikov (1929). In the 1950s and 1960s, LM of the scales was replaced by electron microscopy (EM) with Asmund (1955), Fott (1955), Petersen & Hansen (1956), and Harris & Bradley (1956) among the pioneers.

Today, because this group has many species that are very small and because even large forms must be identified based upon ultrastructural details of the scales, special procedures and precautions are necessary (Kristiansen 1979).

First, at the stage of detection, EM is necessary to be aware of the minute species present. Second, both for detection and identification, examination of dried specimens, preferably by EM, is required to ascertain the structure of the scales and bristles. If there is a sufficient calibration and control with EM, several (but by far the smallest part) of the species, (e.g., *Mallomonas punctifera* and *Synura uvella*) can be identified by LM of scale structure and to some extent a few species can also be identified by means of phase contrast microscopy of living cells.

Let us say that we have now tried as conscientiously as possible to detect whatever species may be present and have tried to identify them correctly, taking all precautions as mentioned.

However, what is correct identification when the species concept is unclear? In the recent surveys (Starmach 1980, 1985), 40% of the Mallomonadaceae cannot be identified; their very existence cannot be proved because scale structure was never described. Some aspects of this problem will be discussed in the following.

Many of the early and even some of the later chrysophycean taxonomists did not describe scale structure of their species. If they did, it was based upon LM, which has now been shown to be inadequate. The increasing use of EM in the 1950s and 1960s made silica scales one of the favorite objects of study. Many species were described on the basis of their scale structure as seen in EM but, on the other hand, often with no LM details. Scale ultrastructure became the foundation for taxonomic descriptions.

Thus, we are now in the absurd position that taxonomy is based on the ultrastructure of scales, but this is only known from some of the species. In a great many cases it is impossible to know to which LM-described species the EM-described species (scales) really correspond (cf. Kristiansen 1979). Species without known ultrastructure of the scales are, in fact, worthless. They cannot be identified, and it cannot be determined if they are really independent species or if they are identical with some of the EM-described species. Nevertheless, they still appear in phytoplankton lists. In the monograph by Asmund and Kristiansen (1986), only the species of *Mallomonas* with known ultrastructure of the scales are recognized (93 species plus some varieties and forms).

However, what are the arguments really that silica scale structure should be taxonomically relevant? The immediate reason might be that the silica scales with their well-defined and clear structure give the impression of exactness, and it appears easy to compare them from cell to cell. However, really, this is a pseudo-objective criterion. It is true that scale structure does not contradict other morphological and cytological details as far as they are known, and it appears that at least some species defined in this way have a high degree of stability and often of ecological integrity, with a specific occurrence pattern or with specific environmental requirements (Kristiansen 1975). Also scale morphology appears a constant character even on the global basis.

However, there are difficulties and exceptions. Many species have a specific succession of several scale types in one individual. What if one type does not develop? Imperfect forms occur where scale structure has not been completely finished, or where some types are lacking (Harris 1970).

Some species such as *Synura petersenii* and *Mallomonas crassisquama,* show a great intraspecific variation in scale structure. How great a variation should be allowed within one taxon? Whereas other species, such as *Synura uvella,* are remarkably uniform in scale structure wherever in the world they are found.

Seasonal variation in scale structure may occur, but it has scarcely been examined, and the importance is not known.

This leads to a further problem. How small a variation or difference in scale structure should be taxonomically significant? In LM, only rather coarse differences are visible. However, with the increasing refinement of EM techniques, more and more delicate structures and minute differences can be visualized. What kind and what degree of difference should justify the establishment of new taxa? The value of very small differences must depend on their coincidence with other characters: morphological, ecological, and physiological. That is, the taxon defined in this way must behave or occur in some individual or independent way. However, very little is known about this, and, in fact, most new taxa are established exclusively on very delicate details in silica scale ultrastructure.

All these aspects must be considered by the taxonomist, and the result will depend on his subjective and personal judgment.

The next question is derived from this. How to decide if species level – or lower levels within the infraspecific hierarchy – should be used? It must be remembered that in ecological investigations, often only the species level is recognized, but varieties or forms are omitted.

An example is the scale variation in *Synura petersenii.* There are many morphological types with all transitions between them. Some of these have been given names: *f. kufferathii, f. bjoerkii, var. glabra = S. glabra, f. macracantha = S. macracantha,* etc.

Similarly, various *Synura spinosa* forms were described based upon the morphology of scales found at the anterior end of the cell: *f. spinosa, f. mollispina, f. longispina,* and *f. curtispina.* Here, the special morphology of the posterior scales of the cell gives the clue why it has been considered reasonable to raise *f. curtispina* and *f. mollispina* to independent species, but not *f. longispina.*

A final problem: Are silica structures, in fact, the only relevant taxonomic criteria on species level? The answer is certainly no, but they are the only characters that have been really investigated, and cytoplasmic and other morphological characters are less or scarcely known.

For comparison, in the genus *Ochromonas,* where (apart from cysts) only cell morphology and cytoplasmic characters are available, at least 69 species have been described according to Starmach (1985). However, most are vaguely characterized, and correct identification is very difficult.

Sexuality and zygote formation is generally accepted as a valuable species character, mainly for theoretical reasons. However, it is only known from a small number of species including several *Mallomonas* and one *Synura,* and if not rare, it is at least not often observed. It seems to require both patience and luck on the part of the observer. However, none of the known examples can add new aspects for species delimitation.

Ecology

All these difficulties just mentioned must be borne in mind when literature containing material for the ecology of scale-bearing Chrysophyceae is used and when ecological investigations are planned or actually carried out.

In most phycological or planktological investigations, Chrysophyceae are not in focus. They are as a rule inadequately treated, and only records of very few characteristic and easily recognized species can be trusted [however, see Roijackers (Chapter 17) and Siver and Chock (Chapter 12)].

Three main difficulties are met with regarding Chrysophyceae in ecological investigations. First, electron microscopy will be necessary for detection and identification. For small species (small *Mallomonas* species, *Paraphysomonas, Spiniferomonas* (= *Chromophysomonas*) (cf. Preisig & Hibberd 1982a,b), EM must be used consistently and routinely for examination of the samples. Second, electron microscopes are seldom available in limnological laboratories.

Third, caution is necessary if only silica structures studied by EM are used. It is not certain that a species registered only on the basis of one single scale observed by EM did in fact occur at the locality in the living condition. The scales may have been transported from somewhere else.

One other main problem is quantitative analysis. On the basis of EM observations alone it is not possible to make any estimate of the cell concentration of the species. If there are several species, a subjective estimate of the quantitative relation between these species may be possible. If, on the basis of EM, a species can be recognized in LM, quantification will be possible.

In what habitats do we then find the scale-bearing Chrysophyceae? In the marine environment very few species are present: mainly species of *Paraphysomonas.* It is rather peculiar, that almost all of these, although originally described from the sea, also have been shown to be widely distributed in fresh water. No morphological differences have been described between marine and freshwater forms of the same species (Thomsen 1975, Preisig & Hibberd 1982a,b).

In salt marshes, several *Mallomonas* species occur, but the main habitat for most species is definitely in fresh water.

Temporarily flooded areas on fields are the habitat of several *Mallomonas* species. Ponds, mainly rather clean, but also slightly eutrophied from the surrounding cultivated country side, are some of the localities favored by Chrysophyceae; others are found in slightly humic woodland ponds. Ponds heavily polluted by cattle or ducks have only very few silica-scaled Chrysophyceae. Lakes (acid, humic, alkaline, clean, or eutrophic) will also contain several species; few species occur in severely polluted lakes.

Some species are widely distributed and occur in many locality types, notably *Synura petersenii* (Kristiansen 1975). Some locality types, especially acid waters, have a very special flora (e.g., *Mallomonas paludosa* and *Synura sphagnicola,* but also *Synura petersenii*).

A general belief tends to be that scale-bearing Chrysophyceae are typical of oligotrophic to mesotrophic, slightly humic waters. However, a highly developed flora of 33 species was found in the strongly eutrophied lake Tystrup Sø in Denmark, although not in high cell numbers (Kristiansen 1985).

Another aspect is the autecology of the individual species; this can be approached from various angles. Occurrence in relation to environmental factors can be compiled from regional investigations where various environmental parameters have been measured (e.g., Cronberg & Kristiansen 1980). Thus, it is possible to construct occurrence spectra for the individual species, in relation to these factors. *Synura petersenii* has a very broad spectrum, but it is also a very variable species. *Synura sphagnicola* has a narrow spectrum and exhibits little morphological variation in scale structure. Some examples can be cited for various ways to construct such spectra, here specifically with regard to pH: *Synura* spectra from Denmark (Kristiansen 1975), Takahashi's (1978) from Japan, and Rosén's (1981) summation diagrams from the Thousand Lakes Investigation in Sweden.

Another approach is based on the seasonal variation in occurrence as related to variation in environmental factors dependent on season, such as light, temperature, and ice-cover (Takahashi 1978), and is treated mathematically by Roijackers (Chapter 17).

A third approach is based on population dynamics and survival strategies of the individual species. Most species exhibit a seasonal occurrence pattern with their main or exclusive occurrence period in spring. The remainder of the year, the population is maintained at very low numbers, or it is absent from the water but present in the sediment as statospores, surviving until next occurrence period. The encystment pattern and frequency need much attention (cf. Sandgren 1983, and Chapter 15).

Sexual cysts are very rarely reported and observed among the Mallomonadaceae, and the actual rates of zygotes in relation to asexual statospores are not known. It is not known how sexuality is related to population dynamics or what the special trigger mechanisms are (Sandgren 1983).

The synecological aspect has only been subject to fragmentary investigations. Which species occur together (see Roijackers, Chapter 17), and in which phytoplankton communities they find their natural habitats has not been consistently examined. A few examples of the development and variation of *Mallomonas* communities are given here:

The species number of silica-scaled Chrysophyceae varies through the year in lake Tystrup Sø (Kristiansen 1985), with a distinct vernal maximum. A second example shows what happened during and after restoration in Lake Trummen in Sweden; the number of silica-scaled species increased from two to three before the restoration to 22 after. Immediately after the restoration some species, notably *Mallomonas eoa,* occurred in great quantities, although most occurred in small numbers (Cronberg 1982).

The role of Chrysophyceae in the dynamics of the phytoplankton has been treated by Siver and Chock (Chapter 12).

For the sake of completeness also the paleoecological viewpoint should be mentioned here briefly. By identifying statospores and scales in sediments and incorporating the knowledge of the ecology of the respective species today, it is possible to characterize the lake at the time when the sediment was formed (Adam & Mahood 1981).

Nygaard (1956) tried in his pioneer work on the subject to use unidentified cysts. However, cysts and scales should be identified to be of any use. Munch (1980) correlated recent human activities at a lake with the mallomonad stratigraphy in the lake sediment based on identified scales.

Distribution

When we try to ascertain the distribution of a species, globally or more locally, we meet with problems similar to those addressed in the section on ecology. We need clear taxonomy, correct identifications, exhaustive examinations, and a relevant mesh of investigations. We are still in the position (almost) that any distribution map records rather the distribution of chrysophycean specialists and their travels than the actual species. Also, solid indication of absence from areas examined by competent specialists is valuable. However, how does one define absence? Occurrence periods may be very short, and the species easily missed (Kristiansen 1963).

If we compare maps showing the distribution of EM examinations for every decade, we shall find the following patterns:

1950 - 1960: Western and central Europe
1960 - 1970: Japan and southeastern Europe
1970 - 1980: North America and USSR
1980 - present: Southern Hemisphere

For references to regional investigations see Kristiansen and Takahashi (1982). Distributions, based on EM investigation, have been compiled by Asmund & Kristiansen (1986).

According to the mentioned principles, the following distribution types present themselves (Kristiansen 1981). Cosmopolitan species are distributed throughout the world and are more or less common. Fifty percent of the adequately known species belong in this category. Some are reported from almost every place where samples have been taken, e.g., *Synura petersenii* and to some extent also *Mallomonas crassisquama.* However, these species have considerable intraspecific variation, so how can we express their global distribution with any certainty? With intensified investigations, many more species will certainly prove to belong in the cosmopolitan group.

Some species (i.e., *Mallomonas lelymene, M. insignis, M. hamata*) have remarkably scattered occurrences and perhaps such rare species have very special environmental requirements. Some distributions appear absurd: *Chrysodidymus* was described by LM from Malaysia (Prowse 1962) but has since been recorded by EM from Northeast America and Greenland (Nygaard 1979, Wujek & Wee 1983) and South America (Dürrschmidt 1982). *Synura lapponica* is almost exclusively recorded from some scattered places in Europe. *Mallomonas portae-ferreae* is a subtropical and tropical species found in southern Europe, Africa, and southeast Asia.

Very remarkable are the species that are distinctly restricted to one continent. The best example is *Mallomonas pseudocoronata,* which to date and inspite of intense searching, has only been recorded from North America (United States and Canada, but not Greenland). It is so big and peculiar that it could not be missed in a sample if it were present, yet it has only been found in several parts of North America where it is common. Roughly 20% of the described species are only known from Europe.

Some species are only reported from one locality or at least from a very restricted area: *M. canina* from Denmark (Kristiansen 1982) and *M. vannigera* from south Sweden (Cronberg & Kristiansen 1980) and westernmost Russia (Balonov 1980). Some other species of this group are very small and inconspicuous and will eventually be found in many more places.

The present research activities in North America (e.g., Wee 1982), the tropics, and in the Southern Hemisphere (e.g., Dürrschmidt 1982 and Chapter 7) will no doubt broaden our views regarding the distribution patterns of well-known species and will result in descriptions of new ones, such as the very peculiar *M. plumosa* from Australia (Croome & Tyler 1983). The lack of information from the tropics is obvious. Light microscopic observations from India (Philipose 1953) and Malaysia (Prowse 1962) have indicated a rich flora, but modern EM- contributions are still few: from Chad (Compère 1975), from Bangladesh (Takahashi & Hayakawa 1979), and from Malaysia (Dürrschmidt & Croome 1985). However, much work is in progress, and it will be interesting to see, if a distinct tropical flora will appear. Investigations in the Southern Hemisphere have started recently, and many new species have already been described (e.g., Dürrschmidt 1982, Croome & Tyler 1983, Chapter 7).

A special phytogeographical region is much in need of research: the cold tropics (high altitude lakes in tropical mountains). Furthermore, it is still unknown whether the Antarctic flora will it be comparable to the Arctic flora.

However, the rules of the game are peculiar. It is well known that the description of a new species from one corner of the world very quickly will provoke records of this species from many other places in other parts of the world. This is also especially the case with new monographs, such as Preisig and Hibberd's (1982a,b, 1983) recent series on *Paraphysomonas* and its allies: identification will suddenly be made possible and half-forgotten, stored away material will be looked upon with fresh eyes.

What are the further perspectives in our studies on ecology and distribution of these organisms? One of the obvious questions is the problem of dispersal. The mechanisms are not known, but one might guess that statospores might play a role. How much desiccation can they tolerate? How many years do they remain viable? Another complex of problems is species formation and evolution. How old are the species? Can any group at present be considered to be undergoing rapid evolution? Are there centers of evolution?

All these questions and many more need further research in the future. Necessary prerequisites are a solid foundation in taxonomy based on a rational species concept, an extensive knowledge of chrysophycean ecology, and a comprehensive view of chrysophycean distribution in all parts of the world.

References

Adam, D.P. & Mahood, A.D. 1981. Chrysophyte cysts as potential environmental indicators. *Geol. Soc. Amer., Bull. Pt. I,* 92: 839–44.

Andersen, R.A., 1985. Synurophyceae, a new class of algae. *Abstr. Second Int. Phycol. Congr. Copenhagen,* p. 5.

Asmund, B. 1955. Electron microscope observations on *Mallomonas caudata* and some remarks on its occurrence in four Danish ponds. *Bot. Tidsskr.* 52: 163–8.

Asmund, B. & Kristiansen, J. 1986. The genus *Mallomonas. Opera Botanica* 85: 1–128.

Balonov, I.M. 1980. O novom dlja flory SSSR vide *Mallomonas vannigera* Asmund. (Chrysophyta). *Biol. Vnutrennich Vod, Inform. Bjull.* 47: 8–15.

Compère, P. 1975. Algues de la region du lac Tchad. Cah. O.R.S.T.O.M., *Sér. Hydrobiol.* 9: 167–92.

Cronberg, G., 1982. Phytoplankton changes in Lake Trummen induced by restoration. *Fol. Limnol. Scand.* 18: 1–119.

Cronberg, G. & Kristiansen, J. 1980. Synuraceae and other Chrysophyceae from central Småland, Sweden. *Bot. Notiser* 133: 595–618.

Croome, R.L. & Tyler, P.A. 1983. *Mallomonas plumosa* (Chrysophyceae), a new species from Australia. *Br. Phycol. J.* 18: 151–8.

Dürrschmidt, M. 1982. Studies on the Chrysophyceae from South Chilean inland waters by means of scanning and transmission electron microscopy. *Arch. Hydrobiol. Suppl.* 63 (*Algol. Stud.* 31): 121–63.

Dürrschmidt, M. & Croome, R., 1985. Mallomonadaceae (Chrysophyceae) from Malaysia and Australia. *Nord. J. Bot.* 5: 285–98.

Fott, B. 1955. Scales of *Mallomonas* observed in the electron microscope. *Preslia* 27: 280–2.
Harris, K. 1970. Imperfect forms and the taxonomy of *Mallomonas*. *J. Gen. Microbiol.* 61: 73–6.
Harris, K. & Bradley, D.E. 1956. Electron microscopy of *Synura* scales. *Discovery*, 17: 329–32.
Iwanov, L. 1899. Beitrag zur Kenntnis der Morphologie und Systematik der Chrysomonaden. *Bull. Acad. Imp. Sci. St. Petersbourg*, 5. *Ser.* 11: 247–62.
Korshikov, A.A. 1929. Studies in the Chrysomonads I. *Arch. f. Protistenk.* 67: 253–90.
Kristiansen, J. 1963. Observations on the structure and ecology of *Synura splendida*. *Bot. Tidsskr.* 58: 281–9.
– 1975. On the occurrence of the species of *Synura*. *Verhandl. Internat. Verein. Limnol.* 19: 2709–15.
– 1979. Problems in classification and identification of Synuraceae (Chrysophyceae). *Schweiz. Zeitschr. Hydrol.* 40: 310–19.
– 1981. Distribution problems in the Synuraceae (Chrysophyceae). *Verh. Internat. Verein. Limnol.* 21: 1444–8.
– 1982. *Mallomonas canina* n.sp. (Chrysophyceae), a new member of sect. Heterospinae. *Nord. J. Bot.* 2: 293–6.
– 1985. Occurrence of scale-bearing Chrysophyceae in a eutrophic Danish lake. *Verh. Internat. Verein. Limnol.* 22: 2826–9.
Kristiansen, J. & Takahashi, E. 1982. Chrysophyceae: Introduction and bibliography. *In:* Rosowski, J.R. & Parker, B.C. [Eds.] *Selected Papers in Phycology*, Vol. II. Phycological Society of America, Lawrence, Kansas, pp. 698–704.
Munch, C.S., 1980. Fossil diatoms and scales of Chrysophyceae in the recent history of Hall Lake, Washington. *Freshwater Biol.* 10: 61–6.
Nygaard, G. 1956. The ancient and recent flora of diatoms and Chrysophyceae in Lake Gribsø. *In:* Berg, K. & I.C. Petersen, [Eds]. Studies on the humic, acid Lake Gribsø. *Fol. Limnol. Scand.* 8: 32–93.
– 1979. Freshwater phytoplankton from the Narssaq area, South Greenland. *Bot. Tidsskr.* 73: 191–238.
Petersen, J.B. & Hansen, J.B. 1956. On the scales of some *Synura* species. *Biol. Medd. Kgl. Dan. Vid. Selsk.* 23(2): 1–27.
Philipose, M.T. 1953. Contributions to our knowledge of Indian Algae. I Chrysophyceae. *Proc. Ind. Acad. Sci.* 37: 232–48.
Preisig, H.R. & Hibberd, D.J. 1982a. Ultrastructure and taxonomy of *Paraphysomonas* (Chrysophyceae) and related genera. I. *Nord. J. Bot.* 2: 397–420.
– 1982b. Ultrastructure and taxonomy of *Paraphysomonas* (Chrysophyceae) and related genera. II. *Nord. J. Bot.* 2: 601–38.
– 1983. Ultrastructure and taxonomy of *Paraphysomonas* (Chrysophyceae) and related genera. III. *Nord. J. Bot.* 3: 695–723.
Prowse, G.A. 1962. Further Malaysian freshwater Flagellata. *Gardens' Bull., Singapore* 19: 105–45.
Rosén, G. 1981. Phytoplankton indicators and their relation to certain chemical and physical factors. *Limnologica* (Berlin) 13: 263–90.
Sandgren, C. 1983. Survival strategies of chrysophycean flagellates: reproduction and the formation of resistant resting cysts. *In:* Fryxell, G.A. [Ed.] *Survival Strategies of the Algae.* Cambridge University Press, pp. 23–48.
Silva, P.C. 1980. Names of classes and families of living algae. *Regnum Veg.* 103: 1–156.

Starmach, K. 1980. Chrysophyceae-Złotowiciowse. *Flora Słodkowodna Polski.* Vol. 5, Polska Akademía Nauk, Warszawa.
- 1985. Chrysophyceae und Haptophyceae. *In:* H. Ettl, T. Gerloff, H. Heying, and D. Mollenhauer, [Eds.]: *Süsswasserflora von Mitteleuropa,* Vol 1. Gustav Fischer Verlag, Stuttgart.
Takahashi, E. 1978. *Electron Microscopical Studies of the Synuraceae (Chrysophyceae) in Japan.* Tokyo University Press Tokyo.
Takahashi, E. & Hayakawa, T. 1979. The Synuraceae (Chrysophyceae) in Bangladesh. *Phykos* 18; 129–47.
Thomsen, H.A. 1975. An ultrastructural survey of the chrysophycean genus *Paraphysomonas* under natural conditions. *Brit. Phycol. J.* 10: 113–27.
Wee, J.L. 1982. Studies on the Synuraceae (Chrysophyceae) of Iowa. *Bibl. Phycol.* 62: 1–183.
Wujek, D.E. & Wee, J.L. 1983. *Chrysodidymus* in the United States. *Trans. Am. Microsc. Soc.* 102: 77–80.

17

Development and succession of scale-bearing Chrysophyceae in two shallow freshwater bodies near Nijmegen, The Netherlands

R.M.M. ROIJACKERS[1]
Laboratory of Aquatic Ecology, Catholic University
Toernooiveld, 6525 ED Nijmegen, The Netherlands

Introduction

From May 1977 to May 1978 investigations have been carried out on the phytoplankton of a shallow, nutrient-rich freshwater body (pond F of the Oude Waal) in the surroundings of Nijmegen, The Netherlands. The aim of this study was to investigate the possible shading effect of the floating leaves of the dominating nymphaeids on the phytoplankton (Roijackers 1984). As a part of this study, the successively occurring phytoplankton communities were described. As the Chrysophyceae comprised almost all dominating phytoplankton taxa throughout the year regarding biomass as well as number of species, most attention was paid to this group.

Particularly, the scale-bearing Chrysophyceae appeared to be very rich in species. From October 1978 till November 1979 a similar investigation has been carried out in a neighboring pond (pond D of the Oude Waal).

In this chapter, the development and succession of the scale-bearing Chrysophyceae in both ponds will be dealt with.

Study areas

The two ponds form part of the Oude Waal, a backwater of the river Waal, near Nijmegen (Figure 17.1). The Oude Waal was cut off from the Waal in the seventeenth century, and both ponds must originate from that century. The Oude Waal is usually flooded in late winter by the river Waal, due to increased discharge of the river Rhine (rainfall, melting of snow).

Pond F has a maximum depth in summer of 6 meters; pond D has a maximum depth between 1.5 and 2 meters. *Nuphar lutea* (L.) Sm. is the dominating macrophyte in pond F, together with *Nymphoides peltata*

[1]Present address: Laboratory of Hydrobiology, Agricultural University, De Dreijen 12, 6703 BC Wageningen, The Netherlands.

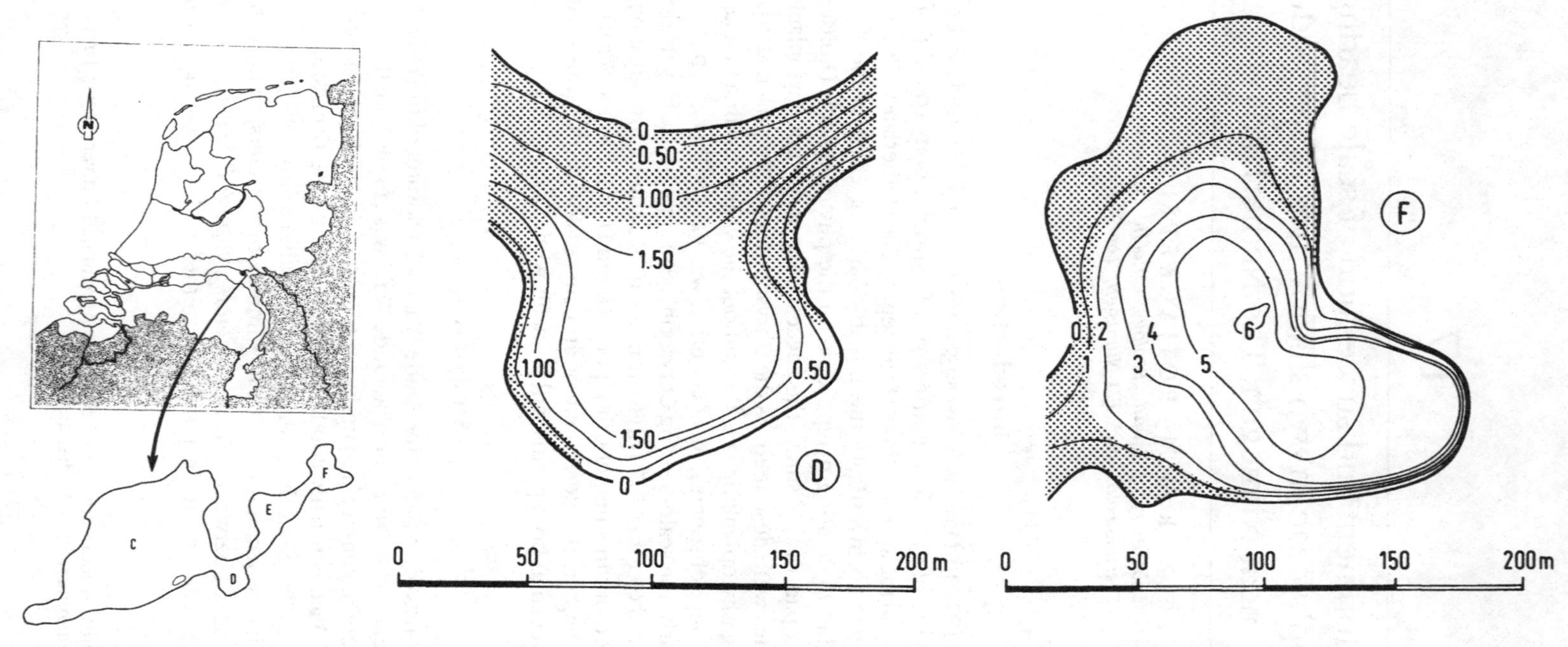

Fig. 17.1. The Oude Waal near Nijmegen, The Netherlands. Left side: the location of the Oude Waal complex with pond D and F. Right side: bathymetric maps of pond D and F with the dotted areas indicating the macrophyte dominated zones (D_2 and F_2) and the open water zones (D_1 and F_1).

(Gmel.) O. Kuntze and *Nymphaea alba* L. Pond D has the same macrophytes, but here *Nymphoides peltata* (Gmel.) O. Kuntze is dominating.

Material and methods

Samples have been taken every week from the upper 50 cm of the ponds. One sample consisted of a mixture of at least 15 1-liter subsamples taken at random throughout the sampling site. Samples were taken at places where nymphaeids occurred (D_2 and F_2) and at places where nymphaeids were absent (D_1 and F_1). At the sampling sites, pH and water temperature have been determined; also samples for chemical analyses have been taken. At the laboratory 1-liter samples were concentrated and studied as soon as possible under the light microscope using phase contrast. One drop of the concentrated material was used for studying the ultrastructure of scales and bristles under the electron microscopes. If necessary, samples have been critically point dried. Photographs and drawings have been made. A Philips EM 201-C transmission electron microscope and a Jeol JSM-U3 scanning electron microscope have been used. A more detailed description of the methods is published by Roijackers (1981). Chemical analyses have been performed according to generally adopted methods.

In order to recognize combinations of scale-bearing Chrysophyceae, the four sampling series have been clustered according to the Patima clustering program of the Computer Center of the Agricultural University, Wageningen. The Patima program performs a hierarchical agglomerative clustering (Boesch 1977). If the development of groups of species and their succession in time is seen as a reaction upon the changing environment, Q-mode studies reveal the affinity between the sampling dates (Sneath and Sokal 1973). All data have been used to compute the Sørensen similarity coefficient (Sørensen 1948). The similarity matrix has been clustered according to Ward (1963). Only species that were present in at least 10% of the samples have been included.

Results

Physicochemical characteristics

Table 17.1 summarizes the most important physical and chemical characteristics of both ponds. Sampling site F_1 has a much greater maximum depth than sampling site D_1 and also the ratio F_1/F_2 area is somewhat higher than the ratio D_1/D_2 area (1.6, respectively, 1.2). So a greater influence of the vegetation at D_2 on D_1 can be expected compared to the situation at F. This is reflected by the higher nutrient load at D_1 compared to F_1 and is expressed in a much higher mean annual phytoplankton biomass at D_1 (41.8 μg chlorophyll *a* per liter, 6.0 mg ash-free dry weight per liter) than at F_1 (17.3 μg chlorophyll *a* per liter, 3.2 mg ash-free dry weight per liter).

Table 17.1. *Physicochemical characteristics (mean annual values) of pond D and F, Oude Waal near Nijmegen*[a]

	Pond D		Pond F	
	D_1	D_2	F_1	F_2
Max. depth (meters)	1.80	1.50	6.00	1.50
area (m^2)	9500	8000	13,000	8000
P as PO_4^{3-} (mg/liter)	0.04	0.04	0.02	0.02
N as NH_4^+ (mg/liter)	0.24	0.24	0.09	0.11
N as NO_3^- (mg/liter)	0.01	0.01	0.01	0.01
N as NO_2^- (mg/liter)	0.01	0.01	0.01	0.01
Ca^{2+} (mg/liter)	48	47	82	80
Mg^{2+} (mg/liter)	14	14	16	15
Cl^- (mg/liter)	76	75	62	62
SO_4^{2-} (mg/liter)	50	49	54	54
SiO_3^{2-} (mg/liter)	3.5	3.5	?	?
HCO_3^- (mg/liter)	?	?	246	245
pH (minimum)	7.1	7.1	7.2	7.4
pH (maximum)	7.9	7.9	8.3	8.4

[a] D_1 and F_1 are the open water zones; D_2 and F_2 are the macrophyte-dominated zones.

The pH of both ponds is neutral to slightly alkaline. The somewhat lower pH values at pond D indicate a higher respiration rate and decomposition activity.

Species composition

All species found in ponds D and F are summarized in Tables 17.2–17.5. Most species have been discussed from a taxonomical viewpoint by Wujek and Van der Veer (1976), Roijackers (1981), and Roijackers and Kessels (1981). The other species will be discussed here. New records for The Netherlands have been indicated with an asterisk.

**Mallomonopsis parvula* Dürrschmidt (1982) (Figure 17.2). This species has already been reported by Roijackers (1981) and Roijackers and Kessels (1981) as *Mallomonopsis* species a.

**Mallomonopsis paxillata* Bradley (1966) (Figure 17.3).

**Mallomonas actinoloma* Asmund and Takahashi (1969) (Figure 17.4).

**Mallomonas alata* Asmund, Cronberg, and Dürrschmidt (1982). Recently the *Mallomonas pumilio* group has been revised by Asmund et al. (1982). In this revision all closely related taxa have been emended, and consequently a new species *M. alata* has been described. *Mallomonas alata* thus includes *Mallomonas* species b as reported by Roijackers and Kessels (1981).

- **Mallomonas areolata* Nygaard em. Asmund (1959) (Figure 17.5).
- **Mallomonas cratis* Harris and Bradley (1960) (Figure 17.6).
- **Mallomonas mangofera* Harris and Bradley (1960) (Figure 17.7).
- **Mallomonas pumilio* var. *munda* Asmund, Cronberg and Dürrschmidt (1982). According to the revision of the *M. pumilio* group by Asmund et al. (1982) this variety is identical with the *M. pumilio* as mentioned by Roijackers (1981) and Roijackers and Kessels (1981).
- **M. punctifera* Korshikov (1941). This species has been reported from The Netherlands as *M. reginae* Teiling (Roijackers 1981, Roijackers and Kessels 1981).
- **Mallomonas striata* Asmund var. *serrata* Harris and Bradley (1960) (Figure 17.8).
- *Paraphysomonas bourrellyi* (Takahashi) Preisig and Hibberd (1982a). In their studies concerning the taxonomy of the genus *Paraphysomonas* Preisig and Hibberd (1982a,b) transferred the type species of the genus *Spiniferomonas* (*Sp. bourrellii* Takahashi) to the genus *Paraphysomonas* as *P. bourrellyi.* The species *P. bourrellyi* thus enclosed *Sp. bourrellii, Sp. conica, Sp. andersonii* (see also Nicholls 1981) and *Spiniferomonas* species a (Roijackers 1981, Roijackers and Kessels 1981).
- **Paraphysomonas takahashii* Cronberg and Kristiansen (1980) (Figures 17.9 and 17.10).
- **Paraphysomonas vacuolata* Thomsen (Thomsen et al. 1981) (Figure 17.11).
- *Chromophysomonas* species c (Figure 17.12). As a consequence of the transference of the type species *Sp. bourrellii* to the genus *Paraphysomonas,* the genus *Chromophysomonas* (Preisig and Hibberd 1982a) has been set up for the remaining species formerly included in the genus *Spiniferomonas.*

  The scales of this taxon are elliptic, with a concentric ring of thickened circles, 1.4–2.4 × 2.2–3.2 μm. The spines are conical with a bifurcate tip, about 5.3 μm long. The diameter of the spine base is about 3.5 μm. The scale structure resembles somewhat that of *Chrysosphaerella coronacircumspina* Wujek et al. (1977). Takahashi and Hayakawa (1979) and also Wee (1982) reported undescribed species that are identical with *Chromophysomonas* species c.

Species development and succession

In Figures 17.13 and 17.14 the seasonal development of the most important taxa has been indicated for ponds F and D respectively. Although a lot of scale-bearing Chrysophyceae tend to live throughout the year, a clear

Table 17.2. *Scale-bearing Chrysophyceae from* D_1[a]

	1978											1979																																		
	Oct.				Nov.				Dec.			Jan.			Feb.		Mar.			Apr.				May					June				Jul.					Aug.				Sept.				Oct.	
	09	16	23	30	06	13	20	27	04	11	18	15	22	29	05	12	12	19	26	02	09	17	23	01	07	14	21	28	05	11	18	25	02	09	16	23	30	06	13	20	27	03	10	17	24	01	08
Mallomonopsis																																															
M. oviformis																		–																													
M. parvula																	–						–																								–
M. paxillata																		–	–																												
M. salina																																													–	–	
Mallomonas																																															
M. acaroides	–	○	–	–	○											–	–		–		○	○	○	–	–		–	–	○	○	○	○	○	–		–		○	○	○	○	–	–	–	–	–	–
M. akrokomos	–	–				×					×	×	×	×	×	+	+	○	–	–	–	–	–		–		–	○	–	–	○	–	–		–		–	–			–						
M. alata																					–																										
M. annulata																	–	–	○	○	–	–				–		–					–														
M. areolata																–																										–					
M. calceolus																																						–									
M. caudata				–		×											–	–																	–	–	–	–	○	○	○	○	○	–	–	–	–
M. crassisquama														×			–		–		–	–		–									–			–	–		–		–			–			
M. cratis																																													–		
M. intermedia																														–	–																
M. mangofera			–																																												
M. monograptus	–	–		–								×	×	×	×	○	○	–	○	–	○	○	○	○	○	–		–			–	○	–	–	–	–	○	○	+	–	–	○	○	○	–	–	–
M. multiunca																						–																									
M. pumilio var. *munda*																			–																												
M. punctifera																																														–	–
M. schwemmlii																												–																			
M. striata var. *serrata*																			–																												
M. tonsurata	–	○		–												–													–	–	○	○	○	○	–	–	○	+	+	–	○	○	○	–	–	–	–
Synura																																															
S. curtispina	–	–															–			–								–					–	○	+	+	○	–	○	○	○	○	○	○	+	○	○
S. echinulata																–							–																								
S. glabra	–	–								×						–	–	–	○	○	–	–	–	–				–		–													–	○	+	○	+
S. petersenii	–	–		–			×	×		×			×			○	–	–	○	○	○	○	–	–	–			–	–		–		–	–	–	–	–	–	–	○	○	○	○	○	○	○	○
S. petersenii f. kufferathii																																													–		
S. petersenii f. praefracta													×			–	–	–	○	○	○	–		–				–					–			–	–	–	○	○	–	–	–		–	–	

S. spinosa																	—	—																					—							
S. uvella	+	+	+	○	○	×	×	×	×	×	×	×	×			—	—	—	—	—	—	—	—	—		—	—		—	—	—	○	—		—			○	—	—	—	—	—	—	—	—
Chrysosphaerella																																														
C. brevispina															—	—	—	—	—	—												—														
C. coronacircumspina																																—	—			—	—	—	—		—	—				
Chromophysomonas																																														
C. cornutus																													—										—	—	—	—	—			
C. trioralis	—	—		—												—		—	—	—							—	—	—	—	○	—	—		—	○	○	—	○	—	—	—	—	—	—	
C. species c																													—				—			—	—									
Paraphysomonas																																														
P. bandaiensis			—																		—	—										—	—													
P. bourrellyi																—	—	—	—				—				—				—					—	—	—			—					
P. butcheri			○	—	—					×									—	—	—	—	—				—	—	—	—	—	○	—					—	—	—	—			—		
P. canistrum				—						×																							—							—						
P. imperforata	—									×								○	—	○	—	—	—				—	—			—	—	—				—		○	—	—		—		—	—
P. imperforata f. no. 2																		—										—	—	—						—	○	—	—							
P. takahashii																	—	○	—	—	—																				—					
P. vacuolata															—	—		—	—											—																
P. vestita	—	○		—							×	×	×	×	○	○	○	+	+	○	○	—	—	—	—	—	—	○	○	○	○	—	○	—		○	○	—	○	—	○	○	—	—	—	

[a] —, rare; ○, common; +, abundant; ×, present, but abundancy not determined.

Table 17.3. *Scale-bearing Chrysophyceae from* D_2[a]

	1978											1979																																			
	Oct.				Nov.				Dec.			Jan.			Feb.		Mar.			Apr.				May					June				Jul.					Aug.				Sept.				Oct.	
	09	16	23	30	06	13	20	27	04	11	18	15	22	29	05	12	12	19	26	02	09	17	23	01	07	14	21	28	05	11	18	25	02	09	16	23	30	06	13	20	27	03	10	17	24	01	08
Mallomonopsis																																															
M. oviformis																		–																													
M. parvula																	–						–																							–	–
M. paxillata																		–	–																												
M. salina																																														–	
Mallomonas																																															
M. acaroides	–	O	–	–				×						×	×	–	–		–		O	O	O	–	–	–		–	–	O	O	O	O	–				–	O	–	O		–	–	–	–	–
M. akrokomos			–	–			×			×		×	×	×	×	+	+	O	–	–	–	–	–		–	–	–	O	–	–	–		–							–			–				
M. alata																					–																										
M. annulata								×									–	–	O	O	–	–							–			–															
M. caudata		–										×					–	–																	–	–	–		–	O	O	O	O	–	–	–	
M. crassisquama		–															–		–		–	–		–			–						–				O		–		–	–	–			–	
M. cratis																																									–		–			–	
M. intermedia																															–	–															
M. mangofera								×																																							
M. monograptus	–	–	–	–										×	×	O	O	–	O	–	O	O	O	O	O	–		–			–	O	–	–	–	–	O	–	O	–	–	O	O	O	–	–	–
M. multiunca																						–																									
M. pumilio var. *munda*																			–																												
M. punctifera																																													–	–	–
M. striata var. *serrata*																			–																												
M. tonsurata	–	O	–	–	–												–											–		–	O	O	O	O	–	–	O	O	O	–	O	O	O	–	–	–	–
Synura																																															
S. curtispina	–	–															–			–													–	O	+	+	O	–	O	O	O	O	O	+	+	O	–
S. echinulata																	–						–																								
S. glabra				–							×					–	–	–	O	O	–	–	–	–									–										–	–	O	O	+
S. petersenii	–	–		–			×	×		×		×			×	O	–	–	O	O	O	O	–	–	–			–			–	–	–	–	–				–	O	O	O	O	O	O	O	O
S. petersenii f. praefracta								×								–	–	–	O	O	O	–		–				–						–		–			O	–	–	–	–		–	–	
S. spinosa																		–	–														–														
S. uvella	+	+	+	O	O	×	×	×	×	×	×	×	×		×	–	–	–	–	–	–	–	–	–	–		–	–		–	–	–	O	–		–	–		–	–	–	–	–	–	–	–	–
Chrysosphaerella																																															
C. brevispina																	–	–	–	–	–																										
C. coronacircumspina																																			–		–	–		–		–					

Taxon																																							
Chromophysomonas																																							
C. cornutus																							–	–									–	–	–				
C. trioralis	–			–								–		–	–	–					–	–	–	–	–	–	–	–	○	○	–	○	–	○	–		–		
C. species b						×					–		–	–	–																								
C. species c																											–			–									
Paraphysomonas																																							
P. bandaiensis			–	–													–	–						–															
P. bourrellyi						×				×	–	–	–	–	–				–		–		–		–						–		–	–		–			
P. butcheri			–	–	–										–	–	–	–	–			○	○	–	–	–	–					–	–	–	–		–		
P. imperforata														○	–	○	–	–	–		○	○		–	–	–	–			–		–	–	–		–			–
P. imperforata f. no. 2														–								–	–	○	–	–	–	–	○	○		–							
P. takahashii													–	○	–	–	–					–											–						
P. vacuolata												–		–	–									–	–									–			–		
P. vestita	–	–		–		×	×	×	×	×	○	○	○	+	+	○	○	–	–	–	–	○	○	+	○	–	○	–	○	○	–	○	–	–	–	–		–	
P. species c																												–											

[a]–, rare; ○, common; +, abundant; ×, present, but abundancy not determined.

Table 17.4. *Scale-bearing Chrysophyceae from F_1*[a]

1977

	May 02	May 09	June 26	July 04	July 11	July 19	July 25	Aug. 04	Aug. 08	Sept. 08	Sept. 12	Oct. 03	Oct. 17	Oct. 24	Oct. 31	Nov. 07	Nov. 14	Nov. 28	Dec. 07	Dec. 12	Dec. 19
Mallomonas																					
M. acaroides				○	○	–			–		×		×		×						–
M. actinoloma																					
M. akrokomos			○	○		–						×									○
M. alata																					
M. annulata			–																		
M. areolata																					
M. calceolus																					
M. caudata	–		–			–	○	○	+	×	×	×	×		×		×				–
M. cratis																					
M. monograptus						–	–	–	○						×				×		
M. multiunca																					
M. pumilio var. *munda*																					
M. striata var. *serrata*																					
M. tonsurata	–	–		–	–	○	○	○	○	×											
M. trummensis																					
Synura																					
S. curtispina		○							–						×						
S. glabra																					○
S. petersenii														×							○
S. petersenii f. kufferathii																					
S. petersenii f. praefracta																					–
Chrysosphaerella																					
C. brevispina																					
C. coronacircumspina						–	–	–	○												
Chromophysomonas																					
C. trioralis				–	–	–	–	–	–				×								
Paraphysomonas																					
P. bandaiensis																					
P. bourrellyi		–																			–
P. butcheri																					
P. imperforata																					
P. imperforata f. no. 2									–												
P. vacuolata																					
P. vestita			○	○	–	–			–		×						×				–

1978

	Jan. 02	Jan. 09	Jan. 16	Jan. 23	Jan. 30	Feb. 06	Feb. 27	Mar. 06	Mar. 13	Mar. 20	Mar. 28	Apr. 03	Apr. 10	Apr. 17	Apr. 24
Mallomonas															
M. acaroides	–	–			–				–	–	○	○	○	○	○
M. actinoloma												–			
M. akrokomos	○	○	○	○	○	○	○	○	○	–	–	–	–	–	–
M. alata		–					–	–	–						
M. annulata		–	–		–	–	–	–	–	–	–	–	–		
M. areolata												–			
M. calceolus													–		
M. caudata	–	–	–	–	–	–		–		–					
M. cratis			–							–	–				
M. monograptus	–	–	–	–	–	–	–	○	○	○	○	○	○	○	○
M. multiunca	–	–	–	○	○	–	–	–	○	–	○	–	–	–	
M. pumilio var. *munda*						–		–	–						
M. striata var. *serrata*								–	–		–				
M. tonsurata									–		–	–	–	–	–
M. trummensis							–								
Synura															
S. curtispina															
S. glabra	+	○	○	–	–	○	–	–						○	–
S. petersenii	○	○	○	○	○	–	–	–	○	+	+	+	+	+	+
S. petersenii f. kufferathii												–			
S. petersenii f. praefracta		–	–		–		–	–			–				
Chrysosphaerella															
C. brevispina							–	–	–	–	–	–	–	–	
C. coronacircumspina															
Chromophysomonas															
C. trioralis		–		–				–		–	–	–	–	○	–
Paraphysomonas															
P. bandaiensis															–
P. bourrellyi				–											
P. butcheri		–					–	–							
P. imperforata											–				–
P. imperforata f. no. 2		–	–					–		–					
P. vacuolata						–									
P. vestita	○	○	–	–	–	○	○	○	○	–	–	–	–	–	–

[a] –, rare; ○, common; +, abundant; ×, present, but abundancy not determined.

Table 17.5. *Scale-bearing Chrysophyceae from F_2*[a]

	1977																			1978													
	May		June		July				Aug.			Oct.				Nov.			Dec.	Jan.					Feb.		Mar.				Apr.		
	02	23	06	26	04	11	19	25	04	08	22	03	17	24	31	07	14	28	19	02	09	16	23	30	06	27	06	13	20	28	03	10	17
Mallomonopsis																																	
M. species a																													–				
Mallomonas																																	
M. acaroides					–	○	–	–	–	–	–									–	–		–						–	○	○	○	○
M. akrokomos			○	○	○		–			–									○	○	○	○	○	○	○	○	○	○	–	–	–	–	–
M. alata																										–	–			–			
M. annulata																					–	–	–	–	–	–	–	–	–	–	–	–	
M. areolata																														–	–	–	
M. caudata	–						–	–	○	+	+	×	×						–	–	–	–	–	–	–							–	
M. cratis																														–			
M. monograptus						–	–	–	–	–	–								–	–	–	–	–	–	–	–	–	○	○	○	○	○	○
M. multiunca																				–	–	–	–	○	–	–	–	○	–	○	–	–	–
M. pumilio var. *munda*																											–		–	–			
M. striata var. *serrata*																											–	–		–			
M. tonsurata	–					–	○	○	○	○	–																		–	–	–	–	–
Synura																																	
S. curtispina		○	–																														
S. glabra																			○	+	○	○	–	–	○	–	–						–
S. petersenii						–													○	○	○	○	○	○	○	–	–	○	+	+	+	+	+
S. petersenii f. kufferathii																								–									
S. petersenii f. praefracta																					–	–	–	–		–	–	–		–	–		
S. spinosa																										–							
Chrysosphaerella																																	
C. brevispina																											–	–	–	–	–	–	
C. coronacircumspina							–	–	–	○																							
Chromophysomonas																																	
C. trioralis					–	–	–	–	–	–	–										–	–				–	–	–	–	–		–	–
Paraphysomonas																																	
P. bourrellyi						–															–					–		–					
P. imperforata																					–	–	–							–	–		
P. imperforata f. no. 2																					–	–	–			–				–			
P. takahashii																						–											
P. vestita				○	○	–	–		–	○									–	–	–	–	○	○	–	○	○	○	–	–	–	–	–

[a] –, rare; ○, common; +, abundant; ×, present, but abundancy not determined.

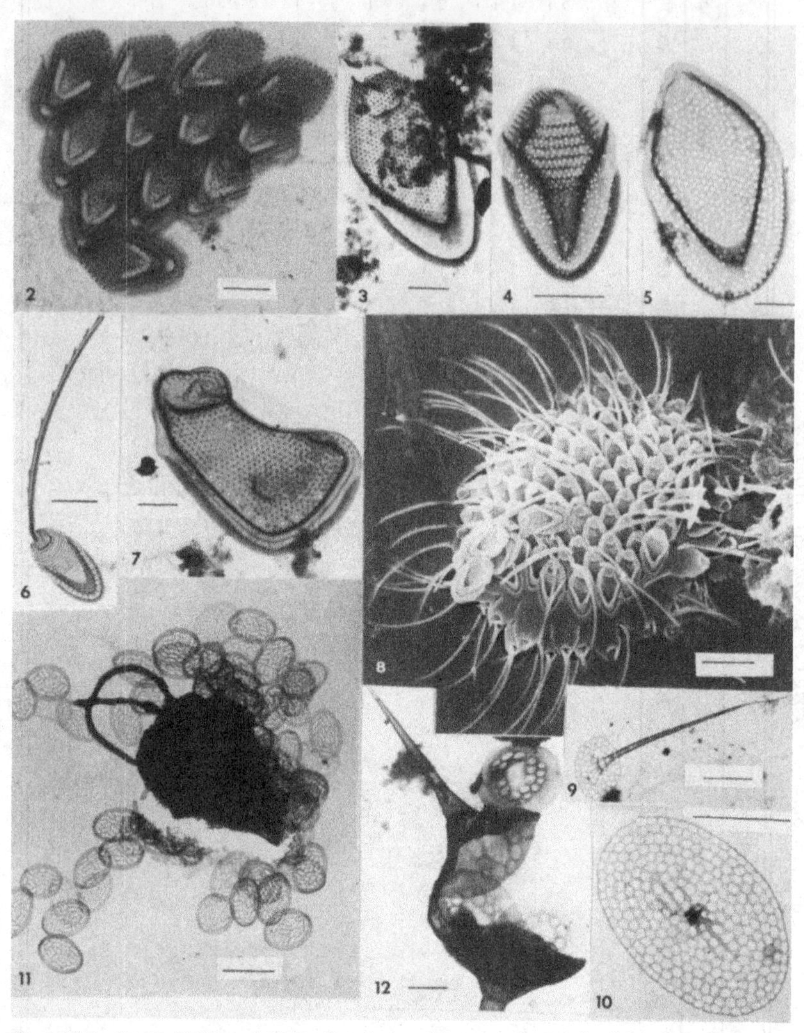

Figs. 17.2–17.12. TEM and SEM of scales and bristles (spines) of some scale-bearing Chrysophyceae from the Oude Waal. Indicated scale is 5 μm in Figures 17.6 and 17.8; in the other figures scale is 1 μm. ***Fig. 17.2.*** *Mallomonopsis parvula* scales. ***Fig. 17.3.*** *Mallomonopsis paxillata* scale. ***Fig. 17.4.*** *Mallomonas actinoloma* scale ***Fig. 17.5.*** *Mallomonas areolata* scale. ***Fig. 17.6.*** *Mallomonas cratis* scale with bristle. ***Fig. 17.7.*** *Mallomonas mangofera* dome-scale. ***Fig. 17.8.*** *Mallomonas striata* var. *serrata* complete cell. ***Fig. 17.9.*** *Paraphysomonas takahashii* short-spined scale. ***Fig. 17.10.*** *Paraphysomonas takahashii* long-spined scale. ***Fig. 17.11.*** *Paraphysomonas vacuolata* complete cell. ***Fig. 17.12.*** *Chromophysomonas* species c scales and spines.

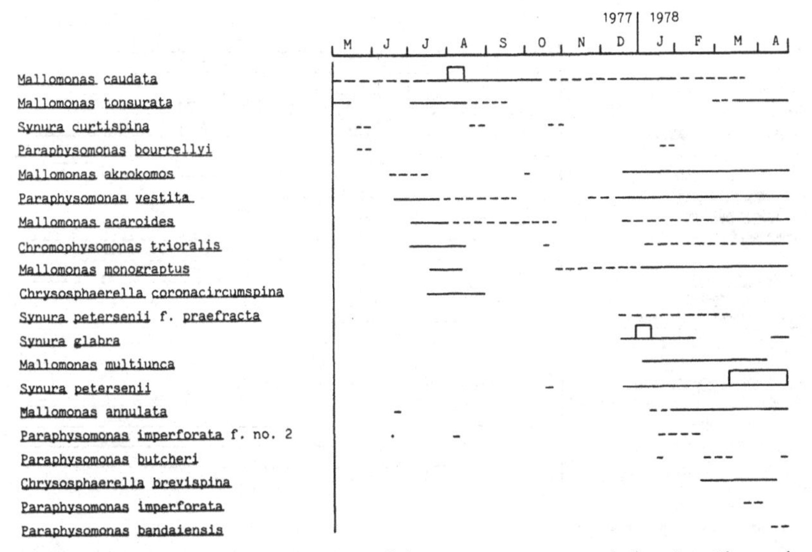

Fig. 17.13. The seasonal development of the most important scale-bearing Chrysophyceae in pond F (F_1) of the Oude Waal. □ taxon abundant to common; —, taxon common to rare.

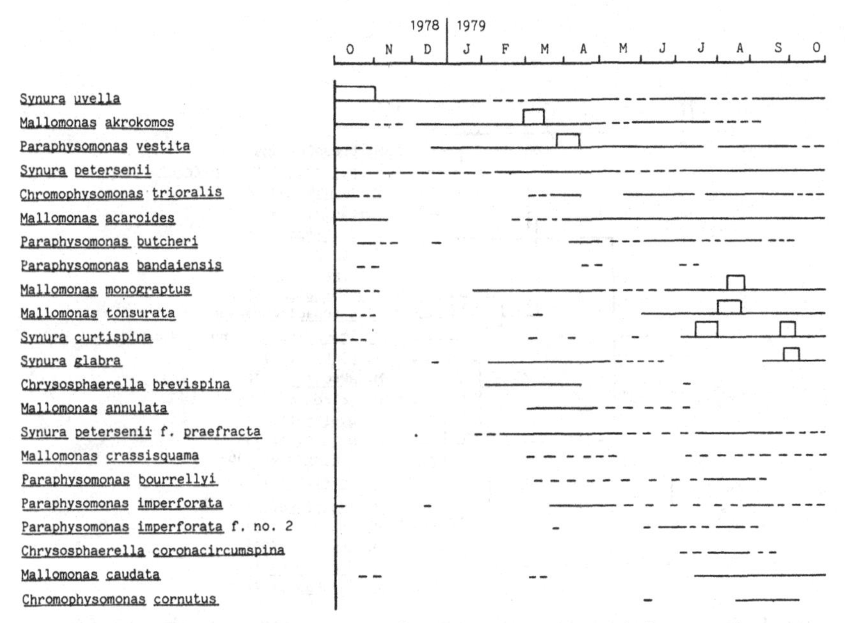

Fig. 17.14. The seasonal development of the most important scale-bearing Chrysophyceae in pond D (D_1) of the Oude Waal. □, taxon abundant to common; —, taxon common to rare.

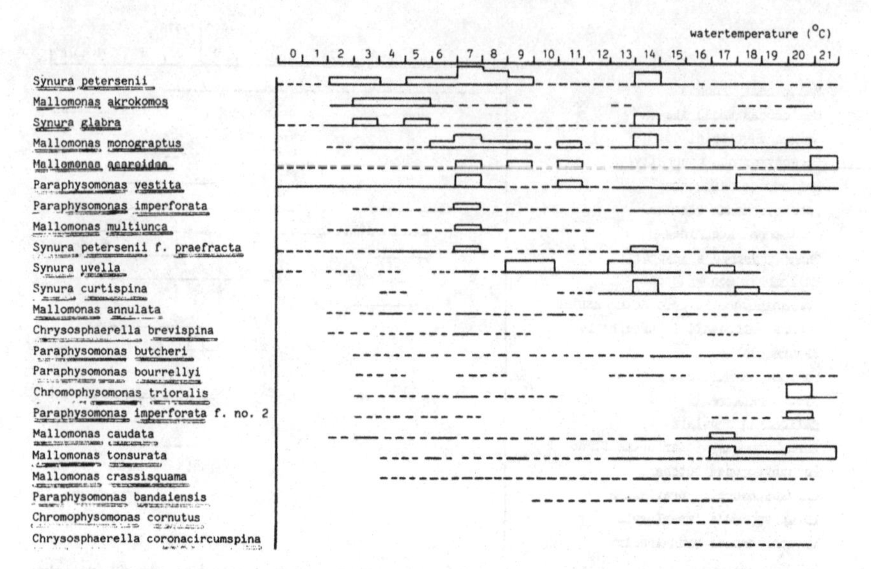

Fig. 17.15. The occurrence of some scale-bearing Chrysophyceae in the Oude Waal as a function of the water temperature. – – –, taxon found in 1–3 of the four sampling sites; ——, taxon present at all the sampling sites (rare); □, indicates the abundancy.

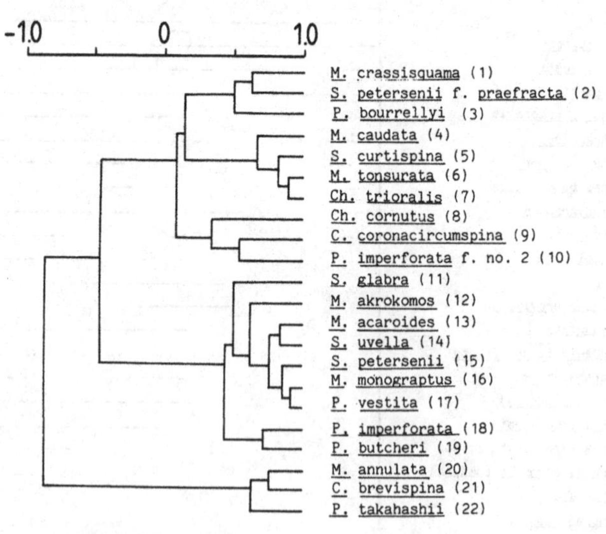

Fig. 17.16. Dendrogram of the most important scale-bearing Chrysophyceae of D_1. The taxa have been numbered.

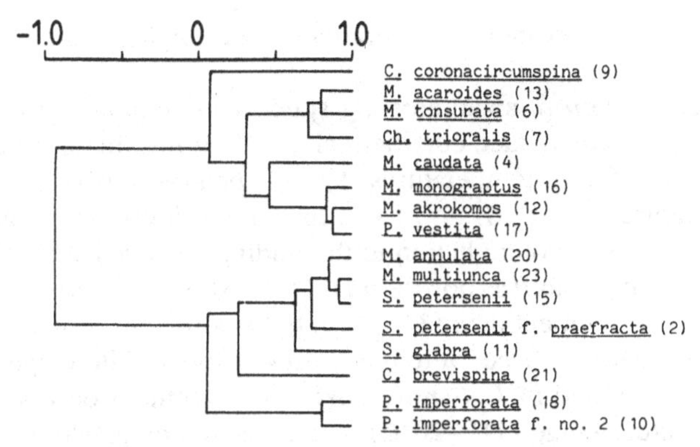

Fig. 17.17. Dendrogram of the most important scale-bearing Chrysophyceae of F_2. The taxa have been numbered in the same way as in Figure 17.16.

species succession can be observed. There are indications that the distribution of the scale-bearing Chrysophyceae is primarily determined by pH (Takahashi 1978) and alkalinity (Roijackers and Kessels in press) and secondarily by the water temperature (Takahashi 1978, Roijackers and Kessels in press). As the alkalinity as well as the pH in both ponds of the Oude Waal are fairly constant throughout the year (Table 17.1), the main parameter that regulates the occurrence of the taxa in the Oude Waal would be the water temperature. In Figure 17.15, the occurrence of 23 taxa has been indicated for a water temperature ranging from 0° to 21°C.

Figure 17.15 is based on Tables 17.2–17.5 and the water temperature data determined at the moment of sampling. All quantitative data of the four sampling places have been mathematically averaged and visualized as indicated in Figure 17.15. From this figure, the water temperature tolerance of some taxa can be clearly seen. The genus *Mallomonopsis* has been omitted from Figure 17.15 because there were too few data. However, from similar investigations by Takahashi (1978), it is known that *M. oviformis* is an oligothermal species (4.5° – 14°C) and that agrees well with my findings (2°C).

The genus *Mallomonas* has been represented in Figure 17.15 by eight species. *Mallomonas akrokomos* seems to be eurythermal, but has a clear prevalence for low temperatures (3° – 5°C). *Mallomonas monograptus* is also eurythermal, but is most common at temperatures above 5°C; the same is true for *M. acaroides*, although Takahashi (1978) decided to classify this species as stenothermal (8.5° – 19°C). *Mallomonas annulata* is classified by Takahashi (1978) as oligothermal (2° – 4°C), but the range must be extended (2° – 11°C and 15° – 18°C) so the species seems to be eurythermal with highest abundancy at lower temperatures. *Mallomonas crassisquama, M. caudata,* and *M. tonsurata* are eurythermal. However,

M. tonsurata has a maximum development at the higher temperatures (17°–21°C).

For the genus *Synura*, an interesting species succession can be observed. All species are, in fact, eurythermal, but their maximum development occurs at different temperatures. *Synura petersenii* occurs at low water temperatures (2°–9°C) and has a maximum development at the range 5°–9°C. This in contradiction to the findings of Takahashi (1978) who classified *S. petersenii* as polythermal (16°–31.5°C). *Synura glabra* occurs also at low temperatures (3°–5°C), but is also abundant at 14°C. *Synura uvella* seems to occur at temperatures in the middle range with maxima at 9°–10°C and 14°C. *Synura curtispina* eventually occurs at the higher temperature range (11°–20°C) with a maximum development at 14°C and 18°C. In the genus *Chrysosphaerella*, a remarkable difference is found between both species involved. Both species are stenothermal. *Chrysosphaerella brevispina* must be classified as oligothermal (2°–9°C), and *C. coronacircumspina* as polythermal (16°–20°C). The findings of *C. brevispina* are in agreement with those of Takahashi (1978), who found this species at temperatures ranging from 2.5° to 10.5°C (see Figure 303 in Takahashi 1978), but unfortunately in the accompanying text the species has been classified as polythermal instead of oligothermal. Although my findings directly point out that *C. coronacircumspina* is polythermal, Preisig and Takahashi (1978) reported the species from a pond in Switzerland occurring at a water temperature of 5.5°C. Most species of the genus *Paraphysomonas* could be classified as eurythermal. *Paraphysomonas vestita* has a maximum development at water temperatures ranging from 18° to 20°C, but also at 7°C and 11°C. Only *P. bandaiensis* is stenothermal (10°–17°C). *Chromophysomonas trioralis* is eurythermal, and *C. cornutus* is stenothermal (14°–18°C).

The combined occurrence of silica-bearing Chrysophyceae

An association is a recurrent group of co-occurrent species (Legendre and Legendre 1978). Because the scale-bearing Chrysophyceae, however, represent only a fraction of the total phytoplankton associations at that moment, the term "combination" is preferred above the term "association" in this context.

Figure 17.16 represents the dendrogram of the 22 taxa from D_1; Figure 17.17 represents the dendrogram of the 16 taxa from F_2. The dendrograms of D_1 and D_2 are quite similar to each other as are the dendrograms of F_1 and F_2. As appears from Figures 17.16 and 17.17, only one clear cluster can be distinguished, namely, the combination of *Mallomonas akrokomos*, *Mallomonas monograptus*, and *Paraphysomonas vestita*. This is a group of species occurring together at ponds D and F from half of December to April, and sharing their ability to develop at water temperatures as low as 5°C.

Another interesting combination is *Mallomonas caudata, Mallomonas tonsurata,* and *Chromophysomonas trioralis,* mostly occurring together from (May) April to September, characterized by their main development at water temperatures higher than 15 °C, but also present at lower temperatures.

Concluding remarks

It is obvious from above-mentioned results that the characteristic succession of silica-bearing Chrysophyceae in the Oude Waal is mainly determined by the water temperature. Although the chemical composition of the water (mainly alkalinity) determines the kind of species that might develop in that water body, it is the water temperature that regulates the development of those species. On the base of cluster analysis, one might detect several species combinations, but these combinations are in most cases coincidences, resulting from the eurythermal behavior of most of the species concerned. Even clustering on the base of Spearman's nonparametric correlation coefficient, which is best suited to these data because they are collected as classes of abundance, did not reveal more than the Sørensen similarity coefficient on binary data.

No differences in species succession and combinations could be detected between samples from the macrophyte zone and the open water zone.

References

Asmund, B., 1959. Elektron microscope observations on *Mallomonas* species and remarks on their occurrence in some Danish ponds and lakes. III. *Dansk Bot. Arkiv,* 18: 1–50.

Asmund, B. & Takahashi, E., 1969. Studies on Chrysophyceae from some ponds and lakes in Alaska VIII. *Mallomonas* species examined with the electron microscope II. *Hydrobiologia* 34: 305–21.

Asmund, B., Cronberg, G., & Dürrschmidt, M., 1982. Revision of the *Mallomonas pumilio* group (Chrysophyceae). *Nord. J. Bot.,* 2: 383–95.

Boesch, D.F., 1977. Application of numerical classification in ecological investigations of water pollution. Special Scientific Report N.T.I.S., U.S.

Bradley, D.E., 1966. Observations on some Chrysomonads from Scotland. *J. Protozool.,* 13: 143–54.

Cronberg, G. & Kristiansen, J., 1980. Synuraceae and other Chrysophyceae from central Småland, Sweden. *Bot. Notiser,* 133: 595–618.

Dürrschmidt, M., 1982. *Mallomonas parvula* sp. nov. and *Mallomonas retifera* sp. nov. (Chrysophyceae, Synuraceae) from South Chile, *Can. J. Bot.,* 60: 651–6.

Harris, K. & Bradley, D.E., 1960. A taxonomic study of *Mallomonas. J. Gen. Microbiol.,* 22: 750–77.

Korshikov, A.A., 1941. On some new or little known flagellates. *Arch. Protistenk.* 95: 22–44.

Legendre, L. & Legendre, P. 1978. Associations. *In:* Sournia, A. [Ed.] *Phytoplankton manual.* UNESCO, New York, pp. 261–72.

Nicholls, K.H., 1981. *Spiniferomonas* (Chrysophyceae) in Ontario lakes including a revision and description of two new species. *Can. J. Bot.*, 57: 557–60.

Preisig, H.R., & Hibberd, D.J., 1982a. Ultrastructure and taxonomy of *Paraphysomonas* (Chrysophyceae) and related genera. *Nord. J. Bot.*, 2: 397–420.

– 1982b. Ultrastructure and taxonomy of *Paraphysomonas* (Chrysophyceae) and related genera. *Nord. J. Bot.*, 2: 601–38.

Preisig, H.R. & Takahashi, E., 1978. *Chrysosphaerella* (*Pseudo-chrysosphaerella*) *solitaria* spec. nova (Chrysophyceae). *Pl. Syst. Evol.*, 129: 135–42.

Roijackers, R.M.M., 1981. Chrysophyceae from freshwater localities near Nijmegen, The Netherlands. *Hydrobiologia,* 76: 179–89.

– 1984. Some structural characteristics of the phytoplankton in the Oude Waal near Nijmegen, The Netherlands. *Verh. Internat. Verein. Limnol.*, 22: 1687–94.

Roijackers, R.M.M. & Kessels, H., 1981. Chrysophyceae from freshwater localities near Nijmegen, The Netherlands. II. *Hydrobiologia,* 80: 231–9.

– in press. Scale-bearing Chrysophyceae from the south-eastern part of The Netherlands. *Nord. J. Bot.*

Sneath, P.H.A. & Sokal, R.R., 1973. *Numerical Taxonomy.* Freeman, San Francisco.

Sørensen, T., 1948. A method of establishing groups of equal amplitude in plant sociology based on similarity of species content and its application to analyses of the vegetation on Danish Commons. *Kgl. Dan. Vid. Selsk. Biol. Skr.*, 5(4): 1–34.

Takahashi, E., 1978. *Electron Microscopical Studies of the Synuraceae* (*Chrysophyceae*) *in Japan. Taxonomy and Ecology.* Tokai University Press., Tokyo.

Takahashi, E. & Hayakawa, T., 1979. The Synuraceae (Chrysophyceae) in Bangladesh. *Phykos,* 18: 129–47.

Thomsen, H., Zimmermann, B., Moestrup, Ø., & Kristiansen, J., 1981. Some new freshwater species of *Paraphysomonas* (Chrysophyceae). *Nord. J. Bot.*, 1: 559–81.

Ward, J.H., 1963. Hierarchical grouping to optimize an objective function. *J. Amer. Statist. Ass.*, 58: 236–44.

Wee, J.L., 1982. Studies on the Synuraceae (Chrysophyceae) of Iowa. *Bibliotheca Phycologica* 62: 1–183.

Wujek, D.E. & Van der Veer, J. 1976. Scaled Chrysophyceae from The Netherlands including a description of a new variety. *Acta Bot. Neerl.*, 25: 179–90.

Wujek, D.E., Gretz, M. & Wujek, M.G., 1977. Studies on Michigan Chrysophyceae. IV. *Michigan Botanist,* 16: 191–4.

18

Chromophysomonas (Chrysophyceae) from twenty-seven localities in the Oslo area

ASBJØRN SKOGSTAD
Department of Biology, Division of Limnology
University of Oslo, P.O. Box 1027
Blindern, N-Oslo 3, Norway

Introduction

Among the smallest members of the family Mallomonadaceae are the species *Chromophysomonas* (formerly *Spiniferomonas*).[1] The spherical, biflagellated cells, covered with siliceous scales and spines, have a diameter of only 3 – 10 μm. The descriptions of the species are principally based on electron microscopic (EM) observations that make this tool necessary for their correct identification. Six species of this genus were originally described by Takahashi and given the name *Spiniferomonas* (Takahashi 1973). The type species of *Spiniferomonas, S. bourrellyi,* was shown not to possess chloroplasts and, therefore, was transferred to the heterotrophic genus *Paraphysomonas* (Preisig & Hibberd 1982). This step invalidated the name *Spiniferomonas,* and the remaining species were renamed *Chromophysomonas* (Preisig & Hibberd 1982). To date eight species are included in this genus. The purpose of this chapter is to report on the occurrence and distribution of these eight species in 27 localities and to add some new systematic and ecological information.

Materials and methods

The main sampling periods were May and August 1979 when all localities were visited. Some of the lakes were revisited in 1980 and 1981 on several occasions.

[1]Note added in proof: Nicholls (1985, *Nord. J. Bot.* 5: 403 – 6.) recently demonstrated the presence of chloroplasts in *Spiniferomonas bourrellii* Takh. and considers *Spiniferomonas* to once again be the valid generic name.

The author wishes to thank Dag Klaveness for valuable comments on the manuscript, Eva Braaten for technical assistance with the SEM, and Sheila Hunt Christiansen for correcting the English grammar.

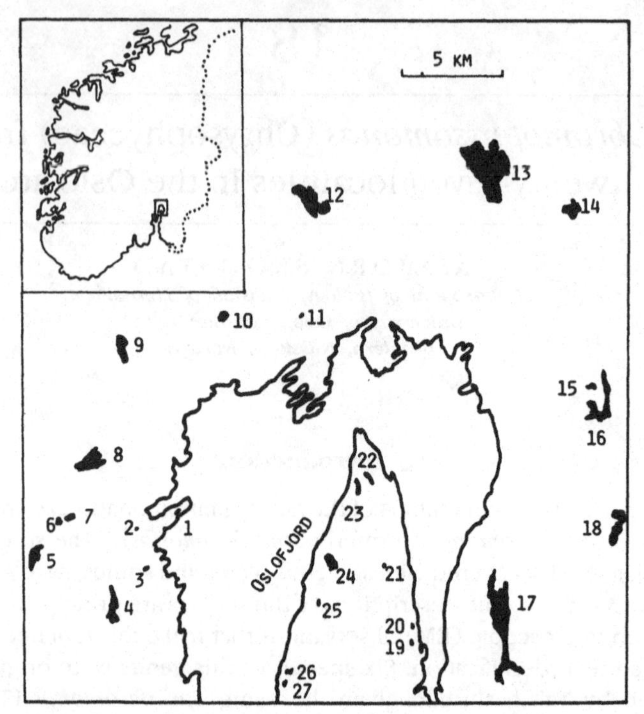

Fig. 18.0. Situation of the ponds and lakes in the Oslo area. The localities are numbered as in Table 18.1.

Investigation area

The localities, 27 ponds and lakes, are situated in Oslo and surrounding district as shown in Figure 18.0. The geology in this area is very complex (Henningsmoen 1960), but can as a simplification be divided in two groups. In this survey, these groups constitute subarea A and subarea B. Localities 1–11, located in subarea A, lie on lime-rich Cambro-Silurian sedimentary rocks. Subarea B, including localities 12–27, has a geology of Precambrian bedrock or Permian intrusives that both yield small amounts of salts to the water. As a consequence, the lakes in subarea A are alkaline and have well-buffered water, while the lakes in subarea B are mainly acidic and poorly buffered. Many localities are influenced by human activity and are heavily polluted (e.g., localities 11, 19, and 22). Others are oligotrophic, clear water lakes; some of them serving as drinking water reservoirs (e.g., localities 8, 13, and 14). Many lakes, especially in subarea B, are humic. The locality types vary from small, shallow ponds to large, deep lakes.

Field methods

Phytoplankton and water chemistry samples were collected by lowering a plastic tube vertically through the water column. In case of thermal stratification, the tube was lowered to a depth corresponding to the epilimnic layer. After closing the tube at the top, it was raised and the contents transferred to polyethylene bottles. From this sample, 100 ml were fixed immediately with Lugol's solution. Concentrated phytoplankton samples were obtained using a 10 μm net and were collected through the same depth as the water samples. A portion of the net sample was fixed immediately with a mixture of glutaraldehyde in a Na-cacodylate buffer to a final concentration of 2% and 0.1 M, respectively, at pH 7.0. All living material was kept cool during transportation to the laboratory.

Laboratory methods: chemistry

Analyses were performed as follows: For pH, potentiometric measurements were taken with a Radiometer pH Meter 29; total P was determined after Golterman, Clymo, & Ohnstad (1978), using a Zeiss PMQ II spectrophotometer. Water color was read against a standardized platinum–cobalt disc in a Lovibond Nessleriser comparator. Calcium was analyzed by atomic absorption using a Perkin Elmer 306.

Phytoplankton observation: light microscopy (LM)

Living material from net samples and centrifuged water samples were examined using bright field, dark field, and phase contrast techniques. The microscope used was a Zeiss Standard RA, fitted with Zeiss Planapo $\times$40/1.0 oil, Zeiss akromat $\times$100/1.27 oil and Zeiss Inko akromat-aplanat condenser, N.A. 1.4. Dry preparations were routinely stained using the Jensen staining method (Petersen & Hansen 1961).

Transmission electron microscopy (TEM). Drops of unfixed net samples or centrifuged water samples were pipetted onto carbon-coated copper grids. Thereafter the material was fixed for 30–60 seconds in OsO_4 vapor, air-dried, washed with distilled water, and redried. Glutaraldehyde-fixed cells were first washed thoroughly with distilled water before being pipetted on the grids. The material was examined directly with a Siemens Elmikroskop 1A or after shadow casting with platinum at an angle of about 30°.

Scanning electron microscopy (SEM). Micrographs printed in this chapter are from glutaraldehyde-fixed material, which after thoroughly washing with distilled water was pipetted on glass-covered specimen stubs and air-dried. The material was coated with gold–palladium in a diode sputter,

Polaron E.5000 before examination with a Jeol JSM-35C scanning microscope. The EM microscopy was done at the EM Laboratory for Bioscience, Oslo.

Results and discussion

As expected, the lakes showed a wide range in chemical composition (Table 18.1). Water chemistry of the two subareas shows some obvious differences. Calcium values are high in subarea A with a correspondingly high pH. Water color is low except for localities 10 and 11. With some exceptions, calcium values in subarea B are low, and the pH is mainly less than 7. Locality 19 has a high phytoplankton biomass, and intensive photosynthesis is responsible for the high pH measured. Many lakes in this area are brown in color due to large bogs in their catchment areas.

The distribution of the *Chromophysomonas* species is given in Table 18.2, and discussed below. New records for Norway are indicated with an asterisk.

Chromophysomonas abei* *(Takahashi) Preisig & Hibberd* *(Figure 18.1)*

Cell size of this species was 3–6 μm. *Chromophysomonas abei* has only one type of scales: simple, oval one-lacuna scales that can be considered the basic scale type in this genus. Observations of single scales make correct determination impossible since other species have similar scales (e.g., *C. trioralis*). The short, delicate spines of *C. abei,* however, cannot be confused with any other known species. Few specimens were observed and only from localities in subarea B. The species may have a preference for humic water. The few reports of *C. abei* indicate a worldwide distribution.

Chromophysomonas alata* *(Takahashi) Preisig & Hibberd* *(Figure 18.2)*

Only a few scales and spines were found and in one locality only. Further distribution of this rare species includes: Japan (Takahashi 1973), USSR (Balonov 1978), and Canada (K.H. Nicholls, 1981, Kling & Kristiansen 1983).

Chromophysomonas bilacunosa *(Takahashi) Preisig & Hibberd* *(Figure 18.3 and 18.4)*

This species proved to be one of the most common members of the genus and was found in both subareas. In addition to the simple one-lacuna

Table 18.1. *Chemical data for the investigated ponds and lakes during May 6–28, 1979*

	Locality	pH	Total P μg P liter^{-1}	Calcium Mg Ca liter^{-1}	Color Mg Pt liter^{-1}
Subarea A	1. ISDAM	7.9	35	40	10
	2. EINEDAMMEN	7.9	21	56	5
	3. ØSTENSTADDAMMEN	7.7	31	34	5
	4. GJELLUMVANNET	7.5	92	19	10
	5. NORDVANNET	8.2	29	23	5
	6. BRENNSRUDVANNET	7.8	7	20	5
	7. HOGSTADVANNET	8.3	30	22	5
	8. SEMSVANNET	7.6	16	11	5
	9. STOVIVANNET	8.5	32	13	5
	10. DÆLIVANNET	8.2	39	25	20
	11. TJERNSRUDTJERN	7.6	136	58	100
Subarea B	12. BOGSTADVANNET	6.8	10	4	10
	13. MARIDALSVANNET	6.6	7	4	5
	14. ALNSJØEN	6.7	7	5	5
	15. ULSRUDVANNET	6.9	14	7	20
	16. NØKLEVANNET	6.5	8	4	5
	17. GJERSJØEN	7.0	23	16	10
	18. SVÆRSVANN	6.0	17	4	30
	19. BLYLAGDAMMEN	8.8	121	14	20
	20. KRYSTALLDAMMEN	7.2	14	14	5
	21. RØERTJERN	6.9	31	8	20
	22. SKOKLEFALLTJERN	6.7	372	19	60
	23. FLASKEBEKKTJERN	5.4	15	3	125
	24. BLEKSLITJERN	6.2	16	4	15
	25. NYBORGDAMMEN	6.2	74	5	40
	26. SPRODAMMEN	6.4	17	4	30
	27. BRÅTADAMMEN	6.8	17	6	15

Table 18.2. *Distribution of* Chromophysomonas *species in 27 ponds and lakes in the Oslo area*

Species	Loc. 1	2	3	4	5	6	7	8	9	10	11	12	13	14	15	16	17	18	19	20	21	22	23	24	25	26	27	Total number
	Subarea A											Subarea B																
Chromophysomonas abei	·	·	·	·	·	·	·	·	·	·	·	·	·	·	·	·	·	+	·	·	+	·	+	·	·	·	·	3
C. alata	·	·	·	·	·	·	·	·	·	·	·	·	·	·	·	·	·	+	·	·	·	·	·	·	·	·	·	1
C. bilacunosa	+	·	·	·	+	+	·	+	·	·	·	+	+	+	·	+	·	+	·	·	+	·	+	·	+	·	·	12
C. crucigera	·	·	·	·	·	·	·	·	·	·	·	·	·	·	·	·	·	+	·	·	·	·	·	·	·	·	·	1
C. cornuta	·	·	·	·	·	·	·	+	·	·	·	+	·	·	·	·	·	+	·	·	·	·	+	·	·	·	·	4
C. serrata	·	·	·	·	·	+	·	·	·	·	·	+	·	·	·	·	·	+	·	·	+	·	·	·	·	·	·	4
C. takahashii	·	·	·	·	·	·	·	·	·	·	·	·	·	·	·	·	·	+	·	·	·	·	·	·	·	·	·	1
C. trioralis	+	+	+	+	+	+	+	+	·	+	·	+	+	+	+	+	·	+	·	+	+	+	+	+	+	+	+	23
Number of species:	2	1	1	1	2	3	1	3	0	1	0	4	2	2	1	2	0	8	0	1	4	1	4	1	2	1	1	

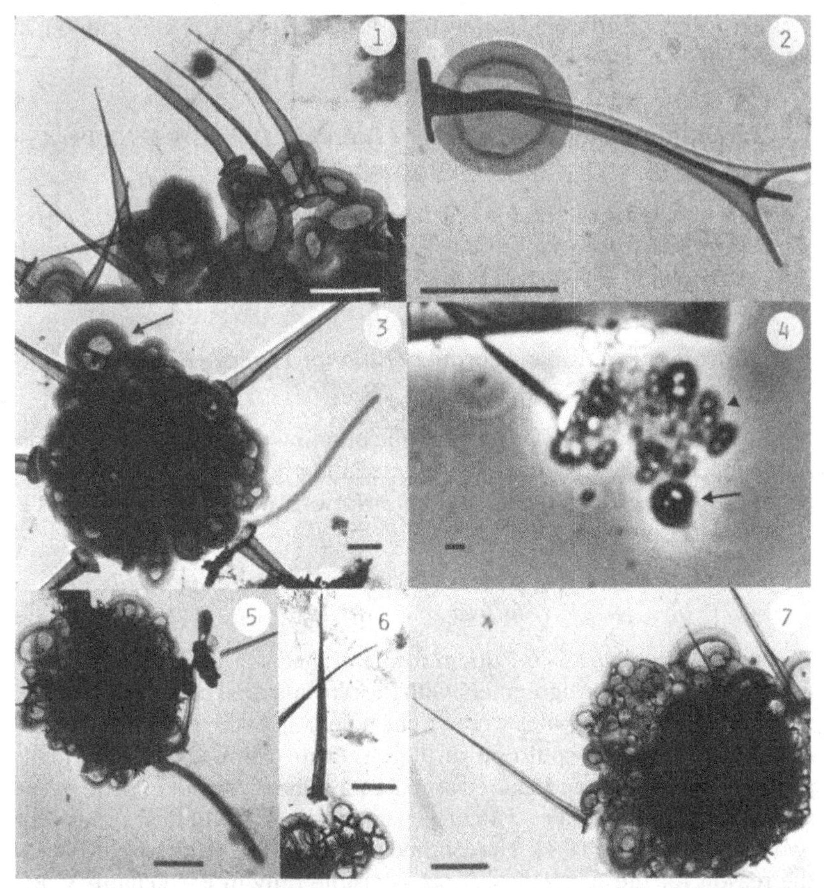

Figs. 18.1–18.2. Chromophysomonas. Scale bar = 1 μm. ***Fig. 18.1.*** *C. abei.* ***Fig. 18.2.*** *c. alata.* ***Fig. 18.3.*** *C. bilacunosa,* with large circular bilacunate scales with rod on median bridge (arrow). ***Fig. 18.4.*** *C. bilacunosa,* LM, Jensen stain, with large circular (arrow) and small elliptical (arrowhead) bilacunate scales. ***Figs. 18.5., 18.6.*** *C. crucigera.* ***Fig. 18.7.*** *C. cornuta.* See text for details.

scales, *C. bilacunosa* possesses small, oval scales with two lacunae and large round or elliptical scales with one lacuna. The bridge separating the two lacunae often had a rod positioned in the center of it. A fourth type of scales, not earlier reported for this species was found on specimens from localities 18 and 21. These are large, round, 1.9–2.3 μm in diameter, with two lacunae (Figure 18.3, arrow). A prominent nodule, 0.5–0.6 μm long, projected from the center of the bridge. These scales are very characteristic and can easily be observed even in LM (Figure 18.4, arrow). A maximum number of 19 spines was found measuring from 4–8.2 μm. Cells carrying the large, round scales with two lacunae, however, had spine lengths of 10–17 μm. This species also proves to have a worldwide distribution. It was earlier reported from Norway by M.P. Nicholls (1981) who found it in an oligotrophic mountain lake.

Chromophysomonas crucigera* *(Takahashi) Preisig & Hibberd (Figures 18.5 and 18.6)*

This rare species was found in one locality only. Other reports of *C. crucigera* are restricted to Japan (Takahashi 1973) and Denmark (Thomsen 1977, Kristiansen 1978).

Chromophysomonas cornuta *(Balonov) Preisig & Hibberd (Figure 18.7)*

Localities from both subareas were inhabited by this species, but few in numbers. To date, *C. cornuta* is reported only from the Northern Hemisphere. It has previously been found in Norway by M.P. Nicholls (1981).

Chromophysomonas serrata *(Nicholls, K.H.) Preisig & Hibberd (Figures 18.8 and 18.9)*

The spherical cells, 4.5–6.5 μm in diameter, had a large, bilobed chloroplast. There was good agreement with the type material concerning spines and scales. The bent spines were so characteristic in the Norwegian material that this species could be distinguished in LM, even in wet-mount preparation. In dried and stained preparations, the two kinds of scales were visible: large, round scales with one lacuna and small, elliptical scales with two lacunae (Figure 18.9). The toothed margin on the spine base, which is the reason for the name *C. serrata,* is visible only in EM (Figure 18.8). Some of the elliptical scales have an erect rod on the median bridge separating the two lacunae. As stated by M.P. Nicholls (1981), these scales have an unipolar distribution limited to the area around the flagellar insertion (M.P. Nicholls Figures 7 and 9). This species is to date reported only from Canada (K.H. Nicholls 1981, Kling & Kristiansen 1983). The organisms illustrated in Figures 7 and 9 in M.P. Nicholls (1981), named as *Spiniferomonas bilacunosa,* are probably *C. serrata.*

Chromophysomonas takahashii* *(Nicholls, K.H.) Preisig & Hibberd (Figure 18.10)*

Only one cell was found, but the spines are so characteristic that confusion with other species is unlikely. The spines were shorter than in the Canadian material (K.H. Nicholls 1981), but otherwise similar. This species has previously been reported only from Canada (K.H. Nicholls 1981, Kling & Kristiansen 1983).

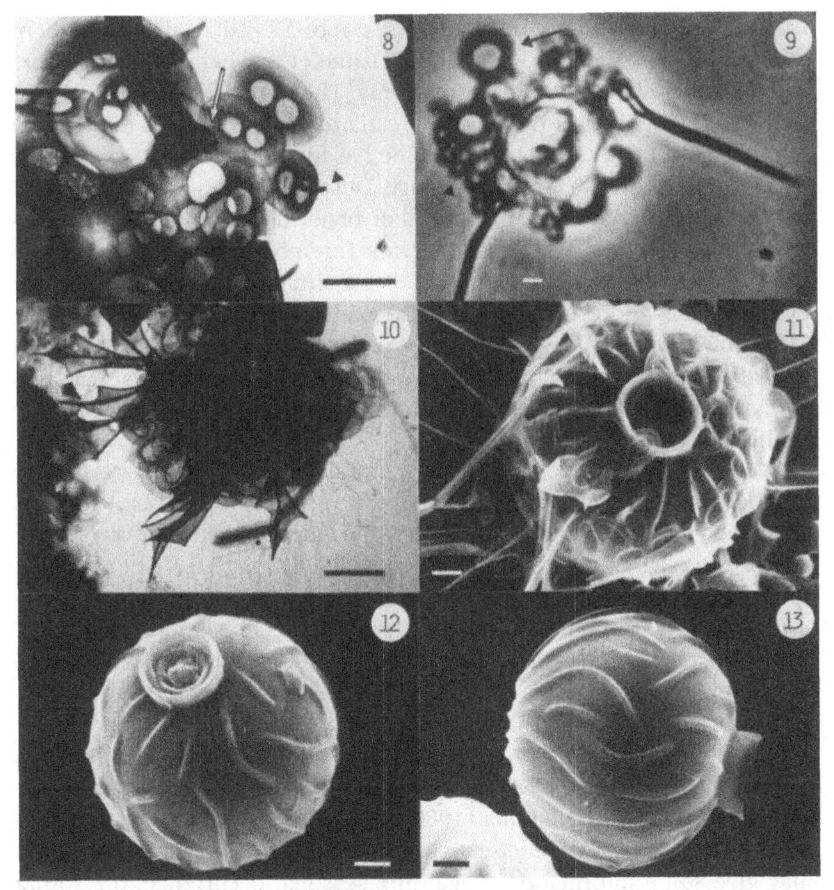

Figs. 18.8–18.13. Chromophysomonas. Scale bar = μm. ***Fig. 1. 18.8. C. serrata,*** elliptical bilacunate scale with erect nodule on median bridge (arrowhead) and toothed margin on spine base (arrow). ***Fig. 18.9. C. serrata,*** LM, Jensen stain, with large single lacuna (arrow) and small elliptical bilacunate scales (arrowhead). ***Figs. 18.11–18.13. C. trioralis*** cysts, SEM. See text for details.

Chromophysomonas trioralis *(Takahashi) Preisig & Hibberd* *(Figures 18.11–18.13)*

This species was found in almost every lake, except four, and sometimes in considerable amounts. In locality 2, the 1979 spring sample showed a large population. Obviously, its growth had taken place under ice or during ice break, since almost only resting stages were found. This, however, made it possible to examine the cyst of this species by means of EM. Figure 18.11 shows an almost mature cyst, but with scales and spines still attached. In Figures 18.12 and 18.13, these are shed and the cyst shows its surface with straight or undulating ridges, running mainly from the pore region and backward. Dimensions of the mature cyst were 7.2–10.5 μm in diameter or 6.3–8.3 × 7.1–9.0 μm if oval shaped. Diameter and height of collar were 2.4–3.0 μm and 0.6–1.0 μm, respectively. The pore opening was 0.5–0.7 μm. In SEM, the cyst of *C. trioralis* greatly resembles the cyst depicted in Figure 2, Plate 5 in Gritten (1977). She has identified her cyst as *Clericia insignis* Deflandre. To my knowledge Deflandre has not described any *Clericia insignis.* He has, on the other hand, described *Archaeomonas insignis* (Deflandre 1933), and this cyst resembles *C. trioralis.* Also *Archaeomonas vermiculosa* Deflandre has a similar ornamentation on the cyst wall (Deflandre 1932, see also Deflandre & Deflandre-Rigaud 1969). Differences, especially in the structure of the collar, make it questionable if *C. trioralis* is identical with any of these *Archaeomonas* species. Scanning electron microscopic investigation of the type materials would solve the problem.

As for the vegetative stage, great variation in spine length and spine number was observed. A number of about 10 was usual, but up to 80 was observed (locality 23). Regarding the spine length, subspecific division by this criterion has been proposed. Balonov (1978) suggested that cells with spine length of 2.9–3.8 μm should be referred to as *Spiniferomonas trioralis* f. *trioralis.* Specimens with spine length up to 12.3 μm were named *S. trioralis* f. *cuspidata.* (The genus name *Spiniferomonas* was used because the paper appeared before the name change.)

The formal division by this criterion can be regarded invalid since spine length varying from 3.7 to 7.0 μm on one and the same organism was observed (locality 12). *Chromophysomonas trioralis* was reported earlier from Norway by M.P. Nicholls (1980).

The ecology of *Chromophysomonas* is obscure, but this investigation may give some information. Regarding seasonal occurrence, the spring seems to be the most important growth season, but all species have been found in summer too. Only *C. cornuta, C. serrata,* and *C. trioralis* were found under ice. Owing to the relatively great differences in water chemistry between the lakes situated in the two subareas, one would also expect to find similar differences in species distribution. From the distribution map, however, this is not so obvious, although subarea B localities seem to be preferred. When other chrysophycean organisms are taken into consideration, this tendency is clearer (Skogstad 1982). An outstanding locality regarding diversity of *Chromophysomonas* species is locality 18, from which all known species were observed. This locality has an especially rich chrysophycean flora (Skogstad 1982).

A comment should be made on reported lack of chloroplasts in *Spiniferomonas bourrellyi* (Preisig & Hibberd 1982). This observation is in contradiction to my own findings that suggest that this organism does have chloroplasts. I feel that this species can be distinguished in LM, and on several occasions this has been verified by EM. This can be done in samples with no other similar species that could confuse the identification, as was the case in the spring samples from locality 20. Cells from this population were seen to possess a chloroplast. Particularly in cyst-producing cells, which augment the cell and chloroplast dimensions, a large, golden and bilobed chloroplast was obvious. Unfortunately, isolation and culturing of these cells has not been obtained, but the author intends to perform this at earliest opportunity.

References

Balonov, I.M. 1978. Electron microscopic study of the genus *Spiniferomonas* Takahashi (Chrysophyta). *Bot. Zh. (Leningrad)* 63: 1639–47 (in Russian).

Deflandre, G. 1932. Note sur les Archaeomonadacées. *Bull. Soc. Bot. Fr.* 79: 346–55.

– 1933. Seconde note sur les Archaeomonadacées. *Bull. Soc. Bot. Fr.* 80: 79–90.

Deflandre, G. & Deflandre-Rigaud, M. 1969. Nannofossiles siliceux I. Archaeomonadaceae. *Fichier Micropaléontologique Gén.-Ser.* 19. CNRS, Paris, pp. I–IX and 119 plates.

Golterman, H.L., Clymo, R.S. & Ohnstad, M.A.M. [Eds.] 1978. *Methods for Physical and Chemical Analysis of Fresh Waters,* 2nd ed. Blackwell Scientific, Oxford.

Gritten, M.M. 1977. On the fine structure of some chrysophycean cysts. *Hydrobiologia* 53: 239–52.

Henningsmoen, G. 1960. Cambro-Silurian deposits of the Oslo region. *In:* Holtedahl, O. [Ed.], *Geologi of Norway.* NGU nr.208. Aschehoug, Oslo, pp. 130–50.

Kling, H.B. & Kristiansen, J. 1983. Scale-bearing Chrysophyceae (Mallomonadaceae) from Central and Northern Canada. *Nord. J. Bot.* 3: 269–90.

Kristiansen, J. 1978. Studies on the Chrysophyceae of Bornholm. II. *Bot. Tidsskr.* 73: 71–85.

Nicholls, K.H. 1981. *Spiniferomonas* (Chrysophyceae) in Ontario lakes including a revision and description of two new species. *Can. J. Bot.* 59: 107–17.

Nicholls, M.P. 1980. *Savalen-Limnologisk undersøkelse.* Hovedfagsoppgave. [Savalen-Limnological investigation. Thesis.] University of Oslo, Oslo, Norway, p. 137.

– 1981. Observations on Chrysophyceae from a Norwegian mountain lake. *Br. J. Phycol.* 16: 373–8.

Petersen, J.B. & Hansen, J.B. 1961. On some neuston organisms III. *Bot. Tidsskr.* 57: 293–305.

Preisig, H.R. & Hibberd, D.J. 1982. Ultrastructure and taxonomy of *Paraphysomonas* (Chrysophyceae) and related genera I. *Nord. J. Bot.* 2: 397–420.

Skogstad, A. 1982. Synuraceae-floraen i 27 lokaliteter i Oslo-området. Hovedfagsoppgave. [The flora of Synuraceae in the Oslo area. Thesis.] University of Oslo, Norway.

Takahashi, E. 1973. Studies on genera *Mallomonas* and *Synura,* and other plankton in freshwater with the electron microscope. VII. New genus *Spiniferomonas* of the Synuraceae (Chrysophyceae). *Bot. Mag. Tokyo* 86: 75–88.

Thomsen, H.A. 1977. External morphology of the choanoflagellate *Salpingoeca gracilis* James-Clark. *J. Mar. Biol. Assoc. U.K.* 57: 629–34.

[illegible]

Hollerbach [illegible]

[illegible] Ocean [illegible]

PART VI

Paleobiology

19

Scale microfossils of Early Cambrian age from Canada

CAROL WAGNER ALLISON
University of Alaska Museum
Fairbanks, Alaska 99701, U.S.A.

AND

JERRY W. HILGERT
Institute of Northern Forestry
U.S.D.A., U.S. Forest Service
308 Tanana Drive
Fairbanks, Alaska 99701, U.S.A.

Introduction

Abundant scale microfossils, originally reported in Allison (1981), occur in limestones of the uppermost Tindir Group and basal beds of the overlying Funnel Creek Limestone. The scale-producing exposures are in west central Yukon Territory, a few kilometers east of the international boundary between Canada and the United States. The dominant fossils in these rocks are blue-green algae, which occur in successive organic-rich layers reflecting the accumulation of alternating algal mat and sediment horizons in a marine subtidal environment. Fossils found in these sediments are best preserved in cherts, which occur as rounded nodules and as thin beds that may be locally broken and transported.

The associated blue-green algal biota (Allison and Awramik, in preparation) contains many taxa closely comparable with previously described mat assemblages from Proterozoic age rocks, particularly with the approximately 900 million year old Bitter Springs biota of Australia (Schopf, 1968, Schopf and Blacic, 1971). The co-occurrence of mineralized spicules with these Tindir Group fossils together with other lines of evidence suggest, however, an Early Cambrian age for the scale fossils.

Materials and methods

Rock samples that have yielded fossil scales include lithologies with nodular or bedded chert. Scales tend to be more abundantly preserved in the latter and are more common near the top of the Tindir limestone, although they occur throughout the upper 20 meters of that unit.

The fossil scales have been studied almost exclusively by light microscopy in thin sections (preparations in which a rock sample is cut and ground thin enough to allow light to pass through). Techniques required to remove the scales from the rock matrix involve the use of acids (HF and HCl) that attack also the scales. The few scales that have survived such processing are in very poor condition and offer much less information than do scales seen in thin section. However, the fact that a few scales did come through the rock dissolving process confirms that at least those types had acid-resistant organic material within, or surrounding, the mineralized scale.

Results

The Tindir scale fossils can be conveniently divided on the basis of morphology into three general types, none of which necessarily reflects a natural grouping of closely related taxa. Over twenty scale morphotypes have been seen, as one to a few randomly occurring scales and/or in clusters of ten to over a hundred. These clusters may reflect an original arrangement around a parent cell. The total size range of the fossil scales is from 5 to over 70 μm; however, the size range of a specific morphotype is typically quite narrow, and most scales fall within the range of 20 to 40 μm. This is larger than modern scales of comparable morphologies, a circumstance that may mirror the gigantism seen in early members of various plant groups.

No representation of fossil cysts has been observed in the biota. This circumstance may relate to the antiquity, the marine paleoenvironment of the mat communities, or a preservational bias against cysts that could have been originally present. Evidence of an organic cell cover has been seen in only one case in which a very thin, pale, crumpled membrane is preserved in a cluster of imperforate scales. We have not observed evidence of bristles, flagella, or cell contents.

The fossil scales, as seen in thin section, are preserved as colorless or hyaline pale to darker brown or occasionally milky pale to dark brown structures. Scales that survived the acid maceration process mentioned above are all medium brown in color. We do not interpret these preserved scale colors to indicate original color (a) because co-occurring blue-green algal and other nonmineralized cells are preserved in the same manner, (b) because degradation of cell membranes and contents commonly results in change to a brownish color, and (c) because the scales could be iron stained.

The three general types of scale fossils are briefly described and illustrated here. They will be more completely described, illustrated and formally named in a forthcoming paper (Allison and Hilgert, in press).

Imperforate Scales

Fossils referred to this group have the most simple morphologies among scales seen in the biota. These scales are more or less oval in outline and range from very slightly to distinctly domed. They tend to occur in discrete clusters of twelve or more scales of a single morphology and narrow size range. In several clusters that indicate original arrangement, the scales all overlap in one direction (Figure 19.1). Similar scales of only one morphotype have been seen in other "arranged" clusters (Figure 19.2) in which overlap is not unidirectional.

Perforate Scales

Scales that have prominent pores (Figures 19.3–19.7) are referred to this general grouping. All included morphotypes are more or less flat. They range in outline from perfectly circular to oval or slightly polygonal. The pores are, in most cases, arranged in a hexagonal pattern and may themselves be round to polygonal in outline. There may be numerous, closely spaced small pores or lesser numbers of larger or more widely spaced pores. However, in scales in which the pores are more or less hexagonally arranged, all of the pores are typically consistent in size and shape.

Several morphotypes included in this general grouping exhibit elaboration of one face of the scale. This may take the form of a prominent central spinelike process that commonly has a barbed tip (Figures 19.6 and 19.7). Another morphotype has a low, slightly inward curved rim on one face of the scale near the margin. When the microscope is focused at the center of a specimen lying perpendicular to the plane of the thin section, this is seen as two projections on the scale (Figure 19.4 and 19.5).

A somewhat different type of perforate scale, in which the pores are radially arranged (Figure 19.8), should perhaps be considered as belonging to a separate subgroup. Scales of this type are clearly perforate, with the pore rows defining their radial nature. All of the radially perforate scales have raised rims around the margin on one side. They tend to be thinner than those with hexagonally arranged pores and tend also to be pale milky brown rather than hyaline brown or colorless. This circumstance could indicate an originally higher organic component than in the hexagonally porous scales, which are commonly colorless or pale hyaline brown.

With a few exceptions, all of the perforate scales have been seen only in clusters containing a single morphotype and narrow size range. Original arrangement of the scales around a parent cell is indicated in a few cases, such as that shown in Figure 19.5, but not to the degree seen among the imperforate scale clusters.

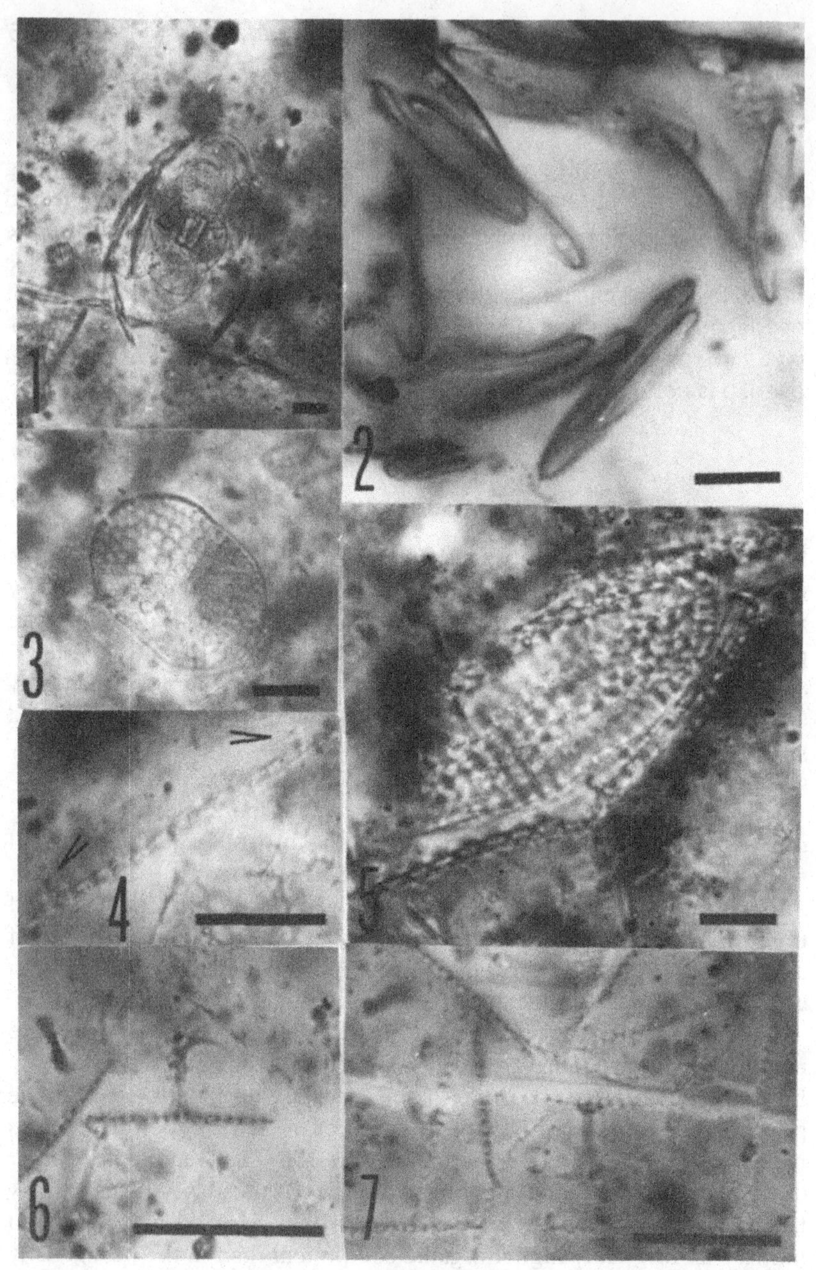

Figs. 19.1–19.7. Scale bar equals 10 μm. ***Fig. 19.1.*** A tight grouping of oval, nonperforate scales indicating original arrangement. ***Fig. 19.2.*** A cluster of shallow, saucerlike, nonperforate scales always occurring in groupings with one scale morphology and size. ***Fig. 19.3.*** A thin, perforate, subquadrate scale with pores arranged in a hexagonal pattern. This type typically occurs singly or in small groupings of similar sized scales. ***Fig. 19.4.*** A perforate scale with a low, slightly inwardly curved rim on one face of the scale. Seen here in side view, the rim appears as two projections near the margin (arrows). ***Fig. 19.5.*** A cluster of porous scales seen in cross section in a configuration suggesting original arrangement around a cell. ***Fig. 19.6 and 19.7.*** Perforate scales with a prominent central spine terminating in a barbed tip. These scales also tend to be found in loose groupings of one morphology.

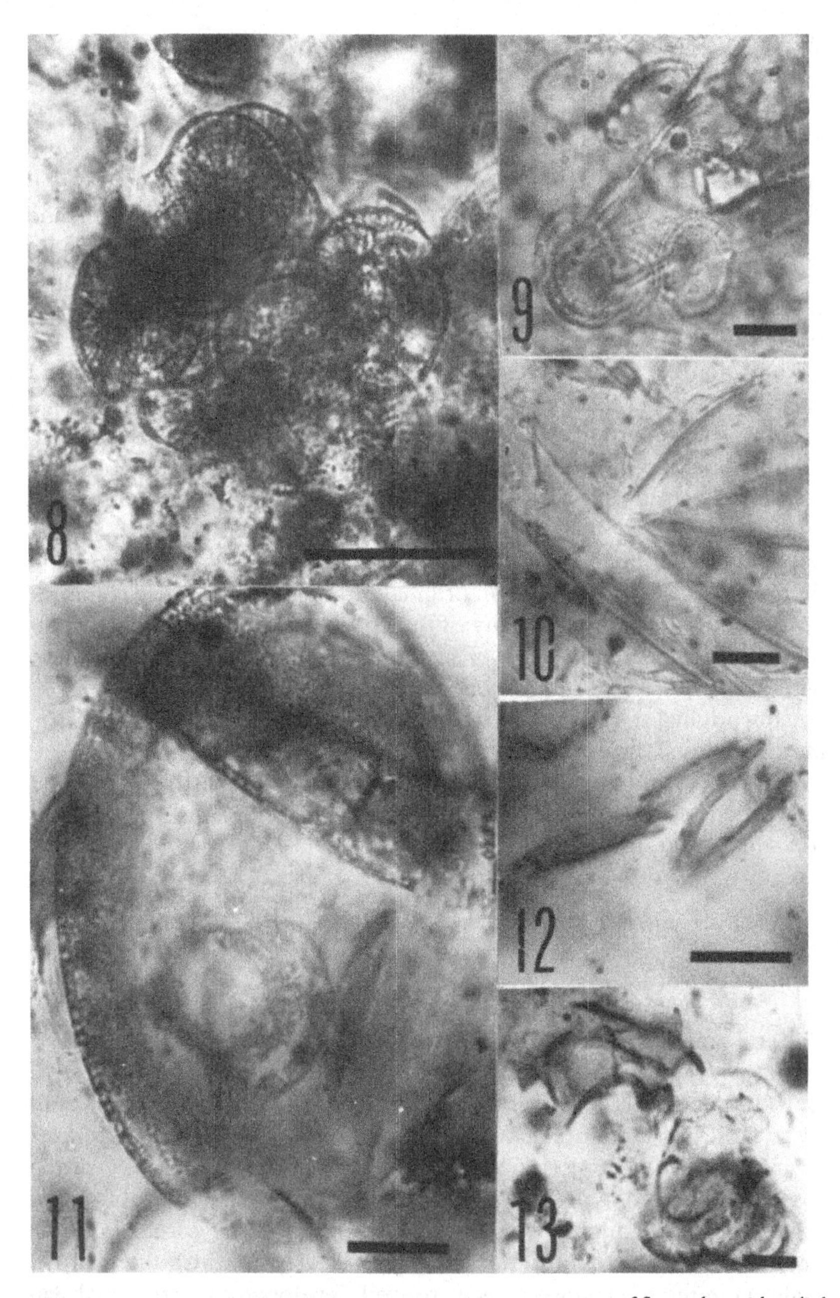

Figs. 19.8–19.13. Scale bar equals 10 μm. ***Fig. 19.8.*** A grouping of flat scales with radial markings. ***Fig. 19.9.*** A cluster of oval, collarlike scales with apparent concentric marginal structure. These occur with porous, needlelike structures and the larger ovals seen in Figure 19.10. ***Fig. 19.10.*** A cluster of collarlike scales seen in side view showing inflated "tops" and a slight, convex peripheral surface. ***Fig. 19.11.*** A large collarlike scale occurring with smaller domed ovals, with much smaller, deeply invaginated scales shown in ***Fig. 19.12,*** and with a tight cluster of thin, small oval rings (not shown). ***Fig. 19.13.*** A cluster of scales with a distinct three-dimensional morphology as seen in cross section, and markedly different diameters of the upper and lower surfaces.

Collarlike Scales

Included here are scales that have a distinct three-dimensional morphology, the basic shape being that of a more or less inclined peripheral band with or without discernible top and/or bottom covering layers. These scales are, in some cases, morphologically reminiscent of coccoliths; others bear resemblance also to modern scales referred to non-Prymnesiophyte protistan groups.

The collarlike scales include a much wider range of morphologies and sizes than has been observed among the imperforate or perforate scale groups. The occurrence of discrete clusters containing more than one morphotype and size (Figure 19.11) is also high among the collarlike scales. It is clearly possible that the forms discussed under one general heading may,in fact, include taxa representing several natural stocks that need not be closely related. Caution is especially advisable in suggesting their phyletic affinity because an original carbonate versus silica composition is not discernible for many of the collarlike scale morphotypes. Although many discrete clusters of scales that very likely reflect association with a single parent cell have been observed, few clusters showing convincing original arrangement have been found among the collarlike scales. Morphotypes shown in Figures 19.9–19.13 are typical of this basic scale type, which includes many additional morphotypes that will, as mentioned above, be illustrated and formally described in a forthcoming paper (Allison and Hilgert, in press).

Discussion

The scale microfossils reported here open a new window on early eukaryote history. Further study is clearly necessary before confident statements can be made about the phyletic affinities of these fossils. Their comparability to scales of modern taxa currently referred to the Chrysophyceae or Prymnesiophyceae (=Haptophyceae) is evident; however, some of the fossil scale morphotypes may instead represent ancestral stock of one or more other modern groups of the Protista, or indeed be unrelated to any known modern or fossil group. In any case, they establish the fact that scale-secreting microorganisms were well developed by Early Cambrian time and almost certainly have an earlier origin. Jost (1968) reported, as Type A, a scalelike structure, from a finely crushed core sample in the approximately one billion year old Nonesuch Shale. If this structure did, in fact, come from those rocks, scale-secreting microorganisms comparable to some in the biota reported here are by far the earliest mineral-secreting organisms known.

References

Allison, C.W. 1981. Siliceous microfossils from the Lower Cambrian of northwest Canada: Possible source for biogenic chert. *Science* 211: 53–5.

Allison, C.W. & Awramik, S.M. in preparation. Organic-walled microfossils from the Early Cambrian of northwest Canada.

Allison, C.W. & Hilgert, J.W. in press. Scale microfossils from the Early Cambrian of northwest Canada. *J. Paleont.*

Jost, M. 1968. Microfossils of problematic systematic position from Precambrian rocks at White Pine, Michigan. *Micropaleont.* 14: 365–8.

Schopf, J.W. 1968. Microflora of the Bitter Springs Formation, Late Precambrian, central Australia. *J. Paleont.* 42: 657–88.

Schopf, J.W. & Blacic, J.M. 1971. New microorganisms from the Bitter Springs Formation (late Precambrian) of the north-central Amadeus Basin, Australia. *J. Paleont.* 45: 925–60.

20

Chrysophycean cysts and scales in lake sediments: a review

GERTRUD CRONBERG
Institute of Limnology
University of Lund, Box 3060
S-220 03 Lund, Sweden

Introduction

A characteristic feature of the chrysophytes is their ability to produce resting stages in the form of endogenously formed silicious cysts. These structures have also been called statospores or stomatocysts. Ehrenberg (1854) was the first to describe organisms that looked very much like cysts (Figures 20.1–20.2). However, the formation of cysts was first described by Cienkowski (1870) for *Chromulina nebulosa* Cienk. (Figure 20.3). Later Doflein (1921) made a careful description of endogenous cyst formation in *Ochromonas granularis* Dofl. Geitler (1935) observed that many cysts of *Dinobryon divergens* Imh. contained two nuclei, and he suggested that these cysts might be sexually produced. Confirmation that sexual reproduction can take place was made by Skuja (1950), who followed isogamous copulation in *Dinobryon borgei* Lemm. and the formation of its sexual cysts. In the same population, Skuja also recorded normal asexual cyst formation, although he observed that the asexually formed cysts were smaller than the sexually produced ones. Sexual reproduction in chrysophytes has also been recorded by Fott (1959), Wawrik (1960), and Kristiansen (1961). Sandgren (1981) studied sexual and asexual cyst formation in the laboratory in six clones of *Dinobryon cylindricum* Imh.

The shape of the cyst is spherical or oval. The cyst wall is made of silica with a smooth or ornamented surface. The cyst is provided with a pore that can be plugged with a silicified stopper (Figure 20.5). The pore may possess a collar or be ornamented in different ways (Figures 20.6–20.8), and the construction of the pore–collar complex is of great taxonomic value (Andrieu 1936, Rampi 1937a, b, Tappan 1980, Sandgren & Carney 1983). The shape and the size of the statospores seem to be constant, but Sandgren (1983a) found that in *Dinobryon cylindricum* the spine and/or collar length of the cysts varied when the *Dinobryon* clones were exposed to different temperatures.

When a cyst germinates, the stopper is dissolved, thus allowing the

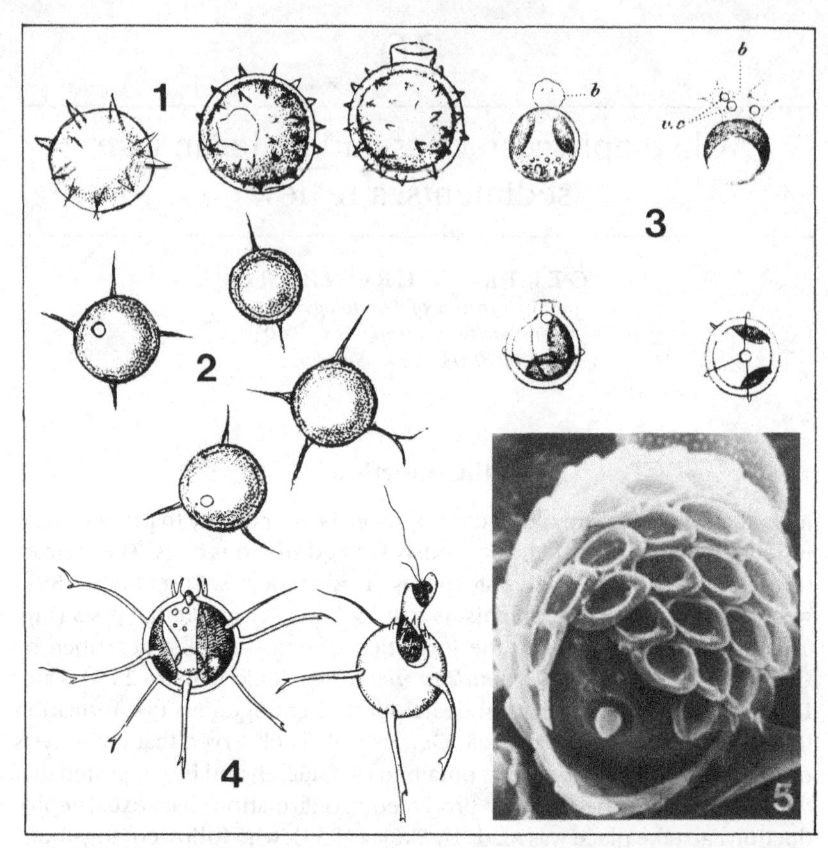

Figs. 20.1–20.5. Fig. 20.1. Chrysophycean cysts? Reproduced from Ehrenberg (1854), Taf IX, Figure 13a–c (no name given) from diatomite, Ceyssat, France. ×750. ***Fig. 20.2.*** *Chaetotyphla volvox,* probably chrysophycean cysts. Reproduced from Ehrenberg (1854), Taf XIV, Figures 93 and 94a,b, from soil samples from Berlin, Germany. ×750. ***Fig. 20.3.*** Development of cysts in *Chromulina nebulosa* Cienk. Reproduced from Cienkowsky (1870), Taf XXIV, Figures 57–61. ***Fig. 20.4.*** Cell emerging from a cyst of *Echinochrysis chodatii* Conr. Reproduced from Conrad (1926), Taf IX, Figures 67–68. ***Fig. 20.5.*** Cyst of *Mallomonas insignis* Penard with the pore plugged. Photo M. Dürrschmidt.

organism to emerge. Pascher (1917) observed ameboid cells emerging from a cyst of the colorless chrysophyte *Leucochrysis pascheri* Bourr. Conrad (1926) depicted the emergence of a single cell from the cyst of *Echinochrysis chodatii* Conr. (Figure 20.4). The emergent cell had the appearance of an *Ochromonas.* However, he also found that the cyst of *Chromulina pascheri* Hof. produced 2–4 zoospores upon germination. Sheath, Hellebust, & Sawa (1975) described the germination of the cysts of *Dinobryon divergens.* They found that a special germination chamber was made from the statospore from which approximately four daughter cells were released.

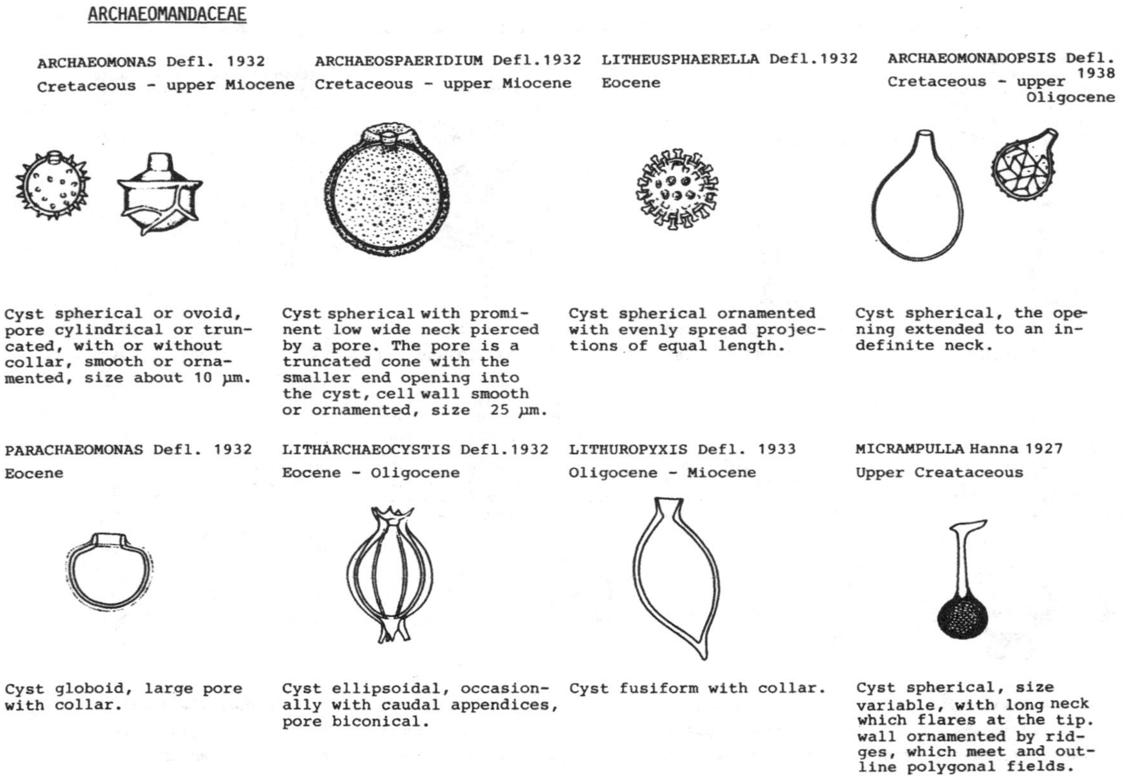

ARCHAEOMANDACEAE

ARCHAEOMONAS Defl. 1932
Cretaceous - upper Miocene

Cyst spherical or ovoid, pore cylindrical or truncated, with or without collar, smooth or ornamented, size about 10 μm.

ARCHAEOSPAERIDIUM Defl.1932
Cretaceous - upper Miocene

Cyst spherical with prominent low wide neck pierced by a pore. The pore is a truncated cone with the smaller end opening into the cyst, cell wall smooth or ornamented, size 25 μm.

LITHEUSPHAERELLA Defl.1932
Eocene

Cyst spherical ornamented with evenly spread projections of equal length.

ARCHAEOMONADOPSIS Defl. 1938
Cretaceous - upper Oligocene

Cyst spherical, the opening extended to an indefinite neck.

PARACHAEOMONAS Defl. 1932
Eocene

Cyst globoid, large pore with collar.

LITHARCHAEOCYSTIS Defl.1932
Eocene - Oligocene

Cyst ellipsoidal, occasionally with caudal appendices, pore biconical.

LITHUROPYXIS Defl. 1933
Oligocene - Miocene

Cyst fusiform with collar.

MICRAMPULLA Hanna 1927
Upper Creataceous

Cyst spherical, size variable, with long neck which flares at the tip. wall ornamented by ridges, which meet and outline polygonal fields.

Fig. 20.6. Cysts belonging to Archaeomonadaceae according to Deflandre (1932a).

CHRYSOSTOMATACEAE

CHRYSOSTOMUM Chodat 1922
emend. Defl.
Tertiary - recent

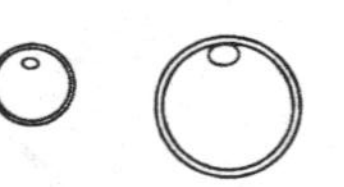

Cyst spherical, ovoid, ellipsoidal with or without collar arising from edge of pore, little or no ornamentation.

CLERICIA Frenguelli 1925
emend. Defl.
Tertiary - recent

Cyst spherical, ovoid or ellipsoidal, collar if present arising from edge of pore, ornamented with short to moderate spines, ridges, granulations, but without alar or single elongate hooklike projections.

CHRYSASTRELLA Chodat 1922
emend. Defl.
Tertiary - recent

Cyst spherical with or without simple collar, smooth with several very long spines.

CARNEGIA Pantocsek 1912
emend. Defl.
Tertiary - recent

Cyst spherical with varying ornamentation, including aliform or recurved projections (stomatocerque).

OUTESIA Frenguelli 1925
Tertiary - recent

Cyst spherical, ovoid or ellipsoidal with a double collar derived from the cyst wall, but without other striking ornamentation.

DEFLANDREIA Frenguelli 1938
Quaternary - recent

Cyst spherical with pore extended into a cylindrical or conical, straight or curved collar.

TRACHELOSTOMUM Freng. 1938
Quaternary - recent

Cyst spherical with long wide, large, cylindrical collar.

AMPHILITHOPYXIS Defl. 1932

Cyst divided in 2 parts, anterior spherical with apical pore, posterior conical with antapical pore, irregular alars connecting the two parts

Fig. 20.7. Cysts belonging to Chrysomonadaceae according to Deflandre (1936).

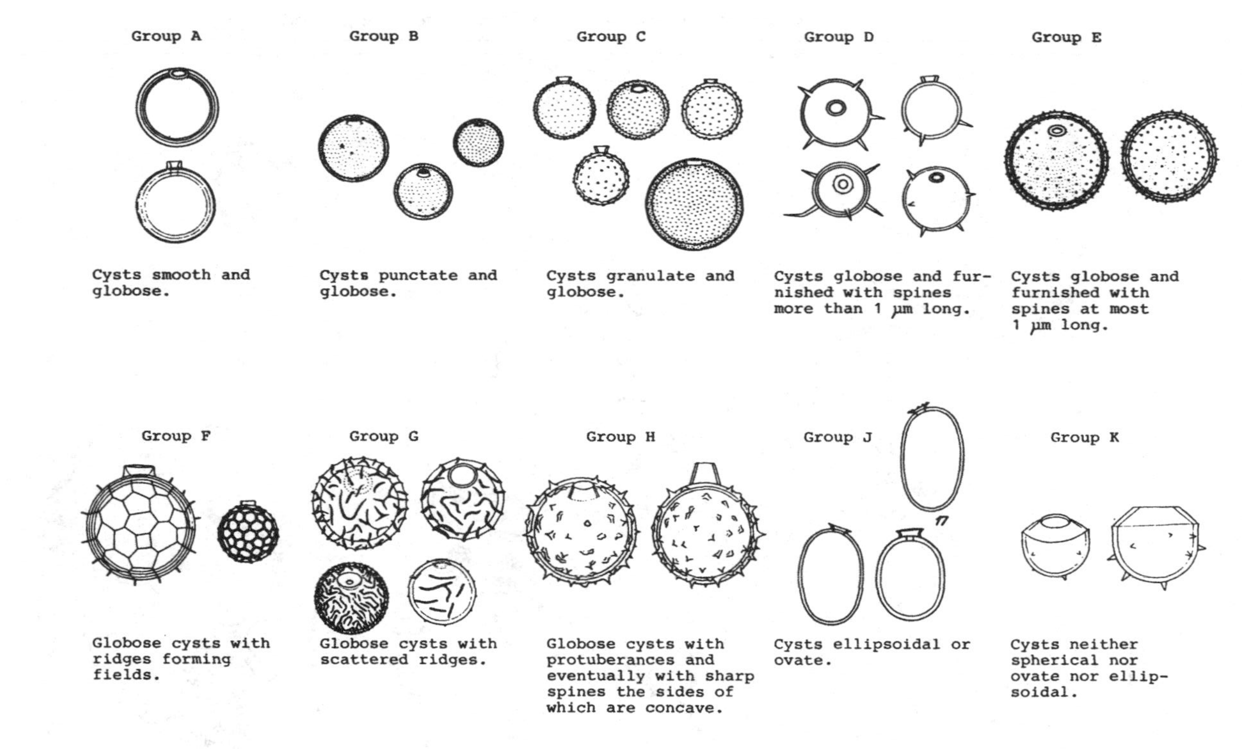

Group A
Cysts smooth and globose.
Group B
Cysts punctate and globose.
Group C
Cysts granulate and globose.
Group D
Cysts globose and furnished with spines more than 1 µm long.
Group E
Cysts globose and furnished with spines at most 1 µm long.
Group F
Globose cysts with ridges forming fields.
Group G
Globose cysts with scattered ridges.
Group H
Globose cysts with protuberances and eventually with sharp spines the sides of which are concave.
Group J
17
Cysts ellipsoidal or ovate.
Group K
Cysts neither spherical nor ovate nor ellipsoidal.

Fig. 20.8. Different types of cysts divided in groups A–K according to Nygaard (1956).

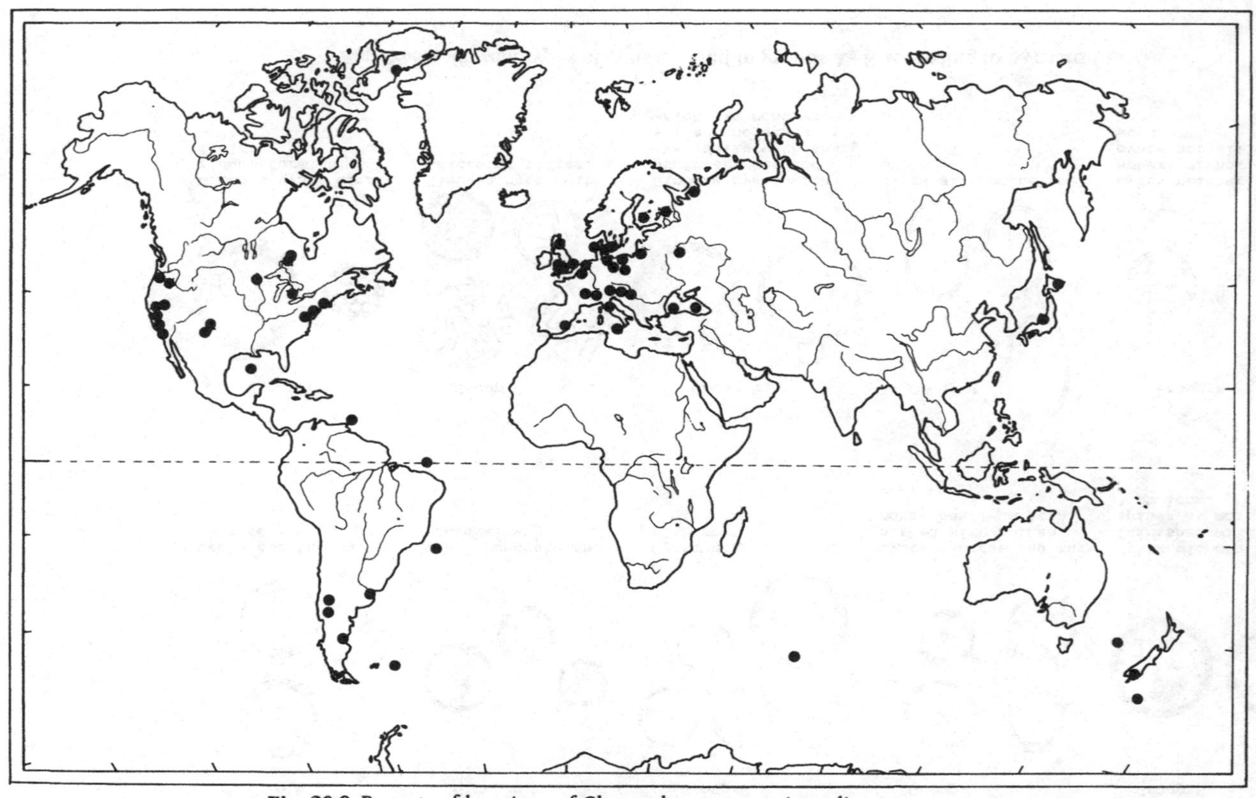

Fig. 20.9. Reports of locations of Chrysophycean cysts in sediments.

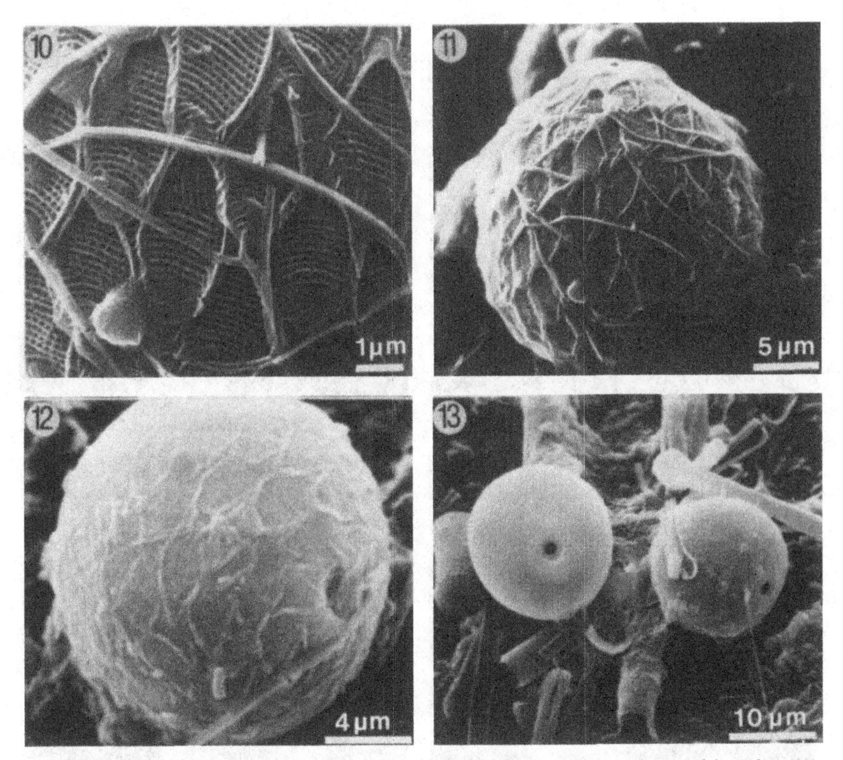

Figs. 20.10–20.13. *Mallomonas striata* Asmund. ***Fig. 20.10.*** Scales and bristles. ***Fig. 20.11.*** Cyst covered with scales; ***Fig. 20.12.*** Cyst covered with partly dissolved scales; ***Fig. 20.13.*** Two cysts. Note raised area surrounding the pore in mature cyst on left.

Chrysophytes belonging to the family Mallomonadaceae have their cells covered with loosely attached silica scales that themselves sometimes possess silica bristles (Figures 20.10, 20.11, 20.14, and 20.16). Iwanoff (1899) was the first to use the silica scales in the species description of *Mallomonas acaroides* Perty emend. Iwanoff. The scales of Mallomonadaceae are minute, and their ultrastructure is characteristic of the species. During the beginning of this century many species of *Synura* and *Mallomonas* were described with light microscopy (LM) (Bourrelly 1957, 1968; Matvienko 1965, Starmach 1980). However, it has been shown that electron microscopy (EM) is necessary for making reliable identifications, and many of these early species descriptions can no longer be used. Fott (1955) and Asmund (1955) were among the first to use transmission electron microscopy (TEM) for identification of *Synura* and *Mallomonas.* To date, about 130 species of Mallomonadaceae have been identified with TEM (Kristiansen 1982).

The chrysophyte cyst is also characteristic of the species, and even within the same genus different species produce cysts of unique size, form, and

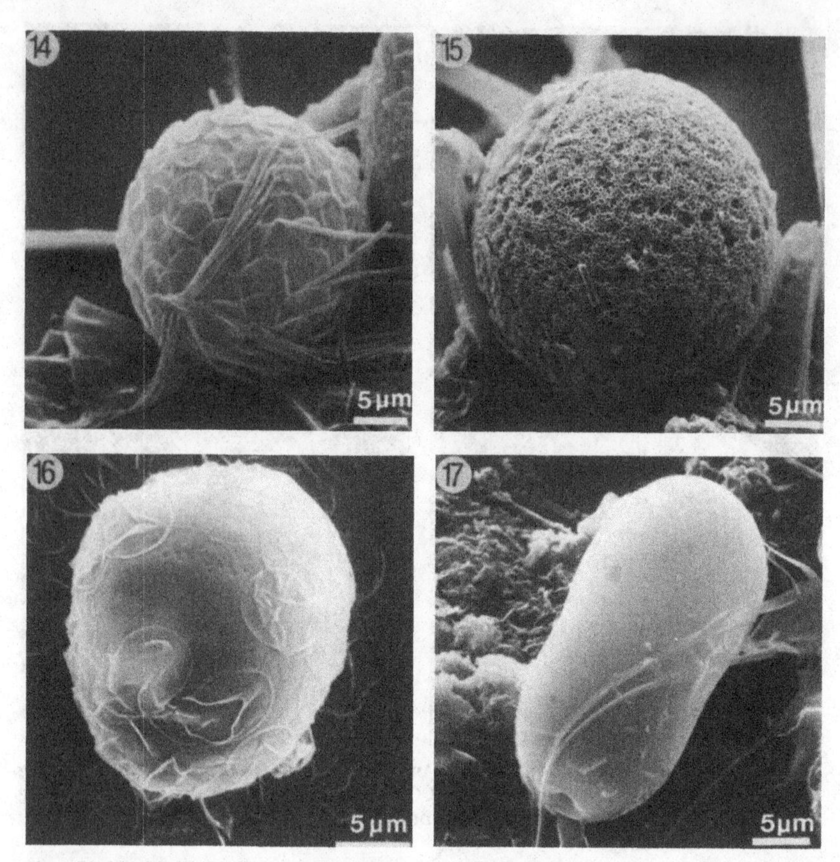

Figs. 20.14–20.17. ***Mallomonas caudata*** Iwanoff. ***Fig. 20.14.*** *M.* ***caudata*** covered in scales with a cyst developed inside. ***Fig. 20.15.*** Cyst. Cell wall has a spongelike appearance. ***Fig. 20.16.*** Cyst of ***Mallomonas vannigera*** Asmund with body scales remaining. ***Fig. 20.17.*** Cyst of *M.* ***vannigera*** smooth surfaced and oval shape

ultrastructure (Figures 20.13, 20.15, and 20.17–20.26). Scanning electron microscopy (SEM) has proved to be an excellent technique for identification of the chrysophyte cysts, just as TEM has for identification of their scales.

Most chrysophytes are found in freshwater habitats, and the majority of them are flagellates. However, there also exists a benthic chrysophyte flora, and chrysophytes have also been reported from marine habitats. Particularly large concentrations of cysts have been found in diatomites from Cretaceous, Miocene, Pliocene, and Quaternary times (Table 20.1 and Figure 20.9) [e.g., Deflandre (1932b, 1933, 1936), Frenguelli (1932), Andrieu (1936, 1937, 1938), Rampi (1937a,b), Tynan (1960, 1971), VanLandingham (1964), Hajos (1968, 1975), Stradner (1971), Cornell (1972), Perch-Nielsen (1975, 1977), Gombos (1977), Schrader (1978) and Adam

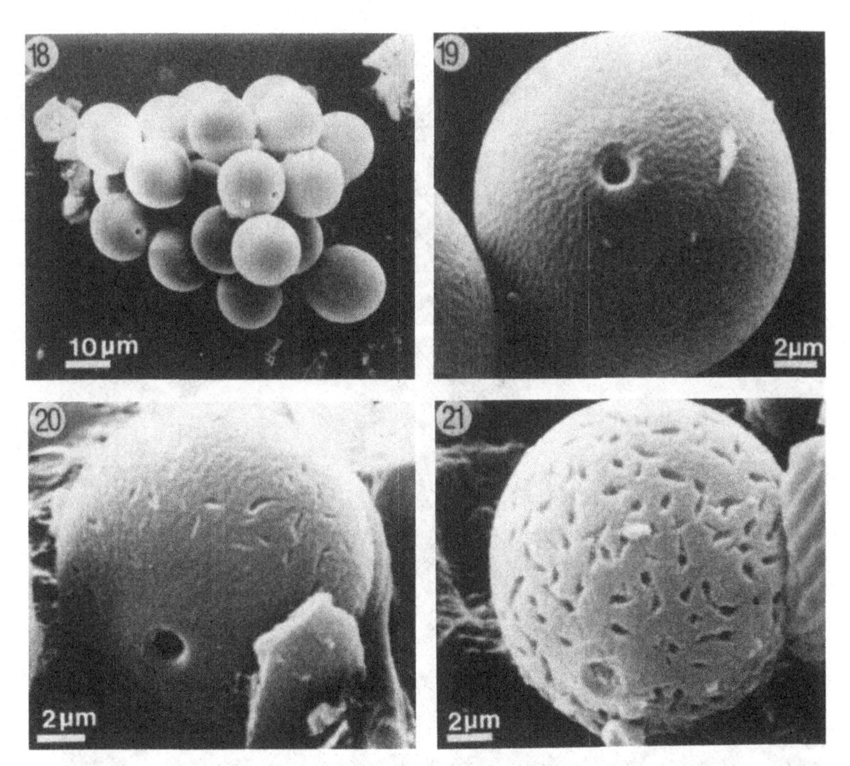

Figs. 20.18–20.21. Fig. 20.18. A whole colony of *Synura* developed into cysts. ***Figs. 20.19–20.21.*** Different stages in the development of *Synura* cysts, all from the same colony.

& Mahood (1981b)]. On the basis of these fossil deposits, chrysophytes are presumed to have had a marine origin (Tappan 1980).

As cysts and scales are made of silica, they are resistent to dissolution and fragmentation. When a cell dies, the disarticulated scales will be deposited on the sediment surface and, after some time, covered by other material. The resting cysts are immobile and will also be deposited on the sediment surface. They can rest in this position for a long time, but when favorable conditions of light, temperature, and nutrient concentrations for vegetative growth of the resting cell are encountered, the plug is dissolved and the organism leaves the cyst (Figure 20.4). The empty cyst will be left on the sediment surface and will eventually be covered and become incorporated into the sediment assemblages.

Because the chrysophytes are primarily planktonic and presumed to have restricted environmental requirements, they can be good indicators of water quality if taxonomic difficulties can be overcome. The chrysophytes often appear seasonally (Bourrelly 1957, Cronberg 1980, 1982), and

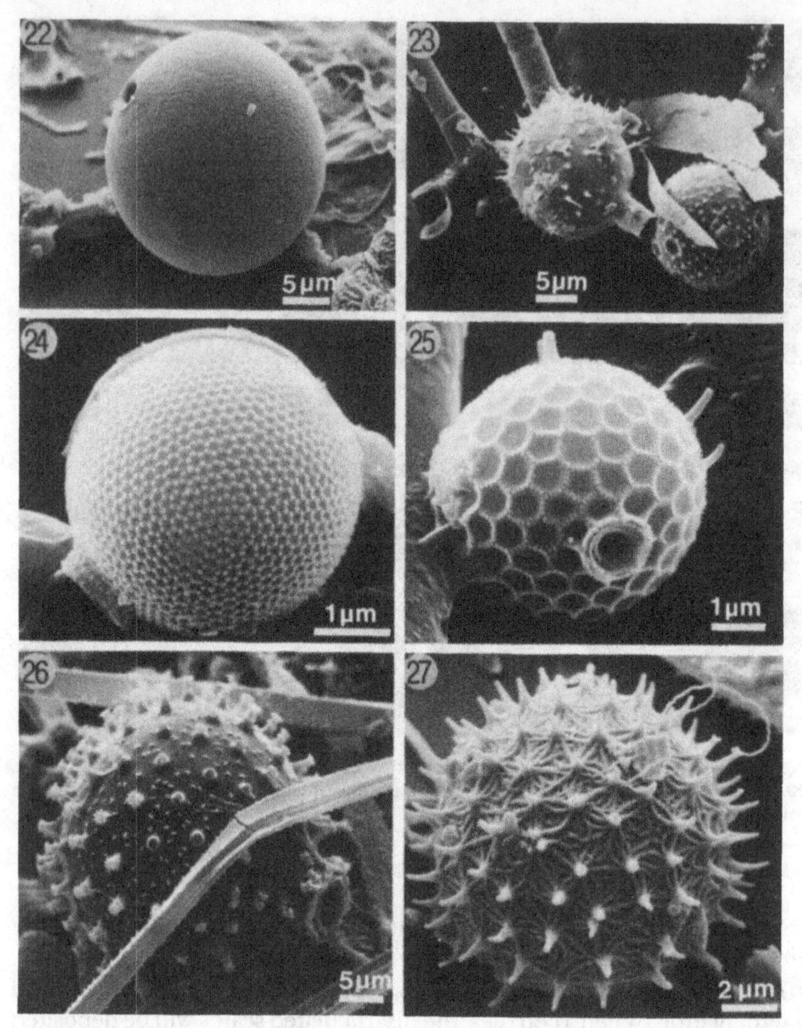

Figs. 20.22–20.27. Unidentified cysts showing different ultrastructure of the cyst wall. Samples collected from lakes in the South Swedish Highlands (Småland) during March–May, 1974–1977.

the scales and cysts may cause distinct layers in the sediment, which can be laminated annually (Tippett 1964, Renberg 1976, Battarbee, Cronberg, & Lowry 1980, Simola, Coard, & O'Sullivan 1981).

In this chapter, I shall review studies that have described chrysophycean cysts and scales in sediments and discuss the potential usefulness of these investigations.

Investigation methods

Sampling techniques

In pioneering investigations, sediments or diatomites were collected from different parts of the world (Table 20.1 and Figure 20.9). No systematic sampling of stratigraphic records was made because the investigators only wanted to get samples rich in diatoms. They found that the diatomite was also rich in other siliceous organisms, including chrysophyte cysts. The geological age of the sediments was estimated on the basis of the diatom content or other plant and animal remains.

With the invention of the first sediment corer at the beginning of the century, a new era for palaeontologists and geologists began. With the corer, sediments could be removed from the bottom of lakes and bogs with their stratigraphy intact. It was possible to remove samples from different layers or at regular intervals, e.g., every 10 cm (Cronberg 1980, Carney 1982), every cm (Smol 1980), or every 0.5 cm (Munch 1980). In some deep lakes the sediment is laminated annually. To obtain intact lamination, the cores were removed from the lake using *in situ* freezing techniques (Renberg 1976, Simola 1979, Battarbee et al. 1980, Simola et al. 1981).

Preparation of scales and cysts from sediment

The first description of fossil chrysophytes was made from diatomite from different parts of the world. The diatomite was sometimes very clean, and samples could be embedded in Canada balsam immediately after simply being rinsed with water.

More often the sediment samples were rich in organic matter and had to be treated with acids (HNO_3, H_2SO_4, or combinations of these) and then thoroughly cleaned with distilled water. Andrieu (1936) cleaned his sample with acids and embedded them with Canada balsam. He commented that the sieving method used for collecting the diatoms was not recommended for chrysophyte cysts because the cysts were lost and only large diatoms were retained. Andrieu especially recommended Canada balsam as a mounting medium for the chrysophyte cysts, and not Hyrax, which was otherwise much used for diatom preparations. He observed that Hyrax gave, in the median section of cysts, parallel lines in the cell wall, suggesting many layers. The double wall observed might in some cases be an artifact of the mounting medium. With Canada balsam, the median section and pore region, important taxonomic features, appeared more clearly, although some details in the cell wall pattern could be lost. However, the cysts were mounted with different media, such as Canada balsam, Hyrax, Styrax, Pleurax, Clearax, Diphane or Caedex.

Nygaard (1956) treated his sediment samples with 30% hydrogen perox-

Table 20.1. *Reports of locations of chrysophycean cysts from sediments*[a]

Locality	Foss	SF	REC	LM	TEM	SEM	Author	Year
Asia								
Japan								
Abashiri	+	−	−	+	−	−	Deflandre	1933
Wembets	+	−	−	+	−	−	Deflandre	1938
Noto Peninsula	+	−	−	+	−	−	Ichikawa et al.	1964
Europe								
Austria								
Limberg	+	−	−	+	+	+	Stradner	1971
Schwarzsee, Kitzbühel	−	−	+	−	−	−	von Sarntheim	1949
Belgium								
Nieuport	−	−	+	+	−	−	Conrad	1926
Foret de Soignes	−	−	+	+	−	−	Conrad	1938
Heyst-op-den-Berg	−	−	+	+	−	−	Conrad	1938
Coq-sur-Mer	−	−	+	+	−	−	Conrad	1938
Sample 547 and 548, H. van Heink collection	+	−	−	+	−	−	Conrad	1940
Denmark								
Fuur, Jylland	+	−	−	+	−	−	Deflandre	1932b
Mors, Jylland	+	−	−	+	−	−	Deflandre	1933
Lake Gribsø	−	+	+	+	+	−	Nygaard	1956
Finland								
Lovojärvi, Southern Finland	−	−	+	−	−	−	Simola	1979

Laukunlampi, North Karelia	–	+	+	–	–	–	Battarbee et al.	1980
Pääjärvi, Southern Finland	–	–	+	–	–	–	Simola et al.	1983
France								
Verneuge, Puy-de-Dome	+	–	–	+	–	–	Andrieu	1937
Vassivière, Puy-de-Dome	+	–	–	+	–	–	Andrieu	1938
Lac Chambon, Puy-de-Dome	+	–	–	+	–	–	Firtion	1944
Menat, Puy-de-Dome	+	–	–	+	–	–	Firtion	1945
Germany								
Diebelsee (moor)	–	+	+	+	–	–	Krieger	1929
Pond near Berlin	+	–	–	+	–	–	Deflandre	1936
Lychen	–	–	+	–	–	–	Conrad	1938
Southeast coast of Schleswig-Holstein	+	+	+	–	–	–	Dahm	1956
Great Britain								
Dalmahoy, Edinburgh	+	–	–	–	–	–	Henderson	1925
Llyn Padarn and Peris, Wales	–	+	+	+	–	–	Elner & Happey-Wood	1978
Strumpshaw broad, Norfolk	–	–	+	–	–	–	Moss	1979
Loe Pool, Cornwall	–	+	+	–	–	–	Simola et al.	1981
Hungary								
Surian	+	–	–	+	–	–	Greguss	1913
Kopacsel	+	–	–	+	–	–	Pantocsek	1913
Karand	+	–	–	+	–	–	Deflandre	1933
Kekko	+	–	–	+	–	–	Deflandre	1933
Borostelek	+	–	–	+	–	–	Deflandre	1938
Nyermegy	+	–	–	+	–	–	Deflandre	1938
Martavorlandes	+	–	–	+	–	–	Hajos	1968
Italy								
Monte Amiata, Toscana	+	–	–	+	–	–	Rampi	1937a

Table 20.1. *(cont.)*

Locality	Foss	SF	REC	LM	TEM	SEM	Author	Year
S:t Fiora, Monte Amiata, Toscana	+	−	−	+	−	−	Rampi	1937b
Marmorita	+	−	−	+	−	−	Deflandre	1938
Isola di Rodi	+	−	−	+	−	−	Rampi	1940b
Rome	+	−	−	+	−	−	Zanon	1947
Licata, Sicilia	+	−	−	+	−	−	Rampi	1948
Sendai, Giappone	+	−	−	+	−	−	Rampi	1948
Lago di Monterosi	+	−	−	+	−	−	Leventhal	1970
Poland								
Imbramowice, Wroclaw	+	−	−	+	−	−	Kaczmarska	1976
Lake Druzno, near the Baltic coast	+	+	−	+	−	−	Przybylowska-Lange	1976
Soviet Union								
Kusnetzk	+	−	−	−	−	−	Deflandre	1932a
Galtgarben-Moore, Königberg	−	+	+	−	−	−	Steinecke	1937
Isenski	+	−	−	+	−	−	Deflandre	1938
Archangelsk	+	−	−	+	−	−	Deflandre	1938
Pyawotschnoje-See	+	+	+	−	−	−	Korde	1961
Nero-See	+	+	+	−	−	−	Korde	1961
Galitsch-See	+	+	+	−	−	−	Korde	1961
Kleine Schartasch	+	+	+	−	−	−	Korde	1961
Uschtschemerowo-See	+	+	+	−	−	−	Korde	1961
Black Sea	+	−	−	+	−	−	Schrader	1978
Spain								
Alicante	+	−	−	+	−	−	Colom	1940
Valencia	+	−	−	+	−	−	Colom	1940

Sweden								
In lime-rich sediments many different localities	+	–	–	–	–	–	Lagerheim	1903
Lake Rudetjärn, Medelpad	–	+	+	–	–	+	Renberg	1976
Lake Växjösjön, South Sweden	–	+	+	–	–	+	Cronberg	1980, 1982
Switzerland								
Zürichsee	–	–	+	–	–	–	Nipkow	1927
Baldeggersee	–	–	+	–	–	–	Nipkow	1927
North America								
Canada								
McKay Lake, Ontario	+	+	+	–	–	–	Tippett	1964
Little Round Lake, Ontario	+	+	+	–	–	–	Tippett	1964
Cape Herschel, Ellesmere Island	–	+	+	–	–	+	Smol	1983
USA								
Moreno shale, CA	+	–	–	+	–	–	Hanna	1927
Maryland	+	–	–	+	–	–	Deflandre	1932b
Poplien	+	–	–	+	–	–	Deflandre	1933
Richmond, VA	+	–	–	+	–	–	Deflandre	1933
San Pedro, CA	+	–	–	+	–	–	Deflandre	1933
Santa Barbara, CA	+	–	–	+	–	–	Deflandre	1933
Atlantic City, NJ	+	–	–	+	–	–	Deflandre	1933
Willington, CT	+	–	–	+	–	–	Conrad	1940
Randle Cliff Beach, Calvert county, MD	+	–	–	+	–	–	Tynan	1960
Yakima and Grant counties, WA								
Kreyenhagen, CA	+	–	–	+	–	–	Rampi	1969
Kellog Creek, CA	+	–	–	–	–	–	Tynan	1971
Moreno, CA	+	–	–	–	–	+	Cornell	1972
Upper Stepney, Botsford, CT	+	–	–	–	–	+	Gritten	1977

Table 20.1. *(cont.)*

Locality	Foss	SF	REC	LM	TEM	SEM	Author	Year
Clear Lake, CA	+	–	–	–	–	+	Mahood & Adam	1979
Upper Echo Lake, CA	–	+	+	–	–	+	Adam & Mahood	1979
Lost Trail Pass Bog, MT	–	+	+	–	–	+	Adam & Mehringer	1980a
Fallen Leaf Lake, Eldorado county, CA	–	–	+	–	–	+	Adam & Mahood	1980a
Alta Morris Lake, Eldorado county, CA	–	–	+	–	–	+	Adam & Mahood	1980b
Castor Pond, Jemez Mountains, NM	–	–	+	–	–	+	Adam & Mehringer	1980b
Haypress Meadows, Eldorado county, CA	–	–	+	–	–	+	Adam	1980b
Flagpole Peak, Eldorado county, CA	+	–	–	–	–	+	Adam	1980c
Fish Lake, Steens Mountains, OR	–	+	+	–	–	+	Adam & Mehringer	1980c
Suzie Lake, Eldorado county, CA	–	–	+	–	–	+	Adam & Mahood	1980c
Kerchner Marsh, MN	–	+	+	–	–	–	Brugham	1980
Lake Aloha, Eldorado county, CA	–	–	+	–	–	+	Adam & Mahood	1981a
Harden Lake Meadow, Yosemite National Park, CA	–	+	+	–	–	+	Adam	1981
Frains Lake, MI	–	–	+	–	–	+	Sandgren & Carney	1983
South America								
Argentina								
Moissac, Cantal	+	–	–	+	–	–	Frenguelli	1925
Various localities	+	–	–	+	–	–	Frenguelli	1932
Platense, Buenos Aires	+	–	–	+	–	–	Frenguelli	1935b
Neuquen	+	–	–	+	–	–	Frenguelli	1936
Various localities	+	–	–	+	–	–	Frenguelli	1938a
Platense, Buenos Aires	+	–	–	+	–	–	Frenguelli	1938b

Barbados								
Springfield	+	–	–	+	–	–	Deflandre	1933
Chile								
Tripoli, Angostura	+	–	–	+	–	–	Frenguelli	1931
Kerguelen island, moor	+	+	+	+	–	–	Andrieu	1936
Australia								
New Zealand								
Oamaru	+	–	–	+	–	–	Deflandre	1933
Oceans								
Southwest Pacific Ocean	+	–	–	–	+	–	Perch-Nielsen	1975
Southwest Pacific Ocean	+	–	–	+	–	+	Hajos	1975
Western South Atlantic Ocean	+	–	–	+	–	–	Perch-Nielsen	1977
South Atlantic Ocean	+	–	–	+	–	–	Gombos	1977
Gulf of Mexico	+	–	–	–	–	–	Jendrzejewski et al.	1978

[a] Foss = fossil cysts deposited from Upper Cretaceous to Pleistocene, SF = subfossil cysts deposited during postglacial time (according to Messikommer 1938), REC = recent cysts deposited from about 1800 AD to present time, LM = light microscopy drawings or micrographs, TEM = transmission electron micrographs, SEM = scanning electron micrographs.

ide (H_2O_2) for 2–3 weeks at 40°–50°C. The samples were then treated for 1 week with a mixture of sulfuric acid and potassium dichromate. They were then rinsed with distilled water and transferred into 96% alcohol, and then finally to xylene or acetone. The cysts were embedded in Sirax, Clearax, or Pleurax.

The method for cleaning diatoms described by Patrick and Reimer (1966) has often been used for cleaning chrysophyte cysts and scales (Leventhal 1970, Gritten 1977, Munch 1980, Carney 1982).

In some cases, the sediment samples have been treated with 10% HCl for removing calcium carbonate and 30% H_2O_2 to remove the organic matter and then rinsed with distilled water. The treatment time with HCl and H_2O_2 depended on the amount of calcium carbonate and organic material in the samples (Kaczmanska 1976, Przybylowska-Lange 1976, Cronberg 1980, Battarbee et al. 1980).

For quantitative analyses, defined volumes of sediments were subsampled and treated with acids or H_2O_2. Defined suspensions of cleaned material were evaporated on cover slips or special evaporation trays (Battarbee 1973).

For the investigation of microlaminated sediments, the sediment core was freeze-dried at −20°C for at least 24 hours. Dried particles were then transferred with adhesive tape (Simola 1979) or carefully removed from individual layers using razor blade edges (Battarbee et al. 1980). Microscopic investigation was made directly on the adhesive tape or on material from microlayers. Individual samples were treated with 30% H_2O_2 and carefully washed with distilled water before being mounted with Microps (Battarbee et al. 1980).

For TEM investigation of scales, samples were treated with H_2O_2 as described above. A drop of the cleaned sample was placed on Formvar-coated grids and dried (Cronberg 1982).

For investigations with SEM, drops of the H_2O_2 or acid-cleaned sample were placed on a round coverslip that was then glued onto a specimen stub. After the drop had dried, the stub was coated with a layer of gold (60%) and palladium (40%) under vacuum (Cronberg 1980).

Literature pertaining to chrysophycean cysts and scales in sediments

Cysts in sediments

For the survey of chrysophyte cysts in sediments, the existing literature can be conveniently divided in four main categories:

1. Description of lake sediments with poor or good stratigraphy, chrysophyte cysts mentioned, but no illustrations given: Lagerheim (1903), Henderson (1925), Steinecke (1937), von Sarntheim (1949), Dahm

(1956), Korde (1961), Perch-Nielsen (1977), Jendrzejewski & Hart (1978), Moss (1979), Brugham (1980), and Smol, Brown, & McNeely (1983).

2. Description of sediments, especially diatomites from different parts of the world, with poor or no stratigraphy, but with good descriptions of chrysophyte cysts including LM drawings, LM micrographs or EM micrographs: Pantocsek (1912, 1913), Greguss (1913), Frenguelli (1925, 1929, 1931, 1932, 1935a,b, 1936, 1938a,b), Hanna (1927, 1934), Krieger (1929), Deflandre (1932a,b,c, 1933, 1934, 1935, 1936, 1938, 1952), Andrieu (1936, 1937, 1938), Rampi (1937a,b, 1940a,b, 1948, 1969), Conrad (1938, 1940), Firtion (1944, 1945), Zanon (1947), Korde (1966), Perch-Nielsen (1975) and Gritten (1977).
3. Description of sediments with good stratigraphy and chrysophyte cysts shown in LM or EM micrographs:
   a.–Investigators following the systematics of Deflandre (1936): [Andrieu (1936, 1937, 1938), Frenguelli (1936, 1938a,b), Rampi (1937a,b, 1940a,b, 1948, 1969), Firtion (1944, 1945), Tynan (1960, 1971), Ichjkawa, Fuji, & Bachmann (1964), VanLandingham (1964), Hajos (1968), Stradner (1971), Cornell (1972), Gombos (1977), and Schrader (1978)].
   b. Investigators following the systematics of Nygaard (1956): [Leventhal (1970), Elner & Happey-Wood (1978), Carney & Sandgren (1983), and Sandgren & Carney (1983)].
   c. Investigators not following the above-mentioned systematics, chrysophyte cysts numbered or identified to species: Kaczmarska (1976), Przybylowska-Lange (1976), Adam & Mahood (1979, 1980a,b,c, 1981a,b), Mahood & Adam (1979), Adam (1980a,b,c, 1981), Adam & Mehringer (1980a,b), Cronberg (1980, 1982), and Smol (1983).
4. Description of sediments with microstratigraphy or annually laminated sediments, with or without chrysophyte cysts pictured with LM or EM: Nipkow (1927), Tippett (1964), Renberg (1976), Simola (1977, 1979), Simola et al. (1981), Simola & Uimonen-Simola (1983).

Important works on chrysophyte cysts and their classification

Ehrenberg (1854) pictured, among others, the alga *Chaetotyphla volvox* (Figure 20.1) from soil samples from Berlin, Germany. This must be one of the first drawings of chrysophyte cysts, although he placed them among the "Polygastern" (i.e., the diatoms).

Pantocsek (1912, 1913) found cysts in lake sediments from Hungary and described the genera *Carnegia* and *Echinopyxis.* However, he placed them in the tribe Diflugioidea (probably Rhizopoda).

Frenguelli (1925, 1929, 1931, 1932, 1935a,b, 1936, 1937, 1938a,b) described about 110 cysts in diatomites from different localities in Argentina. At first he referred the cysts to the genus *Trachelomonas* (Euglenophyta), but after discussions with Deflandre (1935), Frenguelli (1936) changed his opinion and, like Deflandre, classified the cysts as resting stages of chrysophytes.

Deflandre (1932a) described the cyst of *Litharchaeocystis costata*, but did not refer it to any known organism. Later in the same year (Deflandre 1932b), he proposed the family Archaeomonadaceae for marine, fossil, small, more or less ornamented, siliceous spores (Figure 20.6). He concluded that these cysts were related to the living chrysophytes. Deflandre (1936) then divided the cysts into three main groups:

A. Cysts produced by living, identifiable chrysomonads: Chrysomonadinea.
B. Cysts produced by unidentified fossil, marine forms: Archaeomonadaceae (artificial family, Figure 20.6).
C. Cysts produced by unidentified fossil or recent freshwater forms: Chrysostomataceae (artificial family, Figure 20.7).

With Deflandre's system, it was possible to classify the cysts in different artificial genera according to the form and cell wall structure of the specimen. Currently, 16 genera have been described.

Zanon (1947) described cysts from diatomites near Rome, Italy. He proposed a new classification wherein he divided the cysts into two main groups:

A. Cysts with single cell wall
B. Cysts with double or multiple cell wall
   I. Symmetrically built cysts
   II. Asymmetrically built cysts

With this system he described 565 species, distributed in 19 genera. However, this classification system of Zanon has not been followed. In contrast, the systematics of Deflandre has been much used (see above).

Nygaard (1956) investigated chrysophyte cysts in sediments from Lake Gribsø, Denmark. He divided them into ten groups according to size, form, and cell wall structure (Figure 20.8). He also included a key for Lake Gribsø cysts. Nygaard named each statospore as *cysta* plus a descriptive epithet (e.g., *cysta teres* Nygaard). He pointed out that many cysts have a very characteristic appearance and could be used as indicators, if they could be identified.

Nygaard made a careful study of the distribution of cysts in Lake Gribsø sediments. He described and pictured 77 different morphotypes. Sediments from 3000 B.C. to recent times were studied. During the early period 3000 B.C.—400 B.C. the lake was eutrophic and had an estimated pH of 6–7, but it decreased to pH 4–6 around 400 A.D. At the present time, pH varies between 4.6–5.3. Chrysophycean cysts were most common when pH was high around 3000 B.C. This indicates that cyst-producing species thrived better during more eutrophic conditions.

As early as 1927, Nipkow investigated annually laminated sediments deposited between 1890 and 1925 in Lake Zürichsee and between 1904 and 1925 in Lake Baldeggersee, Switzerland. Oligotrophic conditions were recorded during 1890–1895 in Lake Zürichsee and between 1904 and 1908 in Lake Baldeggersee, when *Dinobryon* cysts were deposited. An increase in nutrients then occurred, and in connection with this no more *Dinobryon* cysts were recorded in the sediments. Eutrophication had changed the algal assemblage in the lakes.

Tippett (1964) also found laminated sediments in two lakes in Ontario, Canada. Here the varves consisted of *Dinobryon* cysts, although *Mallomonas* scales were also present.

Fossil cysts have also been investigated from marine environments, such as the Gulf of Mexico (Jendrzejewski & Hart 1978), the southwest Pacific Ocean (Perch-Nielsen 1977), the south Atlantic Ocean (Gombos 1977), and the southwestern Atlantic Ocean (Perch-Nielsen 1977). These marine sediments were rich in chrysophyte cysts, and many species were identical or similar to those species already described by, for example, Deflandre (1932b,c) and Rampi (1940a,b, 1948). The genera *Archaeomonas* and *Archaeosphaeridinium* were particularly well represented.

Leventhal (1970) studied chrysophyte cysts in a sediment core from Lago di Monterosi, Italy. She identified 130 morphotypes and grouped them according to Nygaard's classification system. Most cysts were found in the early sediments representing the oligotrophic period of the lake. Leventhal tried to identify as many cysts as possible, by species, but her identification of different species of, for example, *Chromulina, Ochromonas,* and *Mallomonas* is only tentative.

Elner and Happey-Wood (1978) investigated diatom remains and chrysophyte cysts in the sediments of Llyn Padarn and Llyn Peris, Wales, United Kingdom, from 6000 and 900 years ago, respectively, up to the present day. The investigation included statistical analyses to determine the precision in microfossil counts from the same depth zones and also from different depths. The variations in number of diatoms or cysts taken from the same horizontal layer were small; thus, the differences found in numbers from vertical samples were supposed to reflect true changes in the sediment profiles. In both profiles, a total of 142 cyst morphotypes were recognized, although only four types were frequent and could thus be followed through the cores. In Llyn Padarn most chrysophyte cysts were recorded during the early and oligotrophic period of the lake. Lower numbers of cysts were recorded during the more eutrophic later period, when *Asterionella formosa* was dominating. In Llyn Peris, on the other hand, most cysts were recorded 100–200 years ago, and this coincided with the beginning of copper mining in the surrounding areas. At the same time *Rhizosolenia eriensis* reached maximal development.

The cysts and diatoms in recent sediments of Frains Lake, Michigan have been carefully studied by Carney (1982). He followed the development of cysts and diatoms from 930 A.D. to present the time. Clear changes were recorded in the diatom assemblage composition when European settlement started around the lake. The shift in chrysophycean cysts was not so evident. However, the recent sediments contained a rich flora of chrysophyte cysts.

Carney and Sandgren (1983) closely examined the chrysophyte cysts from Frains Lake with SEM and also counted the most frequent characteristic forms in the sediment core. Some cysts occurred frequently during the presettlement period, but diminished in number or disappeared completely with increased trophy. Some cysts were growing better during the more eutrophic postsettlement period. The absolute accumulation of cysts increased clearly with the settlement and with the beginning of the eutrophication. Each cyst morphotype was carefully studied with SEM, and 46 different morphotypes were described (Sandgren & Carney 1983). Particular attention was paid to the pore – collar complex and cell wall ornamentation. A descriptive terminology for the cyst collar complex was proposed.

Among others, Adam (Table 20.1) investigated sediments from the western part of the United States, especially California, and this part of the world now appears to be the best investigated area with respect to fossil and recent chrysophyte cysts. Adam and his coauthors pictured the cysts with SEM and also numbered them, but no descriptions of them were given. Altogether they have taken approximately 2750 SEM micrographs of cysts from 110 investigated sediment samples (Adam 1981).

In Lake Växjösjön, Sweden, algal remains and chrysophyte cysts were studied in sediments from a 3-meter sediment core extending from 1100 A.D. to the present time (Cronberg 1980, 1982). Six cysts with characteristic structure were found and could be followed through the core. During the oligotrophic period of the lake, prior to 1800 A.D. these six morphotypes were evenly distributed. As the lake became more eutrophic and polluted, the distribution pattern of the cysts changed. Five of these cysts were identified to species: *Mallomonas eoa* Tak., *M. teilingii* Conr., *M. heterospina* Lund, *Chryastrella paradoxa* Chod., *Uroglena lindii* Bourr., and/or *U. soniaca* Conr. *Mallomonas eoa* was most common during the worst pollution period (Cronberg 1982). *Mallomonas heterospina* and *M. teilingii* increased when the lake became slightly eutrophic, but disappeared during the heavy pollution period and then reappeared again when the lake began to recover. *Chrysastrella paradoxa* and the *Uroglena* cysts were only found during the oligotrophic period. This is a clear illustration of the fact that chrysophytes have different environmental demands. Some grow well under eutrophic conditions, while others are more successful under oligotrophic conditions.

In most investigations concerning cysts in sediments, the emphasis has been on the morphological description of the cysts. At least 450 cysts have been found and named without knowing their identity (Nygaard's *nomina tempora*). Only 30 cysts have been identified to species (Table 20.2).

Scales in sediment

For the survey of chrysophyte scales in the sediments the existing literature has been divided into two categories.

1. Description of lake sediments with good stratigraphy, chrysophyte scales *not pictured* with EM: Tippett (1964), Haworth (1983, 1984), and Simola & Uimonen-Simola (1983).
2. Description of lake sediments with good stratigraphy, chrysophyte scales pictured with EM: Battarbee et al. (1980), Munch (1980), and Smol, Charles, and Whitehead (1984a, b).

Important works on chrysophyte scales

As early as 1899, Iwanoff stated that the scales of chrysophytes were species specific, and he used the scale morphology as a taxonomic criterion when he described *M. caudata* Iwan. (Figures 20.14 and 20.15). During the first half of the current century, many scaled chrysophytes were described. These descriptions were all made with help of LM (Bourrelly 1957, Matvienko 1965, Starmach 1980), but then it was shown that EM is necessary for exact and reliable identifications (Asmund 1955, Fott 1955, Petersen & Hansen 1956). Since 1955 many papers have been published with descriptions of scaled chrysophytes (Takahashi 1978, Kristiansen & Takahashi 1982). According to Kristiansen (1982), 130 species distributed in seven genera of Mallomonadaceae have been described with TEM and/or SEM.

Nygaard (1956) found some scales of *Mallomonas* in the uppermost sediments of Lake Gribsø, but not in older material. He believed that the scales were fragile and not well preserved in sediments. Fott (1966), however, found *Mallomonas* scales in lake sediments and stated they could be useful paleolimnological indicators when describing lake history.

In 1980, several papers (Battarbee et al. 1980, Munch 1980, Smol 1980) were published concerning the use of Mallomonadaceae scales as paleological indicators. Sediments rich in scales and diatoms were investigated, and it was found that changes in the lake's development could be interpreted through chrysophyte scales.

Munch (1980) investigated diatoms and scales of Chrysophyceae in sediments from Hall Lake, Washington. She found that before European settlement in the area at the end of the last century, the lake was oligotrophic with a rich development of *Mallomonas caudata* and *M. crassisquama* (Asm.) Fott. After the arrival of the Europeans, the oligotrophic lake was influenced by human activities; *M. caudata* disappeared first, whereas

Table 20.2. *Chrysophycean cysts of defined specific affiliation, studied by EM*

Species	TEM	SEM	Author	Year
Chromophysomonas trioralis Tak.	+	+	Skogstad	1982
Chromulina placentula Belch. & Swale	+	−	Belcher & Swale	1967
Chrysosphaerella coronacircumspina Wujek & Krist.	−	+	Preisig & Takahashi	1978
Dinobryon cylindricum Imhof	+	−	Sandgren	1980a
D. cylindricum Imhof	−	+	Sandgren	1983a
D. divergens Imhof	−	+	Sheath et al.	1975
D. divergens Imhof	+	−	Sandgren	1980b
Mallomonas acaroides var. *striatula* Asm.	−	+	Cronberg	1980
M. akrokomos Ruttn.	−	+	Cronberg	1980
M. caudata Iwanoff	+	−	Sandgren	1980b
M. caudata Iwanoff (Fig. 15)	−	+	Cronberg	This volume
M. crassisquama (Asm.) Fott	−	+	Gretz et al.	1979
M. eoa Tak.	−	+	Cronberg	1973
M. heterospina Lund	+	+	Cronberg	1982
M. insignis Pénard (Fig. 5)	−	+	Dürrschmidt	1984
M. intermedia Kiss.	−	+	Skogstad	1984
M. pumilio var. *pumilio* (Harr. & Bradl.) emend Asm. Cronb. & Dürrs.	−	+	Asmund et al.	1982
M. striata Asm. (Figs. 12–13)	−	+	Cronberg	This volume
M. teilingii (Teil.) Conr.	−	+	Cronberg	1980
M. torquata Asm. & Cronb.	−	+	Asmund & Cronberg	1979
M. vannigera Asm. (Figs. 16–17)	−	+	Cronberg	This volume
Ochromonas sphaerocystis Matv.	−	+	Andersen	1982
O. tuberculata Hibb.	−	+	Hibberd	1977
Paraphysomonas butcheri Penn. & Clark.	−	+	Takahashi	1981
P. imperforata Lucas	−	+	Takahashi	1981
Synura petersenii Korsh.	−	+	Cronberg	1982
Uroglena americana Calk.	+	−	Sandgren	1980b
U. lindiae Bourr.	−	+	Nygaard	1977
U. volvox Ehrenb.	−	+	Kristiansen	1980
U. volvox Ehrenb.	+	−	Sandgren	1980b

Synura lapponica Skuja and *S. echinulata* Korsh. increased. At that time the lake was probably acidified as a result of wood-cutting and pollution from a saw mill. In 1963, a road was constructed around the lake. Due to changes in environmental conditions, *M. crassisquama* disappeared and *S. spinosa* Korsh. and later on *S. petersenii* Korsh. replaced the other more oligotrophic species of *Synura* and *Mallomonas.* In summary, Munch found that the chrysophyte flora changed due to eutrophication.

Later in the same year, Smol (1980) described the changes of the chrysophyte flora in sediments from three Canadian lakes. He studied the stratigraphic distribution of six *Mallomonas* species, and found that they were sensitive to changes in water quality.

Smol et al. (1983) investigated the scales of Mallomonadaceae in sediments of Little Round Lake, Canada. This lake was eutrophicated in connection with settlement by Europeans (about 1850 A.D.). Before this time, scales were well represented in the sediments, but during the years 1850–1945 they were almost completely absent. However, during the last 30 years the scales of Mallomonadaceae have again been recorded in the sediments, indicating a return to oligotrophic conditions, as a result of the onset of meromixis of the lake. In an investigation of 38 lakes in Adirondack, New York, Smol et al. (1984a,b) studied chrysophyte scales from surficial sediments. Thirty taxa belonging to Mallomonadaceae were identified. The distributions of different scales were correlated with environmental factors. They found a strong correlation between pH and different mallomonadacean assemblages. *Mallomonas hindonii* Nich. and *M. hamata* Asm. were indicative of the most acid lakes. Smol et al. (1984a,b) suggested that these two taxa might be used as indicators of acidification.

Haworth (1983, 1984) used chrysophyte scales for the interpretation of the recent history of Blelham Tarn, in the English Lake District. Scales of *Mallomonas reginae* Teil. were reduced in number and *M. caudata* increased in connection with increasing eutrophy. She also compared plankton records (from Dr. Lund's algal records) from the lake with corresponding deposited mallomonadacean assemblages and found that more species were found in the sediments than in the plankton samples. Haworth (1984) indicated that it was difficult to correlate the recorded species in plankton samples with those found in the sediment because of difficulties in the identification of live cells, and the varying dissolution of scales.

Annually laminated sediments were studied by Renberg (1976) from Lake Rudetjärn, Sweden. He found that each lamination consisted of a light- and dark-colored layer. The light-colored layers were mostly of organic material. These layers represented winter and spring, while the darker layers represented summer and autumn. The organic layers contained diatoms and chrysophyte remains, such as scales bristles and cysts. Simola (1977, 1979) reported on many lakes with annually laminated sediments in Finland and also in Cornwall, England (Simola et al. 1981). He concluded that the chrysophytes have maxima in the spring layers in Finnish lakes.

Battarbee et al. (1980) studied *Mallomonas* scales in annually laminated sediments from Laukunlampi, Finland. The white lamina represented the summer period with *Mallomonas crassisquama* dominating in the oligotrophic sediments below 58 cm, and *M. elongata* Rev. in the more eutrophic, uppermost sediments. The dark lamina contained organic material and cysts and represented the autumn and winter.

Discussion and future prospects

Much work has been devoted to describing chrysophyte cysts from both marine and freshwater sediment. However, the cysts' biological affinities are usually unknown. Cyst taxonomy and identification present considerable problems because very little information is available about their variability in size, form, and ultrastructure within a species. We do not know if all cysts deposited on the sediments have reached maturity with, for example, the pore – collar complex and ultrastructure of the cyst wall completely developed. It is likely that incompletely developed cysts are also preserved in the sediments (Figures 20.18 – 20.21). The mature pore – collar complex is of high taxonomic value, but when it is not fully developed, its use in identification work is doubtful. Cysts are evidently much better preserved than scales in the sediments because they have been found from the Upper Cretaceous, while scales have only been recorded from postglacial times (Figures 20.12 and 20.13). However, the cysts are also exposed to dissolution and fragmentation, so finer structural elements such as spines may be rubbed off in the sediments.

The fact that the chrysophytes can produce both asexual and sexual cysts complicates the identifications even more. As said before, the cysts are species specific, but are these two differently produced cysts of each species identical? Skuja (1950) observed that asexual cysts in *Dinobryon borgei* were smaller than the sexual ones. Hundreds of descriptions of cysts have been made with LM, but as EM is necessary for correct identification, most of the older descriptions are now of little use.

Scanning electron microscopy is excellent for studying the outer ultrastructure and form of the cyst (Figures 20.22 – 20.27). The internal structures of the pore and cell wall can only be observed with LM. Therefore, it is important to use both LM and SEM and combine these descriptions. Note that when a cyst has been identified with SEM, it is often possible to recognize it later with LM, because we now know what is characteristic for the specimen and hence what to look for. Furthermore, the counting of cysts from sediments has to be done with LM. In the future both LM and SEM descriptions and micrographs will have to be combined, if we really want to use cysts as paleological indicators. To date many investigators have counted the different cysts in different groups, but then in the end combined them and stated that the maximum number of cysts deposited on the sediments correspond with the oligotrophic period of the lake. However, this procedure may imply an unrealistic simplicity of the system because the cyst-producing species may in fact have completely different ecological demands (Cronberg 1980, 1982, 1986).

The scale-bearing chrysophytes are more straightforwardly identified, and, furthermore, good taxonomic literature exists containing EM descriptions for their identification. About 130 species have been identified with EM. Most research in this field has been concentrated in the temperate zones. Very little information exists from the tropics, subtropics, arctic, and high mountain areas. However, from LM studies we know that there is also a tropical flora (Philipose 1953, Prowse 1962), and EM studies in tropical areas have been initiated. Through the identification of scales from sediments we can gain more information about the palaeolimnological development of lakes. However, until now scales have only been reported from postglacial sediments. Most often cysts can be rediscovered in sediments, and although their scales may be missing, we know that scale-bearing chrysophytes have been present in the plankton earlier. It seems likely that specific conditions must prevail during the sedimentation process to allow the scales to be preserved. Annulated sediments, in particular, contain very well-preserved scales.

What can we do in the future to improve the use of chrysophyte cysts and scales as paleoecological tools? We can collect from old literature the descriptions on chrysophyte cysts with original drawings and micrographs in a flora. We can include data concerning geological age, distribution, and chemical and physical conditions of the sediments. Data of the already described cysts of living chrysophytes can be compiled, including both LM and EM pictures, to create a flora. As much information as possible about the body of water should be included when describing recent cysts, e.g., lake morphology and ecological data such as pH, water color, salinity, conductivity, total phosphorus, and total nitrogen. This should include good descriptions and illustrations, including SEM and LM micrographs or LM drawings. We can use careful stratigraphy, perhaps with in situ freezing techniques, for sediment investigations. Before any further treatment, subsamples from the sediment core should be microscopically inspected to observe if the cysts seem to have viable chloroplasts or if chrysophyte scales are present.

Some fundamental questions concerning cysts might be answered through the study of living chrysophytes, both in nature and in cultures. In particular, the cyst development process and morphological variations within a species must be carefully investigated.

We have to make systematic investigations and study localities with different nutrient concentrations in different climatic regions throughout the whole year.

My opinion is that the chrysophytes thrive in small lakes, ponds, and pools with not too clean water. Sometimes a special spring, summer, or autumn flora exists. Cyst development is often a very quick process ranging from a couple of hours to a number of days. Temporary ponds frequently have a rich chrysophyte flora. Even though the water dries out completely, the chrysophytes return when the pools are filled up again. It is probably a result of their ability to make very persistent resting cysts that they can survive in these habitats (Sandgren 1983b). The trigger for cyst development probably varies from species to species, with temperature, light, nutrient conditions, and competition being potentially critical factors.

How old can a chrysophyte cyst be and still be able to germinate? No one knows. Very few records of emergence from cysts have been observed. An exciting possibility would be to remove seemingly viable cysts from different sediment layers, take them into the laboratory, and trigger germination under the microscope, and then directly study the emergent cells.

References

Adam, D.P. 1980a. Locality data for some chrysophyte cysts. U. S. Geological Survey, Open-File Report 80–651.

– 1980b. Scanning electron micrographs of modern chrysomonad cysts from Haypress Meadows, Eldorado County, California. U. S. Geological Survey, Open-File Report 80–1235.

– 1980c. Scanning electron micrographs of upper Pleistocene chrysomonad cysts from Flagpole Peak, Eldorado County, California. U. S. Geological Survey, Open-File Report 80–1239.

– 1981. Scanning electron micrographs of modern and late Holocene chrysomonad cysts from Harden Lake Meadow, Yosemite National Park, California. U. S. Geological Survey, Open-File Report 81–46.

Adam, D.P. & Mahood, A.D. 1979. Modern and Holocene chrysomonad cysts from Upper Echo Lake, Eldorado County, California. U. S. Geological Survey, Open-File Report 79–1461.

– 1980a. Modern chrysomonad cysts from Fallen Leaf Lake, Eldorado County, California. U. S. Geological Survey, Open-File Report 80–798.

– 1980b. Modern chrysomonad cysts from Alta Morris Lake, Eldorado County, California. U. S. Geological Survey, Open-File Report 80–822.

– Scanning electron micrographs of chrysomonad cysts from Suzie Lake, Eldorado County, California. U. S. Geological Survey, Open-File Report 80–1250.

– 1981a. Scanning electron micrographs of chrysomonad cysts from Lake Aloha, Eldorado County, California. U. S. Geological Survey, Open-File Report 81–45.

– 1981b. Chrysophyte cysts as a potential environmental indicators. *Geol. Soc. Am. Bull. Pt. I,* 92: 839–44.

Adam, D.P. & Mehringer, P.J. Jr. 1980a. Modern and Holocene chrysomonad cysts from Lost Trail Pass Bog, Montana. U. S. Geological Survey, Open-File Report 80-797.

– 1980b Scanning electron micrographs of modern chrysomonad cysts from Castor Pond, Jemez Mountains, New Mexico. U. S. Geological Survey, Open-File Report 80-1231.

– 1980c. Scanning electron micrographs of modern and Holocene chrysomonad cysts from Fish Lake, Steens Mountains, Oregon. U. S. Geological Survey, Open-File Report 80-1249.

Andersen, R.A. 1982. A light and electron microscopical investigation of *Ochromonas sphaerocystis* Matvienko (Chrysophyceae): the statospore, vegetative cell and its peripheral vesicles. *Phycologia* 21(3): 390–8.

Andrieu, B. 1936. Note sur les Chrysostomatacées d'une tourbe de l'Ile Kerguelen. *Bull. Soc. Franc. Microsc.* 5(2): 51–60.

– 1937. Les Chrysostomatacées d'Auvergne. I. Dépot de Verneuge (Puy-de-Dome). *Bull. Soc. Franc. Microsc.* 6(2): 49–58.

– 1938. Les Chrysostomatacées d'Auvergne. II. Dépot de Vassivière (Puy-de-Dome). *Bull. Soc. Franc. Microsc.* 7(3): 96–100.

Asmund, B. 1955. Elektron microscope observations on *Mallomonas caudata* and some remarks on its occurrence in four Danish ponds. *Bot. Tidsskr.* 52: 163–8.

Asmund, B. & Cronberg, G. 1979. Two new taxa of *Mallomonas* (Chrysophyceae). *Bot. Notiser* 132: 409–18.

Asmund, B., Cronberg, G. & Dürrschmidt, M. 1982. Revision of the *Mallomonas pumilio* group (Chrysophyceae). *Nord. J. Bot.* 2: 383–95.

Battarbee, R.W. 1973. A new method for the estimation of absolute microfossil numbers, with reference especially to diatoms. *Limnol. Oceanogr.* 18: 647–53.

Battarbee, R.W., Cronberg, G. & Lowry, S. 1980. Observations on the occurrence of scales and bristles of *Mallomonas* spp. (Chrysophyceae) in the micro-laminated sediments of a small lake in Finnish north Karelia. *Hydrobiologia* 71: 225–32.

Belcher, J.H. & Swale, E.M.F. 1967. *Chromulina placentula* sp. nov. (Chrysophyceae), a freshwater nannoplankton flagellate. *Br. Phycol. Bull.* 3(2): 257–67.

Bourrelly, P. 1957. Recherches sur les Chrysophycées. *Rev. Algol., Mém. Hors-Sér.* 1: 1–412.

– 1968. *Les Algues d'Eau Douce,* Vol. II. *Les Algues Jaunes et Brunes,* Boubée Paris.

Brugham, R.B. 1980. Postglacial diatom stratigraphy of Kirchner Marsh, Minnesota. *Quart. Res.* 13: 133–46.

Carney, H.J. 1982. Algal dynamics and trophic interactions in the recent history of Frains Lake, Michigan. *Ecology* 63(6): 1814–16.

Carney, H.J. & Sandgren, C.D. 1983. Chrysophycean cysts: indicators of eutrophication in the recent sediments of Frains Lake, Michigan, U.S.A. *Hydrobiologia* 101: 195–202.

Cienkowsky. 1870. Über Palmellaceen und einige Flagellaten. *Arch. Mikrosk. Anat.* 6: 421–38.

Colom, G. 1940. Arqueomonadineas, Silicoflagelados, discoasteridos: fosiles de España. Ciencias. *Anal. Asoc. Espan. Prog. Ciencias, Madrid* 5: 343–56.

Conrad, W. 1926. Recherches sur les Flagellates de nos eaux saumâtres, 2e partie: Chrysomonadines. *Arch. Protistenk.* 56: 167–231.

– 1938. Notes Protistologiques. VI. Kystes de Chrysomonadines ou Chrysostomatacées?. *Bull. Mus. Roy. Hist. Nat. Belg.* 14(46): 1–6.

– 1940. Notes Protistologiques. XVII. Chrysomonadées fossiles des collections du Musée royal d'Histoire naturelle de Belgique. *Bull. Mus. Roy. Hist. Nat. Belg.* 16(45): 1–15.

Cornell, W.C. 1972. Late Cretaceous chrysomonad cysts. *Palaeogeogr., Palaeoclimatol., Palaeoecol.* 12: 33-47.
Cronberg, G. 1973. Development of cysts in *Mallomonas eoa* examined by scanning electron microscopy. *Hydrobiologia* 43(1-2): 29-38.
- 1980. Cyst development in different species of *Mallomonas* (Chrysophyceae) studied by scanning electron microscopy. *Algolog. Stud.* 25: 421-34.
- 1982. Phytoplankton changes in Lake Trummen induced by restoration. *Folia Limnol. Scand.* 18: 1-119.
- 1986. Blue-green algae, green algae and Chrysophyceae in sediments. *In:* Berglund, B.E. [Ed.] *Handbook on Palaeoecology.* Wiley, Chichester.
Dahm, H.-D. 1956. Diatomeenuntersuchungen zur Geschichte der westlichen Ostsee. *Meyniana* 5: 7-50.
Deflandre, G. 1932a. *Litharcaeocystis costata* nov. gen. spec., Chrysophycée marine fossile. Remarques sur les Chrysostomatacées. *C. R. Acad. Sci.* 194: 1273-5.
- 1932b. Archaeomonadaceae, une famille nouvelle de Protistes fossiles marins à loge siliceuse. *C. R. Acad. Sci.* 194: 1859-61.
- 1932c. Note sur les Archaeomonadacées. *Bull. Soc. Bot. France* 79: 346-55.
- 1933. Seconde note sur les Archaeomonadacées. *Bull. Soc. Bot. France* 80: 79-90.
- 1934. Sur l'abus de l'emploi, en paléontologie, du nom de genre *Trachelomonas,* et sur la nature de quelques ex *"Trachelomonas"* siliceux (Chrysomonadines) tertiaires et quaternaires. *Ann. Protistol.* 4: 151-65.
- 1935. *Trachelomonas,* Archaeomonadacées et Chrysostomatacées. Réponse à une note de J. Frenguelli. *Arch. Protistenk.* 85(2): 306-11.
- 1936. Les Flagellés Fossiles. Apercu Biologique et Paleontologique. Role geologique. *Actual. Sc. & Indust. Expos. Geol., Paris* 355: 8-97.
- 1938. Troisième note sur les Archaeomonadacées. *Soc. Fr. Micr. Bull.* 7: 73-88.
- 1952. Chrysomonadines fossiles. *In:* Grasse, P. [Ed.] *Traite de Zoologie,* Vol. 1; Masson et cie, Paris, 560-70.
Doflein, F. 1921. Mitteilungen über Chrysomonadinen aus dem Schwarzwald. *Zool. Anz.* 53: 153-73.
Dürrschmidt, M. 1984. Studies on scale-bearing Chrysophyceae of the Giessen area, Federal Republic of Germany. *Nord. J. Bot.* 4: 123-43.
Ehrenberg, C.G. 1854. *Microgeologie.* Leopold Voss, Leipzig.
Elner, J.K. & Happey-Wood, C.M. 1978. Diatom and chrysophycean cyst profiles in sediment cores from two linked but contrasting Welsh Lakes. *Br. Phycol. J.* 13: 341-60.
Firtion, F. 1944. Spongilles et Chrysostomatacées d'un lignite de la formation cinéritique du Lac Chambon (Puy-de-Dome). *Bull. Soc. Géol. France* 5(14): 331-46.
- 1945. Sur les Chrysostomatacées du Bassin tertiaire de Menat (Puy-de-Dome). *Bull. Soc. Géol. France* 5(15): 45-51.
Fott, B. 1955. Scales of *Mallomonas* observed in the electron microscope. *Preslia* 27: 280-2.
- 1959. Zur Frage der Sexualität der Chrysomonaden. *Nova. Hedwig.* 1: 115-29.
- 1966. Elektronenmikroskopischer Nachweis von *Mallomonas*-Schuppen in Seeablagerungen. *Int. Revue Ges. Hydrobiol.* 51(5): 787-90.
Frenguelli, G. 1925. Sopra alcuni microrganismi a guscio siliceo. *Roma Soc. Geol. Ital. Boll.* 44(1): 1-8.
- 1929. *Trachelomonas* de los esteros de la region del Ybera en la provincia de corrientes, Argentina. *Rev. Ch. Hist. Nat.* 33: 563-8.

- 1931. Analisis microscopico de una muestra de Tripoli de Angostura (Provincia de Colcaqua, Chile). *Revista Chilene de Hist. Natural* 35: 9-14.
- 1932. Trachelomonadi del Pliocene Argentino. *Memoire Soc. Geol. Ital.* 1: 1-44.
- 1935a. Einige Bemerkungen zu den Archaeomonadaceen. *Arch. Protistenk.* 84: 232-41.
- 1935b. Traquelomonades del Platense de la Costa Atlantica de la Provincia de Buenos Aires. *Not. Mus. Plata* 1: 35-44.
- 1936. Crisostomataceas del Neuquén. *Not. Mus. Plata* 1: 247-75.
- 1938a. *Deflandreia,* nuevo género de Chrisostomataceas. *Not. Mus. Plata* 3: 47-54.
- 1938b. Crisostomataceas Platenses. *Acta Geograph.* 14(12): 149-54.

Geitler, L. 1935. Über zweikernige Cysten von *Dinobryon. Österr. Bot. Zeitsch.* 84(4): 282-6.

Gombos, A.M. Jr 1977. 12. Archaeomonads as Eocene and Oligocene guide fossils in marine sediments. *Initial Reports of the Deep Sea Drilling Project* 36: 689-694.

Greguss, P. 1913. Die Kieselalgen der Meerangen von Surian. *Botanikai Közlemények* 12(5-6): 202-25.

Gretz, M.R., Sommerfeld, M.R. & Wujek, D.E. 1979. Scaled Chrysophyceae of Arizona. A preliminary survey. J. Arizona-Nevada Acad. Sci. 14: 75-80.

Gritten, M.M. 1977. On the fine structure of some chrysophycean cysts. *Hydrobiologia* 53: 239-52.

Hajos, M. 1968. Die Diatomeen der miozänen Ablagerungen des Matravorlandes. *Geolog. Hungarica* 37: 1-269.

- 1975. Late Cretaceous Archaeomonadaceae, Diatomaceae, and Silicoflagellatae from the South Pacific Ocean, Deep Sea Drilling Project, Leg 29, Site 275. *Initial Report of the Deep Sea Drilling Project* 29: 913-1007.

Hanna, G.D. 1927. Cretaceous diatoms from California. *Calif. Acad. Sci., Occ. Paper* 13: 5-44.

- 1934. Additional notes on diatoms from the Cretaceous of California. *J. Paleont.* 8(3): 352-5.

Haworth, E.Y. 1983. Diatom and chrysophyte relict assemblages in the sediments of Blelham Tarn in the English Lake District. *Hydrobiologia* 103: 131-4.

- 1984. 6. Stratigraphic changes in algal remains (diatoms and chrysophytes) in the recent sediments of Blelham Tarn, English Lake District. *In:* Haworth E.Y. & Lund, J.W.G. [Eds.] *Lake Sediments and Environmental History.* Leicester University Press. pp. 165-90.

Henderson, E.B. 1925. Notes on a diatomaceous deposit at Dalmahoy, Edinburgh. *Trans. Bot. Soc. Edinburgh* 29(2): 135-44.

Hibberd, D.J. 1977. Ultrastructure of cyst formation in *Ochromonas tuberculata* (Chrysophyceae). *J. Phycol.* 13: 309-20.

Ichikawa, W., Fuji, N. & Bachmann, A. 1964. Fossil diatoms, pollen grains and spores, silicoflagellates and arachaeomonads in the Miocene Hojuji diatomaceous mudstone Noto Peninsula, Central Japan. *Sci. Rep. Kanazawa Univ.* (Japan) 9(1): 26-118.

Iwanoff, L. 1899. Beitrag zur Kenntniss der Morphologie und Systematik der Chrysomonaden. *Bull. Ac. Imp. Sc. St. Petersburg* 11(4): 247-62.

Jendrzejewski, J.P. & Hart, G.F. 1978. Distribution of siliceous microfossils in surficial bottom sediments of the Gulf of Mexico. *Palynology* 2: 159-166.

Kaczmarska, I. 1976. Diatom analysis of eemian profile in fresh-water deposits at Imbramowice near Wroclaw. *Acta Palaeobotanica* 17(2): 3-34.

Korde, N.W. 1961. Charakteristische Merkmale der Stratifikation der Bodenab-

lagerungen in Seen mit verschiedenartigem Zufluss. *Verh. Internat. Verein. Limnol.* 14:524–32.
– 1966. Algenreste in Seesedimenten. *Ergebn. Limnol.* 3:1–38.
Krieger, W. 1929. Algologisch-Monographische Untersuchungen über das Hochmoor am Diebelsee. *Beiträge zur Naturdenkmalpflege* 13(2): 230–300.
Kristiansen, J. 1961. Sexual reproduction in *Mallomonas caudata. Bot. Tidsskr.* 57: 306–9.
– 1980. Chrysophyceae from some Greek lakes. *Nova Hedwig.* 33: 167–94.
– 1982. Chrysophyceae. *In:* Parker, S.P. [Ed.] *Synopsis and Classification of Living Organisms,* Vols. I–II, McGraw-Hill, New York, pp. 81–6.
Kristiansen, J. & Takahashi, E. 1982. Chrysophyceae: Introduction and bibliography. *In:* Rosowski, J.R. & Parker, B.C. [Eds.] *Selected Papers in Phycology* Vol. II, Phycological Society of America, Inc., pp. 698–704.
Lagerheim, G. 1903. Untersuchungen über fossile Algen I, II. *Geol. Fören. Förhandl. No. 217.* 24(7): 475–500.
Leventhal, E. 1970. The Chrysomonadina. Lago de Monterosi. *Trans. Amer. Phil. Soc.* 60: 123–42.
Mahood, A.D. & Adam, D.P. 1979. Late Pleistocene chrysomonad cysts from core 7, Clear Lake, Lake County, California. U. S. Geological Survey, Open-File Report 79-971.
Matvienko, O.M. 1965. *Zolotisti vodorosti, Chrysophyta.* Viznac. Prisnovod. Vodorostej Ukraj. RSR, Kijiv.
Messikommer, E. 1938. Beitrag zur Kenntnis der fossilen und subfossilen Desmidiaceen. *Hedwigia* 78: 107–201.
Moss, B. 1979. Algal and other fossil evidence for major changes in Strumpshaw Broad, Norfolk, England in the last two centuries. *Br. Phycol. J.* 14: 263–83.
Munch, C.S. 1980. Fossil diatoms and scales of Chrysophyceae in the recent history of Hall Lake, Washington. *Freshwater Biol.* 10: 61–6.
Nipkow, H.F. 1927. Über das Verhalten der Skelette planktischer Kieselalgen im geschichteten Tiefenschlamm des Zürich- und Baldeggersees. Ph.D. thesis, Die Eidgenössische Technische Hochschule, Zürich, Switzerland.
Nygaard, G. 1956. Ancient and recent flora of diatoms and Chrysophyceae in Lake Gribsø. *Folia Limnol. Scand.* 8: 32–94.
– 1977. New or interesting plankton algae. *Kgl. Dan. Vid. Selsk. Biol. Skr.* 21(1): 1–107.
Pantocsek, J. 1912. *A fertö to kovamozat viranga: Bacillariae locus Peisonis.* Pozsony, Wigand K.F. Könyvnyomdeja.
– 1913. Die im Andesittuffe von Kopacsee vorkommenden Bacillarien. *Botanikai Közlemenyek* 12(5–6): 126–37.
Pascher, A. 1917. Flagellaten und Rhizopoden in ihren gegenseitigen Beziehungen. *Arch. Protist.* 38: 1–88.
Patrick, R. & Reimer, C.W. 1966. *The Diatoms of United States Exclusive of Alaska and Hawaii,* Vol. I, Monograph 13. Academy of Natural Sciences of Philadelphia, Philadelphia, PA.
Perch-Nielsen, K. 1975. 24. Late Cretaceous to Pleistocene Archaeomonads, erbridians, endoskeletal dinoflagellates, and other siliceous microfossils from the subantarctic southwest Pacific, DSDP, Leg 29. *Initial Report of the Deep Sea Drilling Project, Washington* 29: 873–907.
– 1977. 35. Tertiary silicoflagellates and other siliceous microfossils from the Western South Atlantic Deep Sea Drilling Project, Leg 39. *Initial Report of the Deep Sea Drilling Project, DSDP,* 39: 863–67.

Petersen, J.B. & Hansen, J.B. 1956. On the scales of some *Synura* species. *Biol. Medd. Dan. Vid. Selsk.* 23: 3–27.
Philipose, M.T. 1953. Contributions to our knowledge of Indian algae I. Chrysophyceae. – On the occurrence of *Mallomonas* Perty, *Synura* Ehr. and *Dinobryon* Ehr. in India. *Proc. Ind. Acad. Sci.* 37: 331–8.
Preisig, H.R. & Takahashi, E. 1978. *Chrysosphaerella (Pseudochrysosphaerella) solitaria, spec. nova (Chrysophyceae). Pl. Syst. Evol.* 129: 135–42.
Prowse, G.A. 1962. Further Malayan freshwater flagellata. *Gardens' Bull., Singapore* 19: 105–45.
Przybylowska-Lange, W. 1976. Diatoms of lake deposits from the Polish Baltic coast. I. Lake Druzno. *Acta Palaeobotanica* 17(2): 35–74.
Rampi, L. 1937a. Note sur les Chrysostomatacées tertiares de Sànta Fiora. *Bull. Soc. Fr. Microsc.* 6(2): 67–75.
– 1937b. Les Diatomées et les Chrysostomatacées d'une tourbe du Monte Amita. *Bull. Soc. Fr. Microsc.* 6(4): 129–36.
– 1940a. Archaemonadacee del Cretaceo Americano. *Atti Soc. Ital. Sci. Nat. Mus. Civico* 79: 60–8.
– 1940b. Diatomee e Crisostomatacee dell'Isola di Rodi. *Nuovo Giornale Botan. Ital., n.s.* 47: 572–8.
– 1948. Su alcune Archaeomonadacee (Crisomonadine fossili marine) nuove od interessanti. *Atti Soc. Ital. Sci. Nat. Mus. Civico* 87: 185–8.
– 1969. Archaemonadacées de la diatomite éocène de Kreyenhagen, Californie. *Cahiers Micropaleontol. Sér. I, No. 14, Arch. Orig. Centr. Docum. C.N.R.S.* 461: 1–11.
Renberg, I. 1976. Annually laminated sediments in Lake Rudetjärn, Medelpad province, northern Sweden. *Geol. Fören. Förhandl.* 98: 355–60.
Sandgren, C.D. 1980a. An ultrastructural investigation of resting cyst formation in *Dinobryon cylindricum* Imhof (Chrysophyceae, Chrysophycota). *Protistologica* 16(2): 259–76.
– 1980b. Resting cyst formation in selected chrysophyte flagellates: An ultrastructural survey including a proposal for the phylogenetic significance of interspecific variations in the encystment process. *Protistologica* 16(2): 289–303.
– 1981. Characteristics of sexual and asexual resting cyst (statospore) formation in *Dinobryon cylindricum* Imhof (Chrysophyta). *J. Phycol.* 17: 199–210.
– 1983a. Morphological variability in populations of chrysophycean resting cysts. I. Genetic (interclonal) and encystment temperature effects on morphology. *J. Phycol.* 19: 64–70.
– 1983b. Survival strategies of chrysophycean flagellates: reproduction and formation of resistant resting cells. *In:* Fryxell, G.A. [Ed.] *Survival Strategies of the Algae.* Cambridge University Press, pp. 23–48.
Sandgren, C.D. & Carney, H.J. 1983. A flora of fossil chrysophycean cysts from the recent sediments of Frains Lake, Michigan, U.S.A. *Nova Hedwigia* 38: 129–63.
Schrader, H.-J. 1978. 41. Quarternary through Neogene history of the Black Sea, deduced from paleoecology of diatoms, silicoflagellates, erbridians and chrysomonads. *Initial Report to the Deep Sea Drilling Project* 42: 789–901.
Sheath, R.G., Hellebust, J.A. & Sawa, T. 1975. The statospore of *Dinobryon divergens* Imhof: Formation and germination in a subarctic lake. *J. Phycol.* 11(2): 131–8.

Simola, H. 1977. Diatom succession in the formation of annually laminated sediment in Lovojärvi, a small eutrophicated lake. *Ann. Bot. Fennici* 14: 143-8.
- 1979. Micro-stratigraphy of sediment laminations deposited in a chemically stratifying eutrophic lake during the years 1913-1976. *Holarctic Ecol.* 2: 160-8.
Simola, H. & Uimonen-Simola, P. 1983. Recent stratigraphy and accumulation of sediment in the deep, oligotrophic Lake Pääjärvi in South Finland. *Hydrobiologia* 103: 287-93.
Simola, H.L.K., Coard M.A. & O'Sullivan, P.E. 1981. Annual laminations in the sediments of Loe Pool, Cornwall. *Nature (London)* 290(5803): 238-241.
Skogstad, A. 1982. Synuraceae-floraen i 27 lokaliteter i Oslo-området. Ph.D. thesis in limnology. University of Oslo, Norway.
- 1984. *Mallomonas intermedia* Kisselew (Mallomonadaceae, Chrysophyceae). Vegetative and cyst-stage studied by light- and electron microscopy. *Nord. J. Bot.* 4: 275-80.
Skuja, H. 1950. Körperbau und Reproduktion bei *Dinobryon borgei* Lemm. *Svensk Bot. Tidskr.* 44(1): 96-107.
Smol, J.P. 1980. Fossil synuracean (Chrysophyceae) scales in lake sediments: a new group of paleoindicators. *Can. J. Bot.* 58: 458-65.
- 1983. Paleophycology of a high arctic lake near Cape Herschel, Ellesmere Island. *Can. J. Bot.* 61: 2195-204.
Smol, J.P., Brown, S.R. & McNeely, R.N. 1983. Cultural disturbances and trophic history of a small meromictic lake from central Canada. *Hydrobiologia* 103: 125-130.
Smol, J.P., Charles, D.F. & Whitehead, D.R. 1984a. Mallomonadacean (Chrysophyceae) assemblages and their relationships with limnological characteristics in 38 Adirondack (N.Y.) lakes. *Can. J. Bot.* 62: 911-23.
- 1984b. Mallomonadacean microfossils provide evidence of recent lake acidification. *Nature (London)* 307: 628-30.
Starmach, K. 1980. Chrysophyta I - Złotowiciowce. *In:* Starmach, K. & Sieminska, J. [Eds.] *Flora Słodkowodna Polski* Vol. 5. Państwowe Wydawnictwo Naukowe, Warszawa, pp. 1-775.
Steinecke, F. 1937. Zur Geschichte der Galtgarben-Moore. *Schr. Phys.-Ökon. Ges. Königsb.* 69(2-4): 289-340.
Stradner, H. 1971. On the ultrastructure of Miocene Archaeomonadaceae (Phytoflagellates) from Limberg, Lower Austria. *Proc. Second Plankt. Conf. Roma 1970,* 2: 1183-99.
Takahashi, E. 1978. *Electron Microscopical Studies of the Synuraceae* (Chrysophyceae) in Japan. Tokai University Press, Tokyo.
- 1981. Floristic study of ice algae in the sea ice of a lagoon, Lake Saroma, Hokkaido, Japan. *Mem. Nat. Inst. Polar Res. Ser. E Biol. Med. Sci.* 34: 49-56.
Tappan, H. 1980. *The Paleobiology of Plant Protists.* Freeman, San Francisco.
Tippett, R. 1964. An investigation into the nature of the layering of deep-water sediments in two eastern Ontario lakes. *Can. J. Bot.* 42: 1693-709.
Tynan, E.J. 1960. The Archaemonadaceae of the Calvert Formation (Miocene) of Maryland. *Micropaleontology* 6(1): 33-9.
- 1971. Geologic occurrence of the Archaeomonads. *Proc. Second Plankt. Conf. Roma 1970,* 2: 1225-30.
VanLandingham, S.L. 1964. Chrysophyta cysts from the Yakima basalt (Miocene) in South-Central Washington. *J. Paleontol.* 38(4): 729-39.

von Sarntheim, R.G. 1949. Moor- und Seeablagerungen aus den Tiroler Alpen in ihrer waldgeschichtlichen Bedeutung. *Österr. Botan. Zeitsch.* 95(1): 73–85.

Wawrik, F. 1960. Sexualität bei *Mallomonas fastigata* var. Kriegeri. *Arch. Protistenk.* 104: 541–4.

Zanon, D.V. 1947. Saggio di sistematica delle Crisostomatacee. *Acta Pontif. Acad. Sci. Roma* 11(4): 43–62.

21

A proposal for the development of standardized nomenclature and terminology for chrysophycean statospores

GERTRUD CRONBERG
Institute of Limnology
University of Lund, Box 3060
S-220 03 Lund, Sweden

AND

CRAIG D. SANDGREN[1]
Department of Biology
University of Texas at Arlington
Arlington, Texas 76019 U.S.A.

Introduction

Statospores are the characteristic siliceous resting cysts of chrysophycean algae. The highly resistant cyst bodies of many statospores are decorated with species-specific ornamentation patterns (e.g., see Cronberg 1980, Sandgren & Carney 1983). These cysts have great potential usefulness in chrysophyte taxonomy and in limnological or paleoecological studies concerned with microalgal assemblages because the vegetative cells of many chrysophytes are small, preserve poorly in water samples, and do not fossilize. The major current deterrent to the increased use of statospores for identifying chrysophyte species is that the statospores of most chrysophytes have not been critically described in sufficient detail to determine their species-specific features. Despite the current lack of precise biological information regarding most statospore morphotypes, however, these characteristic resting cysts can still continue to be developed as an important group of ecological index fossils if sufficient care is taken to provide critical morphological descriptions of new morphotypes. The use of statospores as taxonomic characteristics also obviously depends upon critical morphological descriptions of the statospores. To promote the increased use of chrysophycean statospores in paleolimnology and taxonomic studies, it has, therefore, become advisable to establish a standardized nomen-

[1]Present address: Department of Biological Sciences, University of Wisconsin at Milwaukee, P.O. Box 413, Milwaukee, Wisconsin 53201, U.S.A.

clature and terminology for describing these siliceous cysts. The recently instituted International Statospore Working Group (ISWG) proposes the following nomenclature guidelines as a result of discussions held in our organizational meeting during the First International Chrysophyte Symposium, Grand Forks, North Dakota on August 12, 1983. Present at this founding meeting were D.P. Adam, G. Cronberg, M. Dürrschmidt, J. Kingston, R. Meyer, S. Munch, C.D. Sandgren, P.A. Siver, A. Skogstad, J.P. Smol, and J. Wee. Scientists whose research involves potential consideration of statospores as systematic, ecological, or paleoecological tools are encouraged by the ISWG to comment on these guidelines and make suggestions for their improvement.

Suggested guidelines for statospore descriptions

It is hoped that all statospore morphotypes will eventually be correlated with specific chrysophyte species and that the need for separate consideration of the vegetative cells and the resting cysts will disappear. However, this goal will require many years of research to achieve and may prove impossible for fossilized statospores produced by chrysophyte species now extinct. The ISWG suggests that future published statospore descriptions adhere to a uniform format in order to ensure that all morphotypes are completely described in a manner that will readily facilitate comparisons. It will then be possible to integrate information regarding statospores readily into chrysophyte species descriptions as correlations between vegetative cells and statospore morphotypes become better known. Should the need arise, it will be possible to develop a useful artificial organization for statospores similar to the nonbiological hierarchal schemes that have long been used in micropaleontology for hystrichosphaere dinoflagellate cysts and acritarchs. Published statospore descriptions should, therefore, include the following information: (1) clear identification of the original specimen on which the description is based, (2) a morphological analysis of the statospore that conforms to generally acceptable descriptive terminology, and (3) a discussion of pertinent ecological or stratigraphic information concerning the morphotype (and the chrysophyte species producing the statospore when this is known).

Identification of the original specimen

To ensure that a statospore morphotype is clearly and distinctly defined, it is advisable to provide photographic illustrations of characteristic specimens as well as information regarding the source of the statospore-containing sample. Identification of the chrysophyte species that produces the statospore should also be included if an unambiguous correlation can be made.

High-resolution scanning electron microscope (SEM) micrographs are the most appropriate means of illustrating the fine details of statospore surface ornamentation. Accompanying light micrographs and line drawings are very useful because they can depict internal cyst wall or cytoplasmic structures and because paleoecologists must frequently rely on light microscopical imagery for routine morphotypic identifications and quantitative analyses. The negative numbers of all published micrographs should be included in the publication to facilitate the future exchange of information. We believe it is advisable to designate a single photographed statospore as an "original specimen" that is most illustrative of the morphotype. This original specimen should be photographed in such an orientation that critical details of both the cyst body ornamentation and the collar/pore complex can be seen. Additional SEM or light microscopy (LM) micrographs representing alternative views of the same specimen or representing additional specimens presumed to be of the same morphotype may be included to illustrate environmentally or developmentally induced diversity. This procedure will eliminate nomenclatural difficulties that may arise from situations in which morphological diversity originally perceived to represent natural variation in a single morphotype is later determined to represent several similarly ornamented and co-occurring statospore types.

Each new morphotype should be given a unique number by the describer, and each describer should maintain a personal series of statospore reference numbers for both the morphotypes and the photographic negatives. To prevent possible confusion, these numbers should not be repeated or reassigned. Living statospores with established biological affinities should be given numerical designations equivalent to fossil or unknown morphotypes. This procedure will facilitate the inclusion of all available statospore morphotypes into a single "universal listing" proposed to be maintained by the ISWG as a convenient resource for future statospore research (see later discussion).

The description of the source locality for a statospore morphotype should be brief unless it is not amplified upon in the publication's "Methods" section or in subsequent ecological or stratigraphic discussion. It should include the name of the sampling site (e.g., lake, rock formation, stratigraphic unit); the site's longitude, latitude, and altitude; the date the sample was taken; and the general nature of the sample (e.g., plankton tow, sediment grab sample, sediment core, soil scraping, rock hand sample). If the sample represents a specific stratum in a sediment core then the corresponding sample depth and sediment type should be included.

The biological affinity of a statospore morphotype is an extremely important piece of information. Such a correlation to a living chrysophycean species will suggest that information available concerning the ecology of the living species would be potentially applicable in describing the environment in which the statospore was produced. It is, therefore, of great importance that correlations between statospore morphotypes and living chrysophycean species be made with great care. Ill-considered correlations that prove to be erroneous could create confusion in the growing statospore literature and retard the development of statospores as paleoecological or taxonomic indicators. Assignment of a biological affinity in the description of the statospore morphotype should only be done with absolute certainty.

The statospore morphological description

New statospore descriptions should be based upon both the light (LM) and scanning electron (SEM) micrographs published with the original designation. The primary illustration of the original specimen, as explain above, would appropriately be an SEM micrograph depicting both the collar and the body ornamentation. Some mention of the SEM sample preparation technique should be made because this can influence morphological descriptions. We recommend acid or H_2O_2 cleaning so that the morphology of fossil and living material will be comparable.

A suggested standardized descriptive terminology for statospores will be developed over the next several years by members of the ISWG. This terminology will address the two key aspects of statospore morphology: (1) statospore superficial ornamentation patterns and (2) the morphology of the collar complex surrounding the pore. Several currently available sources of terms applicable for describing surface features of statospores, other types of algal resting spores, or pollen grains can be consulted until the new guidelines are made available (Nygaard 1956, Faegri & Iverson 1975, Sandgren & Carney 1983, Evitt 1985, see also the discussion of acritarch systematics in Tappan 1980). Authors are encouraged to make reasonable comparisons of their newly described morphotypes to forms previously described so that the existing literature on statospores can be incorporated into new publications.

Ecological or stratigraphic information

As much pertinent ecological information as is available should be included as a part of the statospore description. The intent of this information should be: (1) to describe critical chemical and physical parameters of the environment in which the statospore was produced and (2) to delimit the environmental requirements of the living chrysophycean species (when known). Environmental information need not be repeated in the statospore description if it is presented elsewhere in the same publication or has been previously published, but clear reference to all pertinent literature should be made. We expect that this section will be used by authors as a forum for establishing ecological tolerance limits for statospore production and for the occurrence of vegetative chrysophyte species.

The specific environmental parameters that should be included in an ecological description will of course be dependent upon the type of sample from which the statospore morphotype was recovered. The following parameters are considered desirable:

1. For living samples from aquatic habitats (water or superficial sediments): description of the water body, depth of water column, sampling depth, water quality parameters (pH, salinity, temperature, water color, conductivity, alkalinity, total and dissolved concentrations of phosphorus, nitrogen, silica), and a list of the dominant co-occurring algal species.
2. For samples from soft sediment cores: location of the coring site in the basin and the depth of water, level in the core from which the sample was taken, sample age (if available), associated microfossils and geochemical data, sediment type.
3. For samples from rock: the stratigraphic unit and estimated age, the precise stratigraphic position, associated microfossils, and the inferred depositional environment.

Universal statospore listing

It is a goal of the ISWG that each new morphotype described will eventually be assigned a permanent and unique reference number in a computerized statospore listing maintained by the ISWG. This listing will contain key descriptive characteristics of each statospore type as well as information regarding source localities and pertinent publications. It is intended to contain reference to all clearly described statospores, both fossil and living, regardless of whether their biological affinity is known. The utility of this listing will obviously be greatly enhanced if future published statospore descriptions maintain a degree of uniformity with regard to textual content and illustrations. Provisions will be made for inclusion in the universal listing of previously published statospore morphotypes that are adequately illustrated with LM and SEM micrographs or line drawings. It is expected that this listing of statospores will be a valuable aid to investigators pursuing biogeographical, systematic, and paleoecological studies of chrysophycean species.

An example of a statospore description

Identification of the original specimen

Statospore No. 1. Cronberg, G.M.C., as represented by Figure 7 in: Cronberg, G.M.C. and Sandgren, C.D. 1986. *In:* Kristiansen, K. and Andersen, R. [Eds]. *Chrysophytes—Aspects and Problems.* Cambridge University Press, New York.
Negative number of original specimen: 331/31 (Cronberg).

Locality and date: From a sediment depth of 70 cm. in sediment core BP1 from Lake Växjösjön, Småland, Sweden (55° 51′N, 14° 50′E, 161 m. a.b.s.), February, 1972.
Biological affinity: *Mallomonas eoa* Takahashi, 1963.

Statospore description (terminology after Sandgren & Carney 1983)

Specimens were cleaned by the hydrogen peroxide technique (Cronberg 1980). The statospore is widely obovate, 10–12 μm in length and 5–7 μm in width, with the pore (0.3 μm in diameter) situated on the broader anterior end (Figures 21.1–21.7). The cyst body is psilate in young cysts, but is densely microechinate in mature specimens. The collar surrounding the pore is of a characteristic complex type, being composed of an elaborate secondary (or outer) collar surrounding the simple primary collar and pore. The secondary collar is 1.5–2.1 μm in diameter and is composed of a ring of branched and irregular baculate elements at least 1 μm in length that vary greatly among specimens with respect to their individual diameter and occasionally appear acute and finely divided (see Cronberg 1973). The primary collar is very short cylindric, with an erect to slightly obconical outer margin and a swollen convex inner margin that slopes to the pore margin. There is no annulus surrounding the pore. The pore and collar complex is occasionally situated asymmetrically on the cyst (Figures 21.1 and 21.4–21.6), giving the statospore a characteristic appearance that is often possible to distinguish with the light microscope.

Ecological background information

The original locality for this morphotype, Lake Växjösjön, is situated in the town of Växjö in the central South Swedish Highlands, and downstream from the restored Lake Trummen (Cronberg 1982a). The lake area is 0.87 km^2, and the maximum depth is 6.5 meters. Stratigraphic evidence suggests that lake Växjösjön has been oligotrophic for most of its history, but was polluted with sewage from the town during the period of 1880 to 1930 (Battarbee & Digerfeldt 1976, Digerfeldt 1977, Battarbee et al. 1980, Cronberg, 1982b). The sewage water was diverted in 1929, and some signs of recovery have been observed (Cronberg 1982b), but the lake is still very eutrophic. Statospore No. 1 was found at all depths in the sediment core from the surface down to 3 meters. However, it was most abundant at the depth of 0.7 meter, corresponding to the period of worst pollution, 1920–1930 (Cronberg 1980, 1982b).

In the nearby Lake Trummen, vernal blooms of scaled chrysophytes, especially *Mallomonas* species, appeared after the restoration (Cronberg 1982a). The *Mallomonas* population consisted primarily of *M. eoa* (Cron-

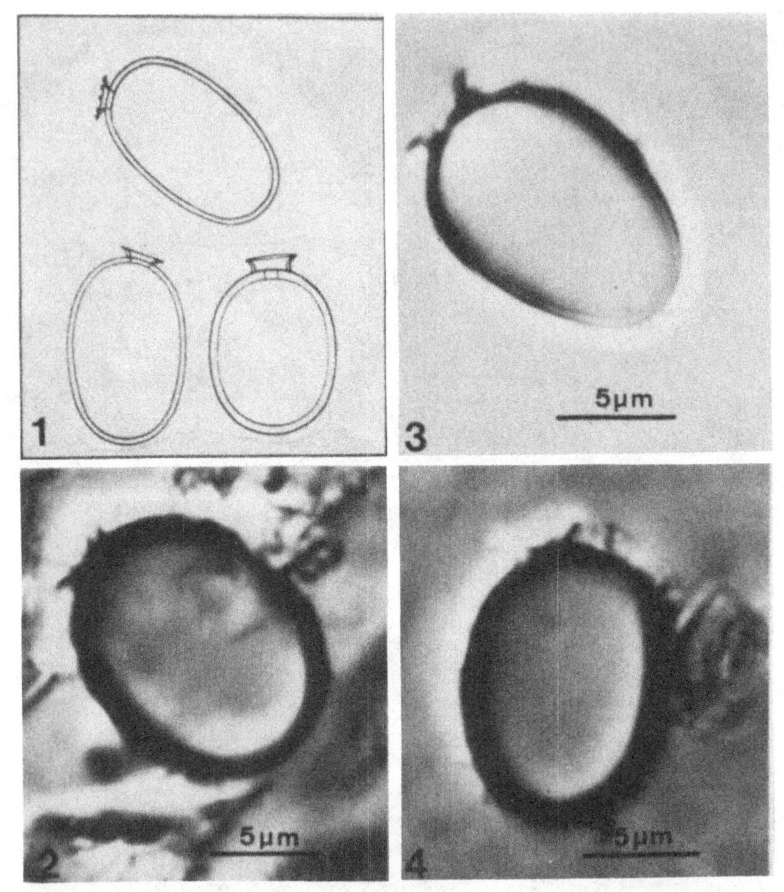

Figs. 21.1–21.4. Fig. 21.1. Light microscope drawing of *cysta teres* from Nygaard (1956). ***Fig. 21.2.*** Light micrograph of *cysta teres* from Nygaard's original slides of Lake Gribsø material. ***Fig. 21.3.*** Light micrograph of the cyst of *Mallomonas eoa* from the plankton of Lake Trummen, Sweden. ***Fig. 21.4.*** Light micrograph of statospore No. 1 Cronberg from the sediments of Lake Växjösjön, Sweden.

berg, 1973). Statospore development was frequently recorded during the end of March and the beginning of April. A comparison of LM and SEM micrographs of *M. eoa* cysts from Lake Trummen and Statospore No. 1 from Lake Växjösjön reveals that they are identical (Figures 21.3–21.7). A summary of the recorded distribution of *Mallomonas eoa* and its statospore is presented (Table 21.1).

Cyst development in Lake Trummen was correlated with changing temperature in the lake. In 1971 when the temperature exceeded 5°C, the entire population of *M. eoa* encysted and disappeared from the plankton in less than 1 week. In the plankton, *M. eoa* was typically associated with other *Mallomonas, Synura,* and *Cryptomonas* spp. Ecological data for Lake

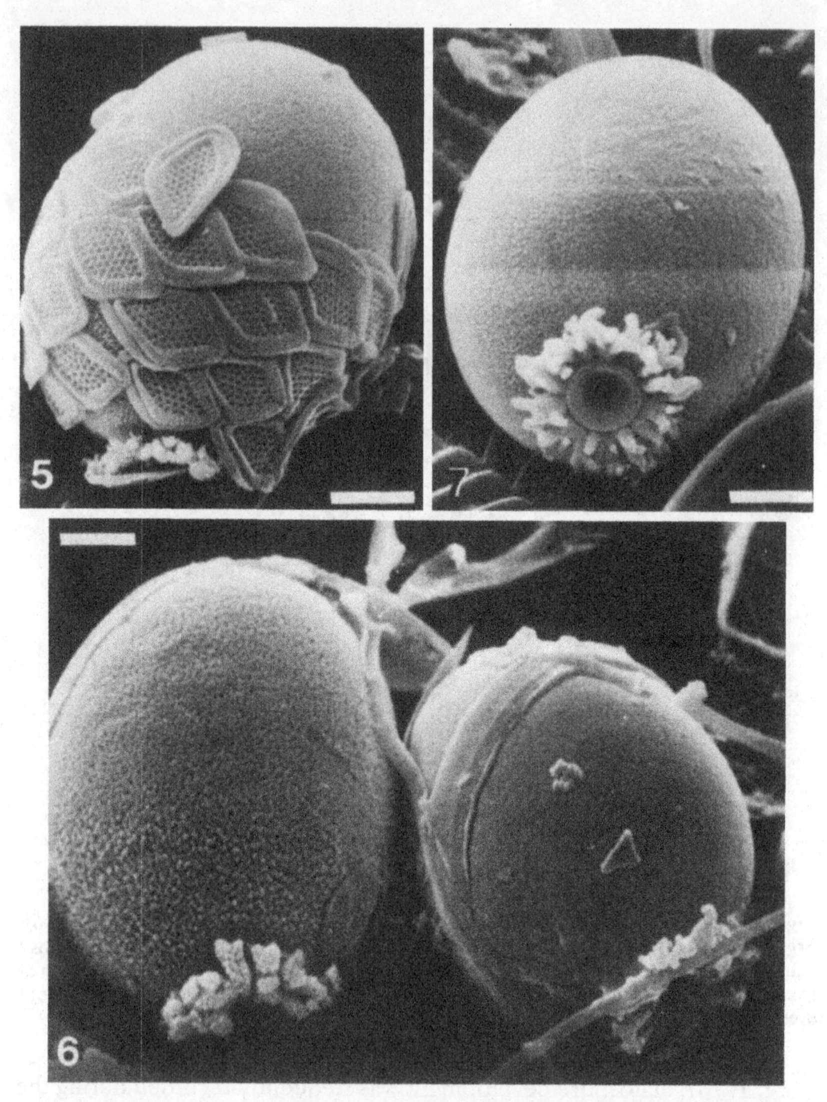

Figs. 21.5–21.7. Scanning electron micrographs of *Mallomonas eoa* and statospore No. 1. Scale = 2 μm. ***Fig. 21.5.*** Statospore of *M. eoa* covered by the layer of characteristic vegetative cell scales; from the plankton of Lake Trummen. ***Fig. 21.6.*** Statospores of *M. eoa* from the plankton of Lake Trummen exhibiting different stages of maturity in ornamentation. ***Fig. 21.7.*** Statospore No. 1 Cronberg (original specimen) from the sediments of Lake Växjösjön, Sweden.

Trummen during the periods when *M. eoa* was most abundant is presented (Table 21.2).

Nygaard (1956) described 77 statospore morphotypes from a sediment core of Lake Gribsø, Denmark. One of his morphotypes, *cysta teres,* strongly resembles Statospore No. 1 as described here. However, when

Table 21.1. *Distribution of* Mallomonas eoa *Takahashi*

The vegetative stage	
Country	**Reference**
Japan	Takahashi 1963, 1978
Alaska	Asmund & Takahashi 1969
Sweden	Cronberg 1973, 1980, 1982; Cronberg & Kristiansen 1980
Russia	Balonov & Kuzmin 1975, Balonov 1978
Norway	Skogstad 1982
Chile	Dürrschmidt 1980
New Zealand	M. Dürrschmidt (personal communication)
The statospore	
Denmark (as *cysta teres*)	Nygaard 1956
Sweden	Cronberg 1973, 1980, 1982a,b
Japan	Takahashi 1978
Russia	Balanov 1978

Nygaard described *cysta teres* in 1956, he published a total of 10 line drawings of this form showing minor structural differences without designating a single drawing as the original or most characteristic form. A recent preliminary SEM reexamination of his sediment samples (C.D. Sandgren, unpublished) suggests that Nygaard's morphotype, as described, probably represents several distinct types that can be distinguished using the high resolution of SEM. A second SEM examination of Nygaard's samples (G. Cronberg, unpublished) confirms that the cysts of *Mallomonas eoa* as defined here were indeed present (Figure 21.2), but their abundance relative to the other submorphotypes of Nygaard's *cysta teres* remains to be determined.

In the sediment core of Lake Gribsø, *cysta teres* was most common at the 4 meter depth, corresponding to 3000 B.C. The pH during that period is estimated to have been 7. During the last 5000 years, Lake Gribsø has shifted from a eutrophic to an acidic, oligotrophic lake with an accompanying decrease in pH to 4–5 (Nygaard 1956).

Conclusions

A proposal for standardization of the format used to describe chrysophycean statospores has been outlined. The suggested procedure involves incorporation of high-resolution scanning electron microscopical imagery into future descriptions and promotes the use of generally acceptable descriptive terminology. Standardized descriptions will facilitate comparisons of statospore morphotypes among studies, will hasten the integration

Table 21.2. *Ecological data from Lake Trummen, 1971–1978, when* Mallomonas eoa *Tak. was most frequent*

Year	1971	1972	1973	1974	1975	1976	1977	1978
Date	March 25	April 10	March 21	March 19	March 18	April 14	May 5	April 18
Temperature (°C)	2.8	6.2	5.6	3.0	2.1	6.0	11.1	8.3
pH	7.23	7.00	6.73	7.28	7.08	7.08	7.28	7.59
Color (mg Pt/liter)	45	20	35	45	50	40	35	45
Conductivity (μS_{20})	167	178	191	208	204	223	248	245
Total P (μg/liter)	132	69	54	50	46	47	43	33
Total N (mg/liter)	1.52	1.53	1.76	1.90	2.21	1.34	2.41	2.11
SiO_2 (mg/liter)	6.1	3.94	4.85	3.15	7.50	0.75	—	5.94

of statospores into descriptions of chrysophycean species, and will encourage the development of statospores as useful paleoecological indicators. A number of biologists and geologists involved in statospore research have formed an International Statospore Working Group to promote the increased critical use of statospores in chrysophyte taxonomy and in paleoecology. The ISWG intends to develop a computerized descriptive listing of properly described statospores that can serve as a resource for future research. The descriptive guidelines developed in this proposal are used to describe the statospore of *Mallomonas eoa* Takahashi as collected from phytoplankton samples and sediment cores of eutrophic lakes in Sweden.

References

Asmund, B. & Takahashi, E. 1969. Studies on Chrysophyceae from some ponds and lakes in Alaska VIII. *Mallomonas* species examined with the electron microscope II. *Hydrobiologia* 34(3-4): 305–21.

Balonov, I.M. 1978. Species of the genus *Mallomonas* Perty (Chrysophyta) in the Volga and its basin. (In Russian) *Trudy A.N.* SSSR 35(38): 75–102.

Balonov, I.M., & Kuzmin, G.V. 1975. Electron microscopic study of the species of the genus *Mallomonas* Perty from the reservoirs of the Volga Cascade. II. (In Russian). *Bot. Zurn.* 60: 1289–96.

Battarbee, R.W. & Digerfeldt, G. 1976. Paleoecological studies of the recent development of Lake Växjösjön. I. Introduction and chronology. *Arch. Hydrobiol.* 77(3): 330–346.

Battarbee, R.W., Digerfeldt, G., Appleby, P. & Oldfield, F. 1980. Paleoecological studies of the recent development of Lake Växjösjön. III. Reassessment of the recent chronology on the basis of modified ^{210}Pb dates. *Arch. Hydrobiol.* 89(4): 440–6.

Cronberg, G. 1973. Development of cysts in *Mallomonas eoa* examined by scanning electron microscopy. *Hydrobiologia* 43: 29–38.

– 1980. Cyst development in different species of *Mallomonas* (Chrysophyceae) studied by scanning electron microscopy. *Algological Studies* 25: 421–34.

– 1982a. Phytoplankton changes in Lake Trummen induced by restoration. *Folia Limnol. Scand.* 18: 1–119.

– 1982b. *Pediastrum* and *Scenedesmus* (Chlorococcales) in sediments of Lake Växjösjön. Sweden. *Algological Studies* 29: 500–7.

Cronberg, G. & Kristiansen, J. 1980. Synuraceae and other Chrysophyceae from central Småland, Sweden. *Bot. Notiser* 133: 595–618.

Digerfeldt, G. 1977. Paleoecological studies of the recent development of Lake Växjösjön. II. Settlement and landscape development. *Arch. Hydrobiol.* 79(4): 465–77.

Dürrschmidt, M. 1980. Studies on the Chrysophyceae from Rio Cruces, Prov. Valdivia, South Chile by scanning and transmission microscopy. *Nova Hedwigia* 33: 353–88.

Evitt, W.R. 1985. *Sporopollenin Dinoflagellate Cysts: their morphology and interpretation.* American Association of Stratigraphic Palynologists Foundation, Dallas, Texas.

Faegri, K. & Iverson, J. 1975. *Textbook of Pollen Analysis.* Hafner, New York.

Nygaard, G. 1956. Ancient and recent flora of diatoms and Chrysophyceae in Lake Gribsø. *Folia Limnol. Scand.* 8: 33–262.

Sandgren, C.D. & Carney, H.J. 1983. A flora of fossil chrysophycean cysts from the recent sediments of Frains Lake, Michigan, U.S.A. *Nova Hedwigia* 38: 129–63.

Skogstad, A. 1982. Synuraceae-floraen i 27 lokaliteter i Osloområdet. Hovedfagsoppgave i Limnologi. University of Oslo, Norway.

Takahashi, E. 1963. Studies on the genera *Mallomonas, Synura* and other plankton in freshwater with the electron microscope. IV. On two new species of *Mallomonas* found in ditches at Tsuruka in northeast Japan. *Bull. Yamagata Univ. Agr. Sci.* 4(2): 169–87.

– 1978. *Electron Microscopical Studies of the Synuraceae (Chrysophyceae) in Japan: Taxonomy and Ecology.* Tokai University Press, Tokyo.

Tappan, H. 1980. *The Paleobiology of Plant Protists.* Freeman, San Francisco.

Index

Note: Numbers in **boldface** indicate pages where illustrations occur.